Nanometer CMOS Sigma-Delta Modulators for Software Defined Radio

Alonso Morgado • Rocío del Río
José M. de la Rosa

Nanometer CMOS Sigma-Delta Modulators for Software Defined Radio

Springer

Alonso Morgado
Instituto de Microelectrónica de Sevilla
Centro Nacional de Microelectrónica
IMSE-CNM (CSIC/Universidad de Sevilla)
Parque Tecnológico de la Cartuja
C/Américo Vespucio s/n
Sevilla, Spain
alonso@imse-cnm.csic.es

Rocío del Río
Instituto de Microelectrónica de Sevilla
Centro Nacional de Microelectrónica
IMSE-CNM (CSIC/Universidad de Sevilla)
Parque Tecnológico de la Cartuja
C/Américo Vespucio s/n
Sevilla, Spain
rocio@imse-cnm.csic.es

José M. de la Rosa
Instituto de Microelectrónica de Sevilla
Centro Nacional de Microelectrónica
IMSE-CNM (CSIC/Universidad de Sevilla)
Parque Tecnológico de la Cartuja
C/Américo Vespucio s/n
Sevilla, Spain
jrosa@imse-cnm.csic.es

ISBN 978-1-4614-0036-3 e-ISBN 978-1-4614-0037-0
DOI 10.1007/978-1-4614-0037-0
Springer New York Dordrecht Heidelberg London

Library of Congress Control Number: 2011933912

Printed on acid-free paper

Springer is part of Springer Science+Business Media (www.springer.com)

A Ania
A Juanan y a Mario
A Visi, José Manuel, María y Jaime
A nuestros padres

Preface

This book represents a contribution to the design of sigma-delta ($\Sigma\Delta$) modulators intended for the A/D conversion in multi-standard multi-mode wireless transceivers, implemented in nanometer CMOS technologies. In these transceivers, ADCs are key parts because they need to operate with a wide spread of their specifications; namely, effective resolution and signal bandwidth. $\Sigma\Delta$ modulators are very suited for the implementation of reconfigurable ADCs in highly integrated transceivers. On the one hand, the key principles of $\Sigma\Delta$ modulators (oversampling and noise shaping) determine the dynamic range of the ADC, so that their adjustment contributes to adapt the converter performance to different specifications with large hardware reuse. On the other, both principles make them robust with respect to non-idealities of an integrated implementation.

In spite of the advantages mentioned above, the design of nanometer CMOS $\Sigma\Delta$ ADCs is not easy, specially when considering adaptability and reconfigurability features. It involves a number of practical issues and trade-offs at both architectural and circuit level that must be taken into account for optimizing performance in terms of power dissipation (device portability and autonomy), silicon area (cost) and time-to-market deployment.

In this context, the work in this book presents innovative solutions for the implementation of flexible $\Sigma\Delta$ modulators intended for the next generation of wireless hand-held mobile terminals, implemented as a SoC in nanometer CMOS processes. Novel adaptive and reconfigurable $\Sigma\Delta$ modulator topologies—based on the combination of resonation, unity signal transfer function, and a new type of cascade with extra inter-stage feedback loops and simplified digital cancellation logic—are presented. These strategies allow to reduce the requirements of the embedded amplifiers in terms of finite DC gain, non-linearity and output swing. This makes them very suited for the implementation of low-voltage low-power wideband ADCs. At the circuit level, different strategies are applied to adapt the performance of the $\Sigma\Delta$ modulators to the different sets of specifications with adaptable power consumption.

A number of architectures, circuit techniques and design procedures presented in this book are demonstrated through the design, implementation and experimental characterization of three IC prototypes. The first one—implemented in a 130-nm CMOS technology—consists of an expandable cascade topology that comprises a 2nd-order front-end stage followed by 1st-order stages, with the last one being switchable and also incorporating multi-bit quantization. The chip reconfigures the modulator loop filter order, the sampling frequency and the number of bits of the internal (back-end) quantizer, and scales the power consumption of internal building blocks in order to adapt the performance to the specifications of 2G/3G standards, considering a direct conversion receiver. Measurement results show a correct operation for GSM/Bluetooth/WCDMA standards, featuring a dynamic range of 86.7/81.0/63.3 dB for signal bandwidths of 200 kHz/1 MHz/4 MHz, respectively. The power consumption is 25.2/25.0/44.5 mW, of which 11.0/10.5/24.8 mW corresponds to the analog part of the circuit.

The second chip—implemented in a 90-nm CMOS process—consists of a two-stage (2–2) topology with 3-level quantization and unity signal transfer function in both stages. The chip reconfigures its loop filter, clock frequency and scales power according to the required specifications for different standards included in B3G wireless telecom, covering: GSM, Bluetooth, GPS, UMTS, DVB-H and WiMAX. Measurement results feature a dynamic range of 78/70/71.5/66/62/52 dB within bandwidths of 100 kHz/500 kHz/1 MHz/2 MHz/4 MHz/10 MHz, while consuming 4.6/5.35/6.2/8/8/11 mW, respectively.

The third chip is a 2-2-2 cascade made up of unity-STF stages. Similarly to the second chip, a 1.2-V 90-nm CMOS technology is employed. The chip can reconfigure its loop filtering order from 6 to 4 or 2 by switching off one or two stages in the cascade, respectively. The quantization in each stage can be selected to 3 or 5 levels. Every stage can work concurrently or as part of a cascade, so this $\Sigma\Delta$ modulator can process up to three standards in parallel. The employed architecture incorporates programmable resonation in the last two stages. The bias currents of main modulator building blocks are adjustable on-chip. Also, the sampling frequency can be adapted to the requirements of each operation mode. Experimental characterization of this chip indicates a correct operation for most of the reconfiguration techniques implemented; namely, adaptation of the in-loop filtering order, programmability of the bias currents in the modulator building blocks, variation of the internal quantization and concurrency.

The work in this book demonstrates the feasibility of using $\Sigma\Delta$ modulators for the efficient implementation of multi-standard telecom systems and shows the way for the practical deployment of the software defined radio paradigm.

Sevilla
March 2011

Alonso Morgado
Rocío del Río
José M. de la Rosa

Acknowledgments

The authors would like to express their gratitude to Prof. Ángel Rodríguez-Vázquez, Prof. Belén Pérez-Verdú, Dr. Fernando Medeiro (IMSE-CNM, CSIC/University of Sevilla), Prof. Maurits Ortmanns (University of Ulm), Dr. Ana Rusu (Royal Institute of Technology, KTH, Stockholm), Dr. Geert Van der Plas and Dr. Julien Ryckaert (IMEC, Leuven), for the review of this text. Specially, the authors wish to thank Dr. Geert Van der Plas and Dr. Julien Ryckaert for their valuable inputs and help in the design of the second chip presented in Chap. 5 of this book, and also to Mr. José G. García-Sánchez, Mr. Luis I. Guerrero-Linares and Mr. Sohail Asghar for their support with the design and layout of the third prototype described in Chap. 6.

This work was supported in part by the Spanish Ministry of Innovation and Science (with support from the European Regional Development Fund) under Contract TEC2007-67247-C02-01/MIC and Contract TEC2010-14285/MIC, and in part by the Regional Ministry of Innovation, Science and Enterprise under Contract TIC-2532.

Table of Contents

List of Abbreviations

1G	First Generation
2G	Second Generation
3G	Third Generation
4G	Fourth Generation
ΣΔ	Sigma-Delta
ΣΔM	Sigma-Delta Modulator
A/D, A-to-D	Analog-to-Digital
AAF	Anti-Aliasing Filter
AC	Alternating Current
ADC	Analog-to-Digital Converter
AMPS	Advanced Mobile Phone System
B3G	Beyond 3G
BE	Backward Euler
BPSK	Binary Phase Shift Keying
BT	Bluetooth
BW	Bandwidth
CDMA	Code Division Multiple Access
CMFB	Common-Mode Feedback
CMOS	Complementary Metal Oxide Semiconductor
CPE	Customer Premises Equipment
CPU	Central Processing Unit
CQFP	Ceramic Quad Flat Pack
CR	Cognitive Radio
CT	Continuous-Time
D/A, D-to-A	Digital-to-Analog
DAC	Digital-to-Analog Converter
dB	Decibels
dBFS	Decibels Full Scale
DC	Direct Current
DCL	Digital Cancellation Logic
DCR	Direct Conversion Receiver

DEM	Dynamic Element Matching
DFF	D Flip-Flop
DNL	Differential Non-Linearity
DOR	Digital Output Rate
DR	Dynamic Range
DRC	Design Rule Checking
DSP	Digital Signal Processing
DT	Discrete-Time
DUT	Device Under Test
DVB-H	Digital Video Broadcasting – Handheld
DWA	Data Weighted Averaging
EDGE	Enhanced Data-rates for Global Evolution
ENOB	Effective Number Of Bits
ESD	Electrostatic Discharge
FDD	Frequency Division Duplex
FDMA	Frequency Division Multiple Access
FE	Forward Euler
FFT	Fast Fourier Transform
FIR	Finite Impulse Response
FM	Frequency Modulation
FoM	Figure of Merit
FS	Full Scale
GB	Gain-Bandwidth Product
GMSK	Gaussian Minimum Shift Keying
GPRS	General Packet Radio Service
GPS	Global Positioning System
GSM	Global System for Mobile-Communications
HiperLAN2	High performance radio Local Area Network
HD	Harmonic Distortion
IBE	In-Band Error
IC	Integrated Circuit
IF	Intermediate Frequency
IIR	Infinite Impulse Response
ILA	Individual Level Averaging
INL	Integral Non-Linearity
IO	Input/Output
LNA	Low-Noise Amplifier
LO	Local Oscillator
LPF	Low-Pass Filter
LSB	Least Significant Bit
LTE	Long Term Evolution
MASH	Multi-Stage Noise Shaping
MiM	Metal-insulator-Metal
MIMO	Multiple Input Multiple Output

MoM	Metal-oxide-Metal
MOS	Metal-Oxide-Semiconductor
MOSFET	Metal-Oxide-Semiconductor Field Effect Transistor
MSB	Most Significant Bit
NMOS	N-channel MOS
NTF	Noise Transfer Function
OFDM	Orthogonal Frequency Division Multiplexing
Opamp	Operational Amplifier
OS	Output Swing
OSR	Oversampling Ratio
OTA	Operational Transconductance Amplifier
PCB	Printed Circuit Board
PDF	Probability Density Function
PDM	Pulse-Density Modulated
PGA	Programmable Gain Amplifier
PLL	Phase-Locked Loop
PMOS	P-channel MOS
PSD	Power Spectral Density
QAM	Quadrature Amplitude Modulation
QFP	Quad Flat Pack
QPSK	Quadrature Phase Shift Keying
QoS	Quality of Service
RAM	Random Access Memory
RF	Radio Frequency
ROM	Read-Only Memory
S/H	Sample-and-Hold
SAW	Surface Acoustic Wave
SC	Switched-Capacitor
SDR	Software Defined Radio
SFDR	Spurious-Free Dynamic Range
SI	Switched-Current
SMASH	Sturdy Multi-Stage Noise Shaping
SNR	Signal-to-Noise Ratio
SNDR	Signal-to-(Noise+Distortion) Ratio
SPR	Serial-to-Parallel Register
SoC	System-on-Chip
SR	Slew-Rate
STF	Signal Transfer Function
TDMA	Time Division Multiple Access
THD	Total Harmonic Distortion
UMTS	Universal Mobile Telecommunications System
USTF	Unity Signal Transfer Function
UWB	Ultra-Wide Band
VLSI	Very Large Scale of Integration

Vpd	Differential peak-to-peak Voltage
Vpp	Peak-to-peak Voltage
WCDMA	Wideband Code Division Multiple Access
WiMAX	Worldwide interoperability for Microwave Access
WLAN	Wireless Local Area Network
WMAN	Wireless Metropolitan Area Network

Chapter 1
Introduction

SINCE THE INFORMATION TECHNOLOGY (IT) AGE BEGAN AROUND 25 YEARS AGO with the commercialization of first personal computers, a number of digital electronic devices (audio/video players, cameras, cell phones, etc) have transformed the lifestyle of billions of people all over the world. The majority of these devices—empowered by microelectronic technology innovations—have been gradually connected to the Internet, thus giving rise to what has been called the first wave of the digital revolution. In the first decade of twenty-first century, Internet-based devices have experienced an explosive growth with the continuous increase of both wired and wireless broadband access services, thus enabling the proliferation of interactive social-networking activities, more and more multimedia contents, as well as new personalized on-line experiences on a single mobile hand-held terminal. All this has lead to the run up of the age of the second wave of the digital revolution [Lim08].

This trend is driven by increasingly sophisticated user demands, prompting mobile terminals towards ubiquitous connectivity, with more and more capabilities, reduced cost and increased battery duration. Moreover, it is expected that new generations of wireless communication technologies will go beyond a simple linear extension of the capabilities of previous ones, not only improving their performance but also enabling the convergence with older standards. The goal is to provide an optimum delivery via the most appropriate network available, allowing smoothly transitions between different networks [Evan00, Hui03, Tach03, Deso06, Frat06, Stee07]. The next generations of wireless handsets must therefore be able to communicate on many radio systems, with different carrier frequencies, bandwidths, multiple access techniques, duplex methods, number of user channels, modulation schemes, data rates, etc [Rosa09].

Although the most advanced smart phones already support a large number of standards and applications, the current addition mechanism of new operating modes and services do not scale. Indeed, each new standard usually requires a dedicated radio frequency (RF) and baseband chipset, which is assembled using ultra-dense packaging techniques, so that the physical volume of every new mobile generation either remains constant or even shrinks. Thus, today's cellphones are marvels of miniaturization but fundamentally, they are still just plain old radios [Rube07]. Even

A. Morgado et al., *Nanometer CMOS Sigma-Delta Modulators for Software Defined Radio*,
DOI 10.1007/978-1-4614-0037-0_1,

though this trend turns into a significant increase in component count and bill of materials (BoM), the final product price can be kept or even reduced, partially thanks to the huge number of terminals sold [Abid07, Gian09]. Indeed, the rate at which new functionalities are introduced in a mobile terminal will soon exceed the rate of miniaturization in packaging. Therefore, addressing this challenge implies redefining the concept of wireless hand-held terminals, evolving from a pure hardware-based to a combination of hardware- and software-based radio [Abid07, Rube07, Gian09, Kits09, Ru09]. As first envisaged by Mitola [Mito95], an ideal software-defined-radio (SDR) should be a universal radio platform which can be programmed to steer any band, tune a channel of any bandwidth, and receive any modulation, all with reasonable constraints, while ensuring the required quality of service as well as guaranteeing privacy and security [Abid07].

In addition to allowing mobile terminals to easily accommodate emerging standards via either software or firmware upgrade [Ru09], the implementation of the SDR concept will enable the so-called cognitive radio (CR) paradigm [Mito95, Cabr06, Rube07]. Essentially, CR-based technology enables wireless networks and hand-held terminals to use the RF spectrum in a dynamic manner—instead of a fixed manner as it is used today. As a result, a more efficient use of the licensed/unlicensed spectrum is achieved, with reduced interferences and/or at a lower power consumption. Indeed, in the near future, SDR-based mobile terminals implementing CR technology will be capable of dynamically sensing their spectral environment and of exploiting the captured information to change their transmission/reception parameters in order to improve the communication link and to reduce the energy consumed on the fly [Rube07]. CR mobile phones must therefore be smart enough to incorporate cognitive (spectrum sensing) capabilities and flexible enough to be dynamically programmed according to the information obtained from their interaction with the environment [Cabr06]. This approach is completely different from the fixed spectrum assignment policy followed by today's mobile telecom systems, in which a large portion (around 85%) of the assigned radio spectrum is used sporadically while the remaining bands are truly busy at any given time. It is important to mention here that reducing power dissipation will also make the use of alternative energy sources possible, thus making solar-cell powered cellphones a reality [Rube07]. By means of embedding additional sensorial capabilities, mobile terminals could also act as healthcare devices, making emergency calls in case they detect problems in the vital signs of the user (patient). Moreover, mobile terminals including cognitive abilities can be used in domotic applications, allowing users to remotely control their home appliances, etc.

However, the current situation in most commercial smart cellphones is far from the universal radio platform defined by the CR/SDR paradigm, where the digitization is close to the antenna and most of the processing is performed by a high-speed general-purpose digital signal processor (DSP). As an illustration, Fig. 1.1 shows the block diagram of an ideal SDR transceiver—as it was originally conceived by Mitola [Mito95]—in which the RF signal coming in from the antenna is directly digitized by an analog-to-digital converter (ADC). Hence, many functions like frequency tuning

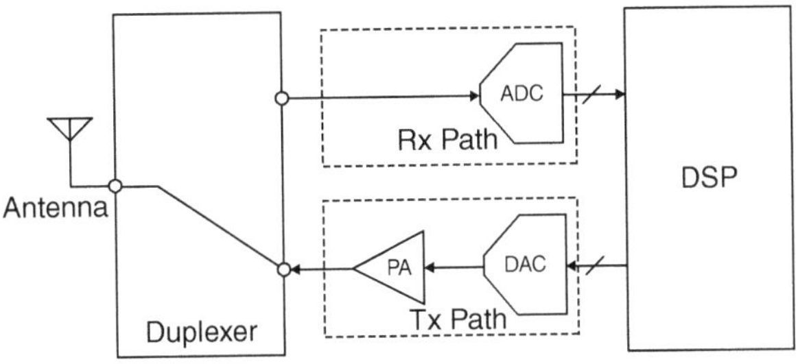

Fig. 1.1 Ideal software-defined radio transceiver as conceived by Mitola. [Mito95]

and translation, filtering, channel selection and demodulation can be implemented by running software on the DSP. A similar reasoning holds in the transmitter side, where a power amplifier (PA) could be combined with a digital-to-analog converter (DAC). Unfortunately, the transceiver in Fig. 1.1 is unrealizable as of today—being even an object of derision by the RF circuit community [Bagh06a, Isma07]—because the required specifications for both the ADC and the DAC are prohibitively stringent. For instance, the receiver would require realizing the A/D conversion of a signal in a frequency band from 800 MHz to more than 10 GHz with an accuracy in excess of 16 bit [Bagh06a]. Hence, a more realistic digital radio transceiver would contain an analog-signal-processing (ASP) section including signal conditioning (frequency translation, amplification and filtering) as illustrated in Fig. 1.2. Although the first steps towards the construction of an SDR are being taken and it will possibly represent the future of universal radio transceivers, the most common situation today consists in a scheme similar to that in Fig. 1.2, where the ASP can be implemented in several ways resulting in different schemes and approaches in the path to SDRs [Abid07, Gian09, Ru09].

The road to future SDR hand-held terminals will require system-on-chip (SoC) transceiver solutions, implemented using mainstream nanometer CMOS processes as the base technologies and embedding digitally-assisted flexible RF and analog front-ends [Ru09]. In this context, the flexibility concept involves the ability for adapting performance metrics and even for dynamically reconfiguring system/circuit topology

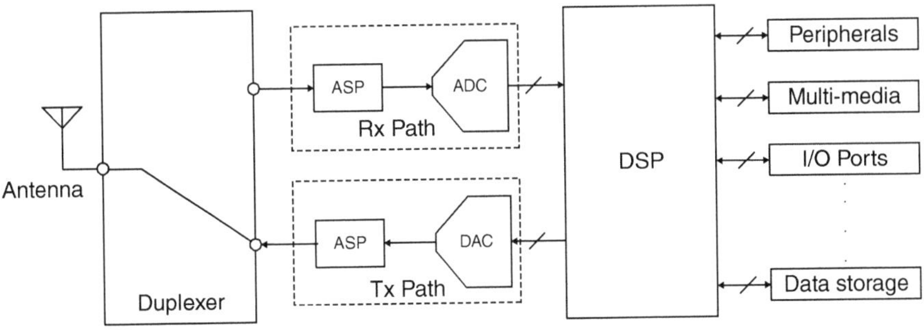

Fig. 1.2 Realistic software-defined radio transceiver

depending on the required transmit/receive parameters [Crom09]. Such universal SoC transceivers are well beyond the state-of-the-art performance. Indeed, most of the reported CMOS multi-mode solutions focus on the selection of a few (two or three) standards or operating modes having similar requirements [Ahol04, Jung04, Dara05, Maed06, Muha06, Behz07, Lin09, Tenb08, Vass08], whereas very little has been really done targeting a really programmable/adaptable analog/RF front-end [Gian09].

In this scenario, this chapter provides an overview of wireless communication systems, their trends and challenges for the implementation of SDR-based transceivers implemented in nanometer CMOS technologies. The chapter is organized as follows. Section 1.1 reviews the different standards and applications that are expected to be covered by the next generation (4G) wireless networks. Section 1.2 discusses different issues related to the design of multi-standard transceiver architectures, their state-of-the-art topologies and the innovations required towards real SDR-based solutions. Section 1.3 moves further down to the circuit level and discusses different design issues of reconfigurable and adaptive building blocks required for the implementation of analog baseband in multi-standard wireless receivers. Especial emphasis is put on the analog-to-digital converter—objective of this book. Finally, Sect. 1.4 gives a survey of the state-of-the-art performance on reconfigurable ADCs.

1.1 Overview of Wireless Standards and Mobile Systems

Although mobile communication is yet a relatively young technology in comparison to other scientific and engineering achievements, its evolution has no precedents. Today, mobile phones are the most used electronics devices in developed countries and the number of mobile subscribers worldwide has grown over 100-fold in the last ten years [Neuv04]. Such is the pervasiveness of mobile communication that it has become a significant contributor to gross national product in the majority of western countries [Evan00], and it is indeed expected to experience the highest growth in the information and communication sector, making ubiquitous communications a reality [Tach03].

Only a few years ago, mobile phones were electronic systems with a single functionality: voice transmission. These systems, based on analog modulation schemes such as AMPS, were called first-generation (1G) cellular or mobile phones. The second generation (2G), based on digital modulation and mainly using the GSM standard, increased their functionality in comparison to 1G enabling the transmission of short message services (SMS), with data rates in the order of tens of kilobits/second (kb/s). In the last few years, with the development of the third generation (3G) of cellular telecommunications, mobile phones have become multimedia electronic devices, combining different functionalities, among others: digital video camera, multimedia players, short-range connectivity (of up to 1 Mb/s using Bluetooth) to other electronic devices and broadband connectivity to the Internet with

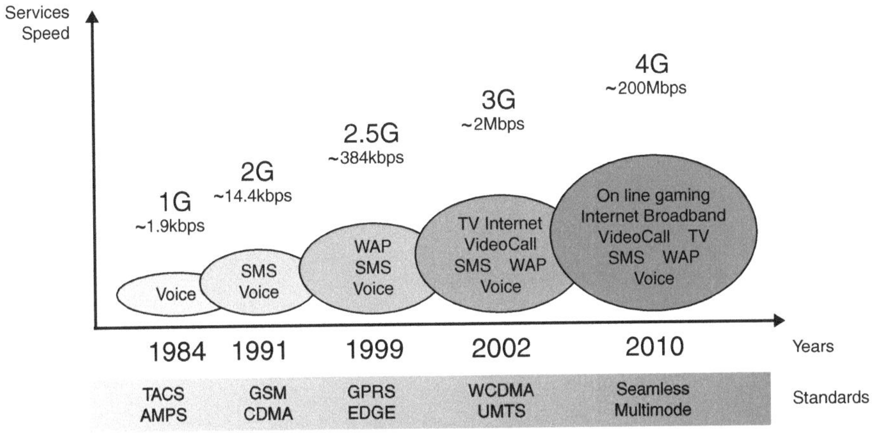

Fig. 1.3 Evolution of wireless applications towards 4G. [Crom09]

data rates of up to 2 Mb/s using the UMTS/WCDMA standard. This trend is foreseen to continue and it is expected that in the next few years the fourth-generation (4G) systems, also named always-best-connected or beyond-3G (B3G) systems, will significantly increase the data rates, exceeding 100 Mb/s point-to-point download transmission speed while mobile and 1 Gb/s when walking and/or stationary. Moreover, as illustrated in Fig. 1.3, 4G systems will foster the convergence of cellular (non-IP-based) wireless networks—like GSM and UMTS (or WCDMA)—with IP-based wireless networks—like WLAN (IEEE 802.11 standard)—providing an optimum data delivery via the most appropriate network available, in a concurrent and always-best-connected mode, and allowing smooth transitions between different networks [Hui03, Gazi05, Deso06, Frat06].

For comparison purposes, Table 1.1 sums up the different generations of wireless telecom by showing a brief description of the services provided by each one, as well as the main characteristics of the standards covered [Li02]. In addition to these standards, other applications like global positioning system (GPS), digital video broadcasting (DVB) and wireless networks covering metropolitan areas (WMAN or WiMAX, based on the IEEE 802.16 standard) and personal areas (WPAN, IEEE 802.15), are expected to be covered by 4G systems. The result is a ubiquitous mobile access to a pool of contents including a huge number of services and applications [Stee07].

1.1.1 Towards 4G Mobile Terminals

As illustrated in Table 1.1, 4G circuits must operate over a variety of standard specifications, including different frequency bands, multiple access techniques, duplex

Table 1.1 Summary of wireless standards and mobile telecom generations

Standard	Description	Access technique	Modulation type	Frequency band Tx/Rx (MHz)	Channel spacing (MHz)	Data rate (Mb/s)
AMPS	Cellular, 1G (analog voice)	FDD	FM	824–849/869–894	0.03	N/A
GSM	Cellular, 2G (digital voice, SMS)	TDMA/FDMA/ FDD	GMSK	890–915/935–960	0.2	0.27
DCS-1800		TDMA	GMSK	1710–1785/1805–1850	0.2	0.27
PCS-1900		TDMA	GMSK	1880–1910/1930–1955	0.2	0.27
IS-95		CDMA	OQPSK	824–849/869–894	1250	1.228
EDGE	Cellular, 2.5G (higher capacity, packetized data)	TDMA	GMSK/ 8-PSK	1850–1910/1930–1990	0.2	0.27
GPRS		TDMA	GMSK	880–915/925–960	0.2	0.27
WCDMA (UMTS)	Cellular, 3G (voice, multimedia, broadband Internet access)	CDMA	QPSK	1920–1980/2110–2170	5	3.84
HSDPA	Cellular, 3.5G (high-speed downlink/uplink packet access)	OFDM/MIMO	QAM	1920–1980/2110–2170	5	14.4
HSUPA		OFDM/MIMO	QAM	1920–1980/2110–2170	5	5.76
Bluetooth (IEEE 802.11 FH)	Wireless Personal Area Networks (WPAN)	CDMA/FH	GFSK	2400–2483	1	1
ZigBee			GFSK	2402–2480	5	0.02, 0.04, 0.250
UWB			SP/FS	3100–10600	500	110–480
IEEE 802.11a	Wireless Local Area Networks (WLAN)	OFDM	BPSK	5000	16.25	6–54
IEEE 802.11b		DSSS	D-BPSK/ D-QPSK	2400	22	1–11
IEEE 802.11g		DSSS	D-BPSK/ D-QPSK	2400	16.25–22	1–54
IEEE 802.11n		OFDM/MIMO	QAM	2400, 5000	22	>100
IEEE 802.16d/e (WiMAX)	Metropolitan Area Networks (WMAN)	OFDM/MIMO	BPSK/QPSK/ QAM	2000–11000	28	>1000 (stationary) >100 (mobile)

methods, number of channels, modulation schemes, data rates, etc. For instance, only considering the GSM standard, there are different operating frequencies, from 890 MHz to 1.9 GHz, depending on the version that is considered; i.e., GSM, EDGE or PCS1900. A similar situation is found in WLAN, where depending on the standard considered (IEEE 802.11 a/b/g/n) the frequency band changes from 2.4 to 5.0 GHz.

The channel bandwidth (BW), which spans from hundreds of kHz to tens or even hundreds of MHz, is directly related to the data rate, as the well-known Shannon's theorem predicts through the expression of the information carrying capacity of a channel as

$$C = \mathrm{BW} \cdot \log_2(1 + \mathrm{SNR}) \tag{1.1}$$

where C is the capacity the capacity in bits per second (b/s), and SNR stands for the signal-to-noise ratio. Therefore, the larger the SNR, the greater the channel capacity will result. Thus, considering a fixed value for BW, Shannon's theorem imposes a requirement for the circuits targeting a set of standard specifications about how strong the information signal must be in comparison to the system noise and also to interference signals in order to achieve the required data rate. In the case of 4G systems, data rates vary from hundreds of kb/s (for instance 270 kb/s in GSM) to hundreds of Mb/s (over 100 Mb/s in WLAN IEEE 802.11n or WiMAX IEEE 802.16d/e).

Channel capacity—C in (1.1)—represents an ideal upper limit on the information carrying capacity of a channel, which has not been reached by practical radios yet, although new modulation and data transmission schemes are enabling a more and more efficient use of bandwidth. Particularly, there are two key technologies that constitute the bases for increasingly data rates of cellular communications and wireless networks. One advance is the orthogonal frequency division multiplexing (OFDM) modulation technique, whereas the other consists of using a multiple-input multiple-output (MIMO) transmission scheme [Stee07]. Indeed, as shown in Table 1.1, these techniques are extensively used by high-speed wireless local area networks (WLAN) and metropolitan (WiMAX) communication standards included in 4G [Boyd06].

The increasing number of standards and applications to be included in 4G systems requires addressing the design of the circuits and systems involved from a different perspective. An evolution from a hardware-based to a software-based concept of wireless systems is needed in order to ensure the required quality of service, reconfigurability, adaptability, as well as privacy and security [Evan00]. Adaptability must be understood in this context so that the system should be aware about the battery level, the power level of the incoming signal or the presence of near interferers in the communication channel, in order to switch between operation modes and/or dynamically adapt to the operating conditions. This will be a major challenge due to the varying bit rates, channel characteristics, bandwidth allocation, and hand-over support among heterogeneous wireless networks. Reconfigurability must be conceived at different levels of the 4G network hierarchy, from protocol stacks to base stations and terminals [Vars01].

Table 1.2 Input-referred receiver requirements for different wireless standards

Specification/Standard	GSM	Bluetooth	UMTS	WLAN
Sensitivity (dBm)	−102	−70	−117	−65
Maximum signal (dBm)	−15	−20	−25	−30
Interferer level (dBm)	−49	−39	−46	−45
Maximum out-band blocker (dBm)	0	−10	−15	0
Maximum in-band blocker (dBm)	−23	–	−44	−30

1.1.2 Circuits and Systems for 4G: Challenges and Innovations

The most efficient implementation of 4G wireless systems implies the use of the so-called multi-standard transceivers, as will be detailed in next section. These multi-standard transceivers must fulfill the performance requirements of each standard separately and in some cases, concurrently [Li02]. For instance, wireless hand-held terminals may use simultaneously Bluetooth and GSM standards in a voice transmission using Bluetooth headphones. At the same time, the mobile terminal can be used to check the e-mail via WLAN/UMTS data network.

The huge number of standard specifications translate into many circuit-level specifications. Unfortunately, documents on the wireless corresponding standard do not give explicit recommendations for the physical realization of transceivers. Instead, a black-box approach is assumed and a set of evaluation tests are outlined to validate the receiver performance in terms of three basic aspects: sensitivity, selectivity and linearity[1]. As an illustration, Table 1.2 sums up the input-referred receiver requirements from these tests for some 4G standards.

Receiver requirements are mapped onto building-block (circuit) specifications (gain, dynamic range, linearity and noise figure) in an iterative synthesis process, generally referred to as receiver planning [Crol97], which will be briefly described in Chap. 4. This process is usually accompanied by a level diagram that shows how the different signals (wanted signals and interferes) evolve along the receiver chain. As an illustration, Table 1.3 shows the specifications[2] for an analog-to-digital converter—main objective of this work—to satisfy the requirements of some wireless standards shown in Table 1.1. It is assumed that the receiver is implemented using a direct conversion architecture[3].

[1] The performance of a radio receiver is mainly characterized by two figures: sensitivity and selectivity. The former measures the ability to detect signals in the absence of any interference other than noise, while the latter characterizes the capacity of the receiver to discriminate between the desired signal and large adjacent-channel interferers. A detailed description of these and other receiver metrics—beyond the scope of this book—can be found in [Raza98, Lee04].

[2] The performance of an ADC can be quantified in terms of two main specifications: effective resolution (measured in bits) and bandwidth of the digitized signal. These terms will explained in detail in Chap. 2.

[3] This is the most common situation in practice, as will be detailed in Sect. 1.2.

Table 1.3 ADC specifications for different wireless standards

Specification/Standard	GSM	Bluetooth	GPS	UMTS	DVB-H	WiMAX
Effective resolution (bits)	12–13	11–12	10–11	9–10	8–9	8–9
Signal bandwidth[a] (MHz)	0.1	0.5	1	2	4	10

[a]A direct conversion or zero-IF receiver would require two ADCs with the enclosed conversion bandwidths, whereas a low-IF receiver would require two ADCs with twice that bandwidths

In order to satisfy the above-mentioned system-level requirements, flexible[4] analog, mixed-signal and RF CMOS circuits and systems are needed. Moreover, the implementation of the SDR paradigm demands for innovative scalable/reusable and programmable solutions at different abstraction levels in the transceiver hierarchy, as well as for keeping the required performance with reduced cost and time-to-market deployment. All these issues translate into a number of significant challenges that can be summarized in the following key points:

- **Incorporation of flexibility capabilities**: As stated above, new wireless standards are complementing (rather than replacing) existing ones. Therefore, as stated above, in order to operate over a variety of different specifications, multi-standard transceivers based on reconfigurable/adaptive building blocks are required. Moreover, SDR-based transceivers are expected to be tuned at run-time to a variable set of receive/transmit parameters [Gian09]. An alternative solution could be using wideband transceivers, although these topologies do not properly filter undesired RF incoming signals and are thus prone to interferences [Sand08]. This limitation is particularly critical for SDR-based transceivers aimed at covering arbitrary frequency bands, where the definition of in-band and out-of-band interferences—and their corresponding RF band-selection filters—may be fuzzy [Sand08, Ru09]. Indeed, the analog front-end in SDR-based transceivers must provide a flexibility comparable to that of the digital back-end, by simplifying the analog pre-processing prior to the ADC as much as possible (see Fig. 1.2) [Bagh06a].
- **Integration in nanometer CMOS technologies**: The integration of more and more parts of telecom systems using nanometer CMOS processes reduces the fabrication cost in comparison to multi-chip platform solutions based on different technologies. Increased integration levels and reduction of fabrication costs pushes the use of standard CMOS technologies whenever possible. However, the benefits of technology scaling for the embedded DSP turns into a number of performance limitations for analog circuit design. In addition to the reduced supply voltage, the degradation of transistor transconductance and output resistance worsens its intrinsic gain. Moreover, threshold voltage variations increase with shrinking technologies [Malo09]. Thus, in the case of 4G circuits, analog circuits must be flexible, but must be also designed to compensate for process variations, with reduced supply voltage (for instance, 1 V in typical 90- and 65-nm CMOS

[4] The term flexibility used in this work is defined as the ability of circuits and systems to reconfigure their topology and to adapt their performance to different specifications.

processes) and consequently dynamic range [Kits09]. Last but not least, the required precision (and resolution) of RF/analog circuits could be severally reduced as a consequence of thermal noise increase with technology scaling, as well as the bad accuracy and linearity of passive elements—particularly problematic when using minimum sizes. All these issues demand for innovative circuit techniques in order to benefit from technology scaling, turning the mentioned limitations into benefits.

- **Need of proper CAD tools and design methodologies**: For a multi-standard hand-held device to be efficient, it should be less expensive than the simple compound of all separate RF chips (normally one per standard) while keeping good performance, lower power dissipation (better portability) and less silicon area (less cost and price) [Giel05]. In order to achieve these objectives, a proper system planning and partitioning should be adopted, supported by proper CAD tools and design methodologies to optimize the system design and to shorten the time-to-marked. In addition, a systematic exploration of different alternative implementations, in terms of architecture selection, reconfiguration strategies and circuit-level implementation is needed in order to optimize the design of multi-standard wireless telecom transceivers [Crol97, Li02].

This book addresses some of the aforementioned challenges, focusing on the systematic design of reconfigurable ADCs for the next generation of SDR-based transceivers. In this context, the rest of this introductory chapter makes a survey of trends and challenges in the cutting-edge performance of the circuits and systems required for the implementation of multi-standard hand-held terminals, putting emphasis on wireless transceiver architectures and required ADCs—main objective of this work.

1.2 Multi-Standard Wireless Transceivers

As discussed above, the wireless communications universe covered by 4G systems calls for the need of developing digital transceiver chips, capable of operating at different modes with multi-standard support features [Li02]. However, as previously stated, the situation today in most commercial mobile terminals is far from this universal radio platform. Indeed, although the number of RF components in mobile handsets has been reduced over the years, each new operation mode and/or application requires its own radio/baseband chipset, which is normally incorporated in the newer generation of the phone by using advanced packaging technologies so that the physical size of the final product is reduced [Bagh06a, Abid07, Isma07]. Nevertheless, the rate at which new applications are introduced together with the need of power/cost-efficient architectures are placing SDRs into the picture. However, we are still far from real and practical SDR-based transceivers as the ones conceptually depicted in Fig. 1.1 and Fig. 1.2.

The majority of reported multi-standard and/or multi-mode transceivers use a single down/up conversion scheme as conceptually depicted in Fig. 1.4 [Li02, Bran05, Mak07]. This architecture eliminates the need for both intermediate frequency (IF)

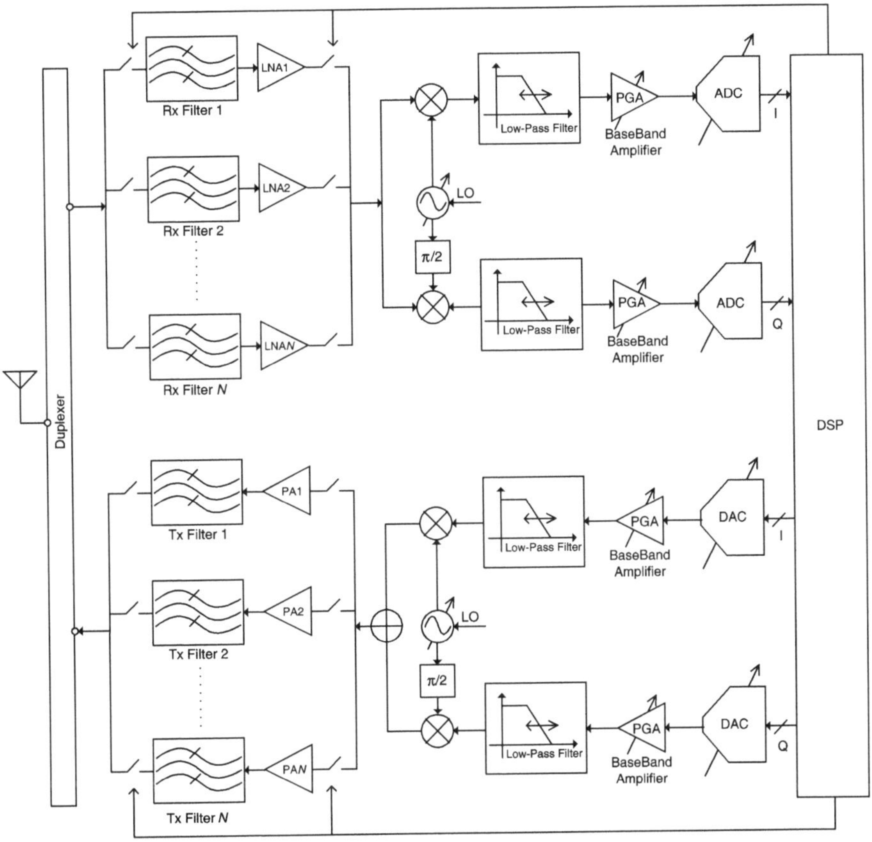

Fig. 1.4 Conventional multi-standard RF transceiver architecture

and image-reject (IR) filtering and requires only a single oscillator and mixer, what makes it very suited for multi-standard applications because it increases the level of integration and facilitates hardware sharing. Moreover, the most common situation is that, in order to cope with the requirements of different standards, separate (switchable) RF front-end paths (normally one per standard) are used, whereas a single, digitally-programmed baseband section (from the mixer to the ADC) is shared by all standards [Bran05].

However, as previously stated, the trend is towards a maximum hardware reuse, by designing as many reconfigurable and adaptive transceiver building blocks as possible [Mak07]. Ideally, an optimized multi-standard transceiver should conceptually look as in Fig. 1.5. This kind of transceivers exploits the benefits of embedding digitally-assisted reconfigurability into the analog/RF front-end building blocks, thus allowing to be adapted to a number of specifications [Bagh06a, Gian09, Ru09]. However, they are relatively complex because of the large number of analog components (and their associated tuning circuitry) which are still needed, what makes these schemes sensitive to the effect of technology parasitics and process variations, as well as more difficult to upgrade by software—one of the benefits of implementing the SDR

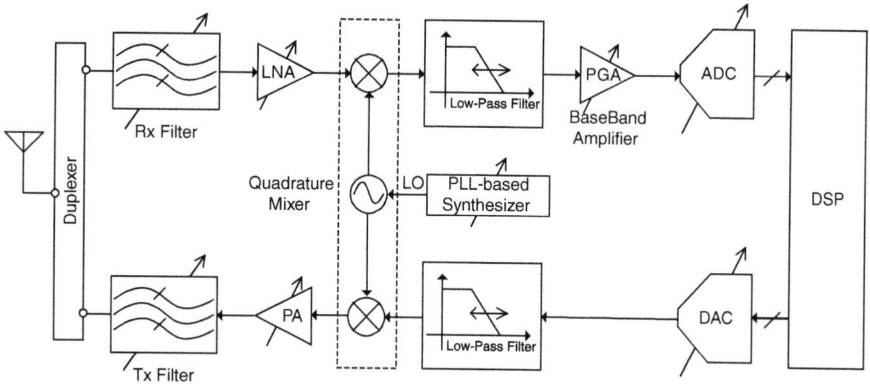

Fig. 1.5 Ideal multi-standard RF transceiver

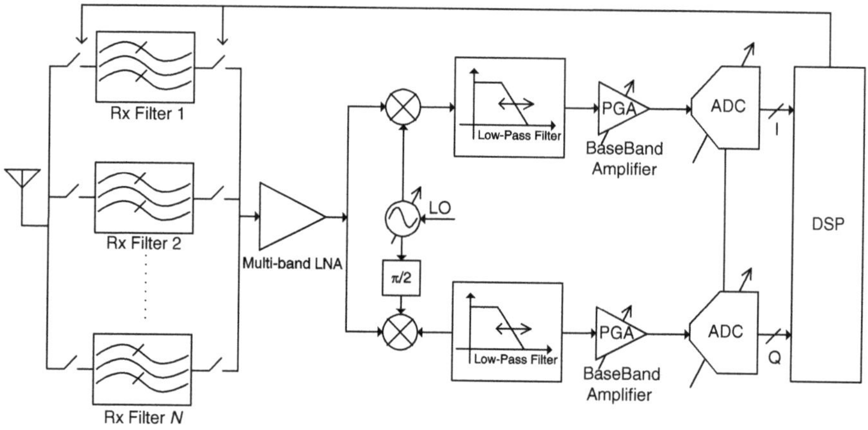

Fig. 1.6 Multi-standard RF receiver sharing a multi-band LNA

paradigm. Moreover, the use of transceiver architectures like the one in Fig. 1.5 for 4G systems present additional disadvantages to be solved. On the one hand, 4G multi-standard transceivers must be able to support different standards concurrently. On the other hand, an important limitation to circuit reuse is the need for multiple off-chip RF SAW filters, which forces to use separate RF front-end paths—made up of a filter and a LNA for each standard—as shown in Fig. 1.4. Another possible implementation is shown in Fig. 1.6, which consists of using separate filters and a single multiple-input multi-band LNA [Bran05].

For the sake of completeness, Table 1.4 shows a summary of the state of the art in CMOS multi-standard RF receivers reported so far[5]. Their main performance characteristics are summarized including, among others, noise figure (NF), third-order

[5] Data have been mainly collected from papers published in the IEEE Journal of Solid-State Circuits and in the IEEE International Solid-State Circuits Conference.

Table 1.4 Summary of the state of the art in CMOS multi-standard wireless receivers

Ref.	Standard	Architecture	NF (dB)	IIP_3 (dBm)	Gain (dB)	Tech.	Supply (V)	Power (mW)
[Agne06]	DCS1800	Low-IF	5.2	−7.5	13.5–28.5	0.13 μm	1.2	20–24
	UMTS	Zero-IF	5.6	0	14.5–29.5			
	802.11b–g	Zero-IF	5.8	−4.8	8.4–23.4			
[Ahol04]	802.11a/b/g	Super-heterodyne	4.9–5.6	−26/−1	8–85	0.18 μm	1.8	207–212
[Bagh06b]	GSM/802.11g	SDR/Zero-IF	5–5.5	−3.5/−4	–	90 nm	2.5/1	25–72
[Cran07]	Cellular/WAN	SDR/Zero-IF	4–8	−9	10–90	0.13 μm	1.2	74.4–144
[Dara05]	Bluetooth	Low-IF	5.5	−8	–	0.35 μm	3	195
	802.11b	Zero-IF						
[Gian09]	GSM/EDGE	Low-IF	2.6–2.4	−15/−5.5	–	45 nm	1.1	60–90
	DVB-H	Zero-IF	2.3	−13/−3				70–100
	WCDMA	Zero-IF	2.4	−18/−6				68–95
	MIMO WiMAX	Zero-IF	3.8	−18/−6				160–210
	MIMO 802.11n	Zero-IF	3.8–6.5	−16/−6				160–230
[Inge10]	–	Zero-IF	2.4	−4.7/−17	73	40 nm	1.1	33–99
	DVB-H		2.6	9/−4.6	77			
	WCDMA		4	−6.3/−11	75			
	LTE		3.4	−3.9/−11	76			
	WiMAX		3.8	−0.7/−12	72			
	–		7.1	−7.5/−15	73			
[Jung04]	Bluetooth	Zero-IF	–	−21	–	0.25 μm	2.7	135
	802.11b		–					175.5
[Maed06]	802.11a/HiperLAN2	Zero-IF	4.4	−2.1	74	0.18 μm	1.8	108
[Rao05]	802.11a/b/g	Super-heterodyne	4.7–5.1	−11.8/−1	7–33	0.18 μm	1.8	8.1

intermodulation intercept point (IIP3) and gain. The receiver architecture used, as well as the standards that are covered, are also included in the table for comparison purposes. Note that most of the reported receivers support multi-mode operation, like for instance WLAN IEEE 802.11a/b/g [Ahol04, Rao05, Maed06], or a couple of standards like Bluetooth and WLAN [Jung04, Dara05]. However, very few integrated circuits (ICs) have been reported that intend to cover more than two different communication standards (cellular and wireless area networks for instance) [Agne06, Gian09, Inge10] and, to the best of the authors' knowledge, no fully-integrated solution has been published in open literature covering the major 4G standards (GSM, UMTS, WLAN and WiMAX)[6]. This is partly due to the huge challenges involved in the design of reconfigurable and adaptive analog/RF front-end circuits.

Indeed, the efficient implementation of CMOS multi-standard wireless transceivers demands for reconfigurable building blocks capable to operate over different co-existing communication protocols, signal conditions and battery status, while minimizing their power dissipation (portability) and silicon area (cost). An overview of the different circuit-level techniques and state-of-the-art performance in CMOS RF front-end circuits is beyond the scope of this work. Instead, the following sections focus on the analog baseband section, putting especial emphasis on the ADC—main objective of the work in this book.

1.3 Flexible CMOS Analog Baseband Circuits for 4G Telecom

The baseband section of a wireless transceiver provides the necessary adaptation, in terms of sensitivity and selectivity, between the RF section (or alternatively the IF section in other types of receiver architectures like superheterodyne or low-IF topologies [Mak07]) and the DSP. Indeed, the term baseband emphasizes the fact that no frequency shifting of the information-bearing signal transferred to or delivered from the DSP block is accomplished at this point and three basic operations are implemented: amplification, filtering and data conversion [Delg03]. These operations are realized with the following building blocks: programmable gain amplifiers (PGAs), low-pass filters (LPFs), ADCs and digital-to-analog converters (DACs). For instance, in the receiver path, PGAs increase the signal amplitude while LPFs reduce the power of out-of-band signals in order to adapt the required dynamic range for the useful signal that is digitized and processed by the DSP.

Multi-standard multi-mode baseband sections must contain building-block circuits which should be very flexible and capable of adapting their performance to the different standard requirements with scalable power consumption [Ryyn06]. This adaptation is performed through a digital control (by the DSP) of biasing, gain programmability, circuit- and stage-level reconfiguration, block power down/up, etc

[6] Note that the transceiver presented in [Inge10] seems to cover the specifications of the major 4G standards from the multi-band LNA to the ADC.

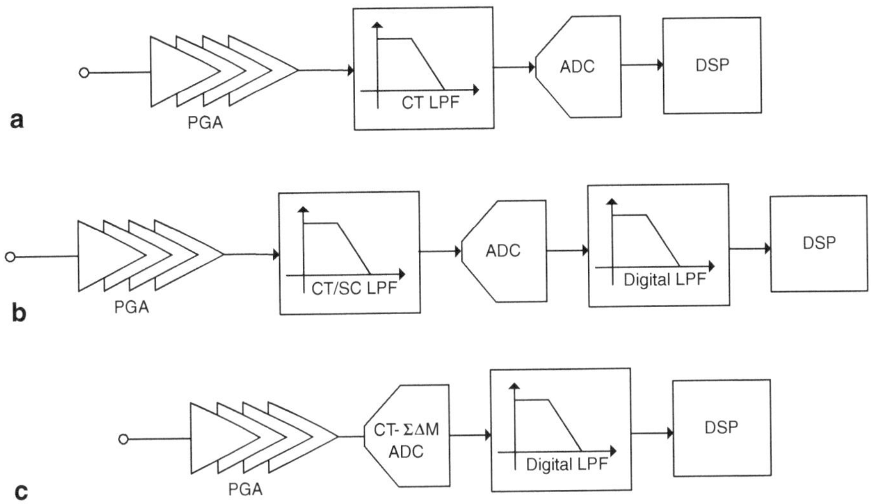

Fig. 1.7 Channel selection approaches: **a** Analog, **b** Mixed-signal, **c** Digital

[Basc06, Gian07]. Moreover, the design of the multi-standard baseband section implies a number of trade-offs between LPF selectivity and circuit complexity [Basc06]. Normally, higher filter selectivity results in a lower number of amplification and filtering stages [Delg03]. Besides, the received channel selection can be implemented with an analog or a digital filter or both. The way in which filtering is shared between the analog and the digital side comprises a number of design trade-offs and challenges. On the one hand, fully digital filtering is desired because it makes the whole transceiver more robust and programmable to different operation modes. On the other hand, the more filtering is translated to the digital side, the more demanding the requirements for the data converters [Ryyn06].

1.3.1 Channel Selection

The main functionality of baseband filters is to perform channel selection, also known as channel isolation. As discussed above, this operation can be made either in the analog domain or in the digital domain or it can be a combination of both. These three main filtering distribution schemes are shown in Fig. 1.7. The way in which this is done directly affects the requirements of the filter and the ADC. For instance, depending on the channel bandwidth, which can vary from hundreds of kHz (200 kHz in GSM) to tens or even hundreds of MHz (100 MHz in IEEE 802.11n), it may be easier to realize an analog filter and PGA than an ADC with comparable dynamic range or vice versa [Delg03].

Figure 1.7a shows a possible implementation of the fully-analog channel selection strategy. This is one of the most common situations in practice, employing continuous-time (CT) filtering in front of the ADC. The role of the filter is two-fold. On the one hand, it provides the required selectivity. On the other, it prevents from aliasing in the ADC sampling process [Holl01]. One of the main issues associated to using CT filters like that in Fig. 1.7a is that their frequency response strongly depends on the absolute values of the passive elements (i.e., capacitors and resistors) and the transconductances involved. These elements are prone to statistical variations that may deviate their nominal values ranging from 10 to 20% in nanometer CMOS technologies. Therefore, in order to compensate for this problem, tuning schemes are normally used [Tsiv92].

An alternative approach that relaxes the problem of using pure CT filters is shown in Fig. 1.7b. In this scheme, switched-capacitor (SC) techniques are combined with CT techniques in order to implement the channel-selection filter. A low-order (typically less than 3) CT anti-aliasing filter (AAF) is followed by an SC filter that samples the incoming signal and rejects adjacent channels [Chan97]. However, this scheme is limited in high-speed applications because of the high sampling frequency required, which usually increases the power dissipation of the implementation.

The problems associated to the schemes in Fig. 1.7a–b can be partially alleviated if a fully-digital filtering scheme is used, as illustrated in Fig. 1.7c. In this case, the channel selection is performed completely in the digital domain, with the subsequent benefits in terms of scalability and programmability. Indeed, if the ADC is based on sigma-delta ($\Sigma\Delta$) modulation[7] some of the functionalities—like amplification and filtering—can be embedded in the ADC, and only a simple AAF filter (typically a 2nd-order passive filter) is needed. Moreover, in case the ADC is implemented using CT circuits, the AAF operation is implicitly implemented by the loop filter of the $\Sigma\Delta$ ADC [Cher00a][8] and the analog baseband can be simplified to just a single building block, the ADC.

1.3.2 Programmable Baseband Filtering

The main design issue associated to multi-mode multi-standard filters is that tunability of the corner frequency must be controlled over many different bandwidths—from kHz to hundreds of MHz (see Table 1.1). Thus, the corner frequency control is

[7] Fundamentals, basic architectures and state-of-the-art performance of $\Sigma\Delta$ modulators will be described in detail in Chap. 2.

[8] Indeed, the use of $\Sigma\Delta$ modulation techniques may benefit from CT circuits to implement CMOS data converters operating within the GHz band [Ryck09] and, what is more important, many radio functions like blocker-rejection filtering, frequency-mixing process, channel-selection, etc, can be embedded within the $\Sigma\Delta$ converter architecture, thus simplifying that of the transceiver (in terms of circuit complexity and number of building blocks), as well as allowing to be adaptable and digitally programmable with high degree of robustness. However, the price to pay is very demanding ADC specifications of linearity, clock jitter and circuit element tolerances and mismatches, apart from their multiple trade-offs with power dissipation [Kim09].

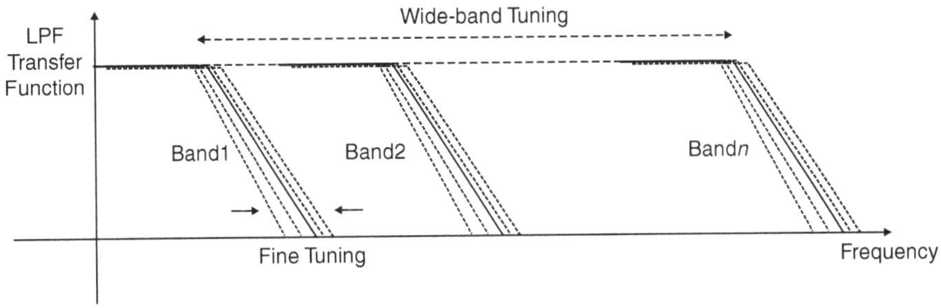

Fig. 1.8 Illustrating the baseband filtering tuning required before the ADC in a multistandard wireless receiver

implemented in practice by discontinuous bands (instead of a continuous tuning) as illustrated in Fig. 1.8 [Ryyn06].

Actually, the most common circuit techniques used to implement programmable/reconfigurable baseband channel-selection filters are active-RC and Gm-C circuit techniques [Basc06, Ryyn06]. Active-RC filters have the advantages of better linearity and larger signal swing, whereas Gm-C filters usually can operate at higher frequencies with less power consumption. The cut-off frequency tuning can be implemented by using switchable paths for rough tuning and element (capacitor, resistor and/or transconductor) variation for fine tuning. The latter implies a trade-off among frequency tuning range, noise and silicon area (cost) [Tsiv92, Ryyn06].

An important observation can be derived from the discussion above and, particularly, from the circuit partitioning conceptually shown in Fig. 1.7. The circuit nature and position of the ADC plays a critical role in the performance of the whole analog baseband in terms of channel selection (selectivity), dynamic range (sensitivity) and power consumption. These figures determine the complete performance of wireless transceivers. This motivates exploring new techniques for the implementation of reconfigurable/adaptive ADCs as the main objective of this book.

1.4 Reconfigurable ADCs for SDR-Based Mobile Terminals

As stated above, one of the most challenging parts of multi-standard mobile terminals is the analog-to-digital interface[9], because of the varying sampling rates and resolutions required to handle the wide range of signals corresponding to each individual operation mode (see for instance Table 1.3). Moreover, the position of the ADC

[9] The analog-to-digital interface is made up of the ADC in the receiver and the DAC in the transmitter. Ideally, these blocks perform complementary data conversions from analog-to-digital and from digital-to-analog sides. However the requirements imposed by many wireless standards to the former are much more tougher to meet [Isma07], making its design much more complicated and critical than in the case of multi-standard DACs, which normally rely on reconfigurable current-steering architectures [Basc06].

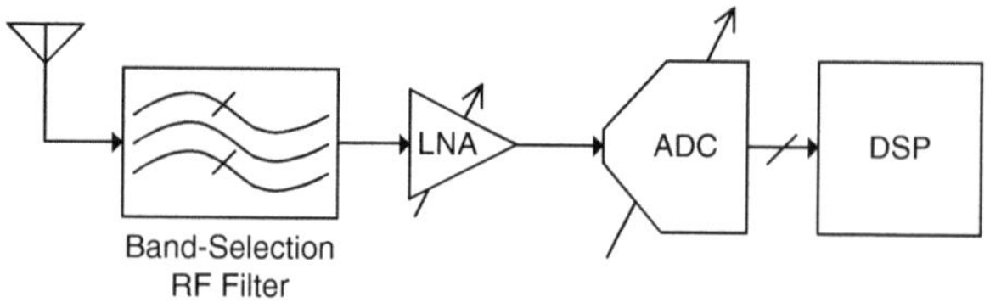

Fig. 1.9 Conceptual scheme of RF digitization

in a receiver has a direct influence on the overall performance in terms of circuit complexity, programmability, power consumption and cost. Thus, depending on the receiver architecture, the ADC has to digitize either intermediate-frequency (IF) or baseband signals [Shi02].

Indeed, a good (and a priori direct) alternative may be pushing the analog-to-digital interface as close as possible to the antenna—an approach commonly referred to as RF digitization, which is conceptually illustrated in Fig. 1.9. As discussed above, this is the evident implementation of an SDR transceiver because the digital-intensive architecture emulates perfectly a true software transceiver, thus incorporating software-upgradable tunability, adaptability, reconfigurability and the capability to simultaneously process multiple channels. Although there have been some attempts to implement such an approach—normally implemented in SiGe (BiCMOS) technologies [Cher00b, Chal07, Than07]—the resulting specifications for the data converters lead to unfeasible power-hungry solutions for practical mobile handsets in which reduced power consumption is a must. However, a number of CMOS $\Sigma\Delta$ ICs have been recently reported based on digitizing only the desired channel, which may result very promising candidates for the implementation of simplified mainly-digital RF transceivers [Yama08, Ryck09, Lu10][10]. In spite of the potential benefits of RF digitization, due to the still very demanding power consumption, the majority of reported multi-standard CMOS ADCs perform a baseband (BB) A/D conversion, whereas a few ICs has been reported using either an IF A/D conversion (also named band-pass ADC) and/or an IF-to-BB A/D conversion [Jant97, Salo03, Veld03, Silv07].

This work focuses on baseband ADCs implemented using $\Sigma\Delta$ modulation techniques. Compared to other ADC techniques, $\Sigma\Delta$ modulators ($\Sigma\Delta$Ms) are very suited for the implementation of reconfigurable ADCs in highly integrated multi-standard direct-conversion transceivers using low-cost, digitally-oriented nanometer CMOS technologies [Mill03, Veld03, Chri07, Ouzo07, Putt07, Bos10, Crom10]. As will be detailed in Chap. 2, this type of ADCs combines redundant temporal data (oversampling) to reduce the quantization noise and filtering (noise shaping) to push this

[10] A good example of the RF-to-digital conversion approach has been recently published in [Ryck09]. This ADC, consisting on a band-pass CMOS CT $\Sigma\Delta$ converter, uses the so-called undersampling technique [Nade08] to digitize 2.4-GHz centered signals using a sampling frequency of 3 GHz. The operation frequency is tuned by varying tank capacitors. Recently, the concept of IF digitization using band-pass SC $\Sigma\Delta$ ADCs has been also successfully demonstrated in [Yama08] and [Lu10], the former being able to program the carrier frequency from DC to 1/3 of the sampling frequency, f_s.

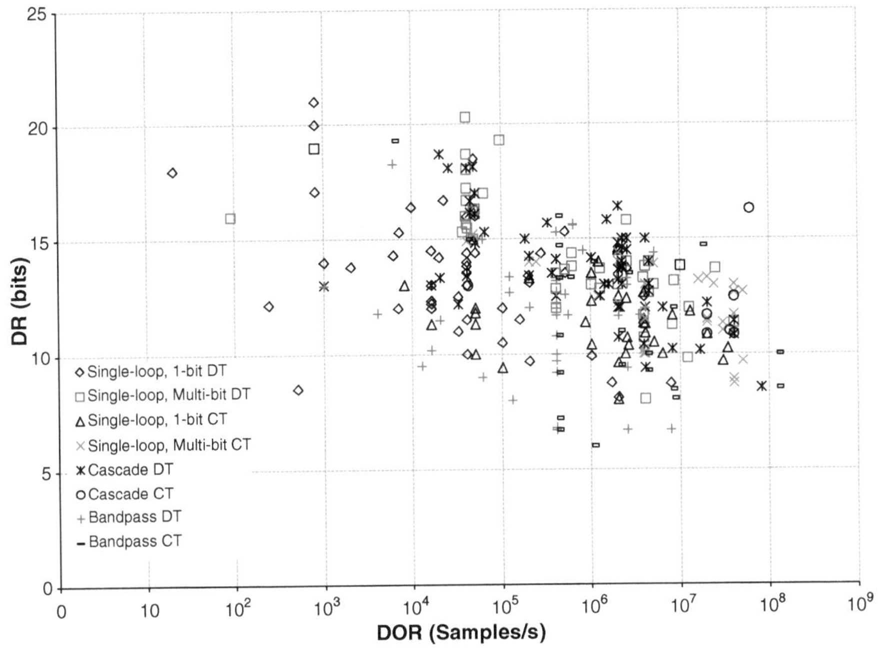

Fig. 1.10 State-of-the-art CMOS ΣΔMs in the dynamic range-vs.-digital output rate plane

noise out of the signal band [Nors97a, Mede99, Geer02, Schr05b, Ortm06]. On the one hand, the use of these analog signal-processing strategies results in high-performance, robust ADCs, which have lower sensitivity to circuitry imperfections than Nyquist-rate ADCs[11], thus making easier to include reconfigurability and programmability functions without significant performance degradation. On the other hand, ΣΔMs trade analog accuracy for signal processing, thus facilitating their integration in nanometer (analog hostile) CMOS technologies, more suited to implement fast digital circuits than precise analog functions.

These reasons have prompted the use of ΣΔMs in a number of applications—going from instrumentation to communications [Nors97a, Mede99, Geer02, Schr05b, Ortm06]. As an illustration, Fig. 1.10 places the state-of-the-art ΣΔ ADCs—which will be described in Chap. 2—in the resolution-bandwidth plane[12]. It can be noted that a wide conversion region is covered, ranging over seven decades in frequency and more than 15 bit in effective resolution. This feature motivates the use

---

[11] According to the Nyquist theorem, the minimum value of the sampling frequency, f_s—often referred to as Nyquist frequency and represented by $f_N \equiv 2\text{BW}$—must be twice the signal bandwidth; i.e., $f_N = 2\text{BW}$. Based on this criterion, those ADCs with a sampling frequency, $f_s = f_N$ are called Nyquist-rate ADCs, while if $f_s > f_N$, the resulting ADCs are known as oversampling ADCs, and $\text{OSR} \equiv f_s/f_N$ is defined as the oversampling ratio.

[12] The data in this figure are classified according to the main types of ΣΔMs. This will be described in detail in Chap. 2.

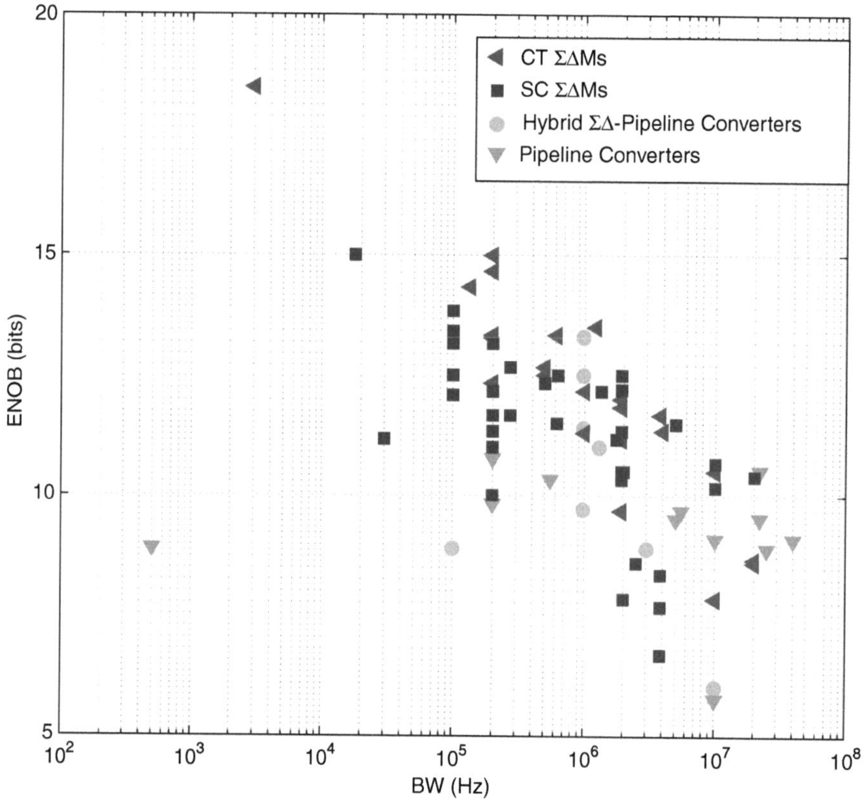

Fig. 1.11 State-of-the-art reconfigurable ADCs for multi-standard applications

of this data conversion technique for the implementation of reconfigurable ADCs in multi-standard telecom systems, where a number of different set of specifications; i.e., effective resolutions and signal bandwidths, must be covered by a single IC.

The above-mentioned advantages have motivated the use of ΣΔMs in the majority of reconfigurable ADCs intended for multi-standard wireless telecom reported so far [Burg01, Dezz03, Salo03, Aria06, Chri07, Ouzo07, Bos10, Crom10, Chri10]. Fig. 1.11 depicts the resolution-bandwidth plane of state-of-the-art reconfigurable ADCs employed for multi-mode multi-standard applications. Fig. 1.11 shows that reconfigurable ADCs based on ΣΔ modulation are widely used and that they achieve the highest resolutions over a range of 3 decades in signal bandwidths. Note from the data in Fig. 1.11 that there is a range of more than 8 bits in the effective number of bits (ENOB) covered by the reconfigurable multi-mode ΣΔMs as will be detailed in this work.

This type of reconfigurable ADC can easily implement a combination of different architectural- and circuit-level strategies to increase the programmability and adaptability of the ADC performance to a wide number of specifications with large hardware reuse and power consumption scalability.

The number of applications is continuously increasing and it is expected that the conversion region covered by reconfigurable ADCs will be larger and larger according to the requirements demanded by next generations of wireless telecom systems.

Chapter 2
$\Sigma\Delta$ ADCs: Basic Concepts, Topologies and State of the Art

SIGMA-DELTA MODULATION HAS DEMONSTRATED TO BE VERY SUITED FOR the implementation of Analog-to-Digital (A/D) interfaces in many different electronic systems, covering a large number of applications from instrumentation to telecom [Nors97a, Mede99, Geer02, Schr05b, Ortm06]. This type of A/D converters (ADCs), composed of a low-resolution quantizer embedded in a negative feedback loop, uses oversampling (a sampling frequency much larger than the Nyquist frequency) to reduce the quantization error and $\Sigma\Delta$ modulation to push this noise out of the signal band [Inos62]. The combined use of redundant temporal data (oversampling) and filtering (noise shaping) results in high-resolution, robust ADCs which have lower sensitivity to circuit parasitics and tolerances, and are more suited for their implementation in modern standard CMOS technologies [Rodr03, Schr05b].

The aim of this chapter is to introduce the foundations of $\Sigma\Delta$ ADCs. The fundamentals of sampling and quantization processes—inherent to the A/D conversion—are revised in Sect. 2.1 for the purpose of juxtaposing the operation principles of $\Sigma\Delta$ modulation (oversampling and quantization noise shaping). Sect. 2.2 presents the basic scheme of a $\Sigma\Delta$ ADC, together with its ideal behavior and a definition of its most important performance metrics. A classification of practical implementations of $\Sigma\Delta$ modulators is then presented in Sect. 2.3. Sect. 2.4 describes single-loop $\Sigma\Delta$ architectures, discussing stability issues and practical topologies to implement high-order loops. Cascade $\Sigma\Delta$ architectures are presented in Sect. 2.5 as an architectural alternative to implement unconditionally-stable high-order $\Sigma\Delta$ modulators. Sect. 2.6 revises $\Sigma\Delta$ modulators with multi-bit embedded quantizers, analyzing their pros and providing techniques to circumvent their impact on the modulator linearity, such as dual-quantization or dynamic element matching techniques. Finally, the state of the art in $\Sigma\Delta$ ADCs is reviewed from integrated circuit implementations reported in open literature.

A. Morgado et al., *Nanometer CMOS Sigma-Delta Modulators for Software Defined Radio*,
DOI 10.1007/978-1-4614-0037-0_2,

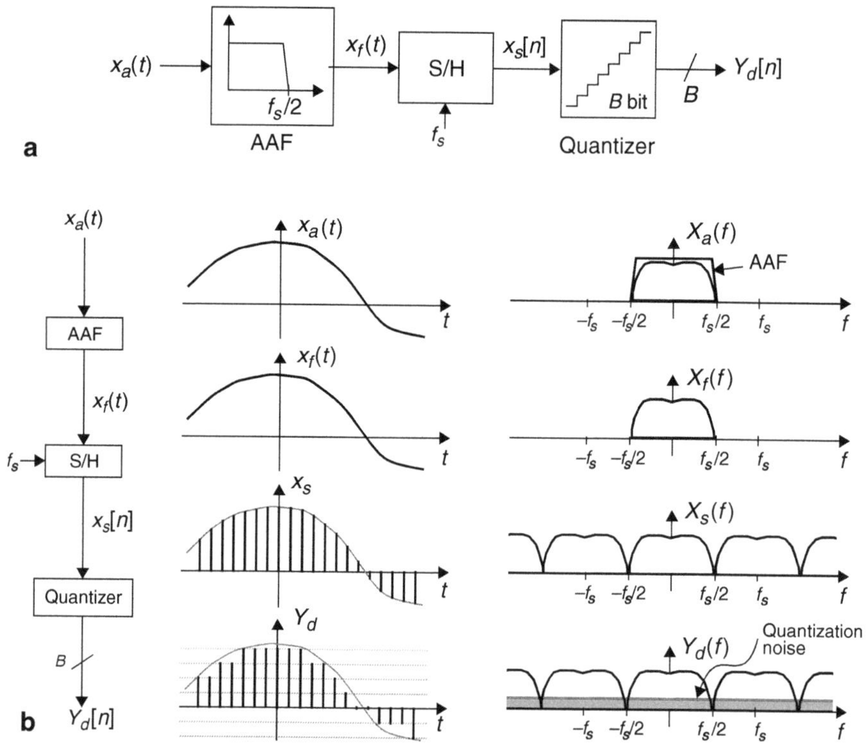

Fig. 2.1 Analog-to-digital converter: **a** generic scheme, **b** signal processing involved

2.1 Fundamentals of the A/D Conversion

An ADC is a system that transforms signals which are continuous in time and amplitude (analog signals) into signals which are discrete in time and amplitude (digital signals). Fig. 2.1a shows the generic scheme of an ADC intended for the conversion of low-pass signals that includes the following blocks: an anti-aliasing filter (AAF), a sampling-and-hold circuit (S/H) and a quantizer. The signal processing involved in the operation of the ADC blocks is illustrated in Fig. 2.1b both in time and frequency domains. First, the analog input signal $x_a(t)$ of the ADC passes through the AAF in order to remove spectral components above one half of the sampling frequency f_s of the S/H. Otherwise, according to the Nyquist theorem, out-of-band components would be folded back into the signal band during the subsequent sampling process, therefore corrupting the signal information. The resulting band-limited signal $x_f[t]$ is then sampled at a rate f_s by the S/H, yielding a discrete-time signal $x_s[n] = x_f[nT_s]$, where $T_s = 1/f_s$. Finally, the quantizer maps the range of amplitudes of $x_s[n]$ into a discrete set of levels using B bits; i.e., each sample of the continuous-valued input is coded onto the closer discrete-valued level out of the 2^B levels that cover the variation interval of the input signal. This process yields the converter digital output $Y_d[n]$.

The fundamental operations involved in the A/D conversion are sampling and quantization, as illustrated in Fig. 2.1. On the one hand, the sampling process performs the continuous-to-discrete conversion of the input signal in the time domain. On the other, the quantization process performs the continuous-to-discrete conversion of the input signal in amplitude. These two transformations inherently impose limitations to the performance of an ADC, even if its implemented using ideal components.

2.1.1 Sampling

Sampling imposes a limit on the bandwidth of the analog input signal. According to the Nyquist theorem, the minimum frequency f_s required for sampling a signal with no loss of information is twice the signal bandwidth, BW; i.e., $f_N = 2\text{BW}$, also called the Nyquist frequency. Based on this criterion, the ADCs in which the analog input signal is sampled at minimum rate ($f_s = f_N$) are called Nyquist ADCs. Fig. 2.1b illustrated the signal processing involved assuming a Nyquist ADC. Since the input signal bandwidth equals $f_s/2$, aliasing will occur if $x_a(t)$ contains frequency components above $f_s/2$. High-order AAFs are therefore required in Nyquist ADCs in order to implement a sharp transition band and remove the out-of-band components with no significant attenuation of the signal band.

2.1.2 Quantization

Quantization also limits the performance of an ideal ADC, since the process itself of mapping continuous-valued levels into a set of discrete levels degrades the quality of the input signal. In the process an error is generated, called quantization error.

This process is illustrated in Fig. 2.2 for the case of a 3-bit ideal quantizer ($B = 3$). As the input signal changes from x_{min} to x_{max}, it is 'rounded' to one out of the eight (2^B) discrete levels. For a B-bit quantizer, the separation between adjacent levels is defined by the quantization step Δ as

$$\Delta = \frac{X_{\text{FS}}}{(2^B - 1)} \qquad (2.1)$$

with X_{FS} being the quantizer full scale.

The quantizer operation can be therefore described mathematically by a linear model

$$y = g_q x + e(x) \qquad (2.2)$$

where g_q stands for the quantizer gain—the slope of the line intersecting the code transitions—and $e(x)$ stands for the quantization error. This error is a non-linear

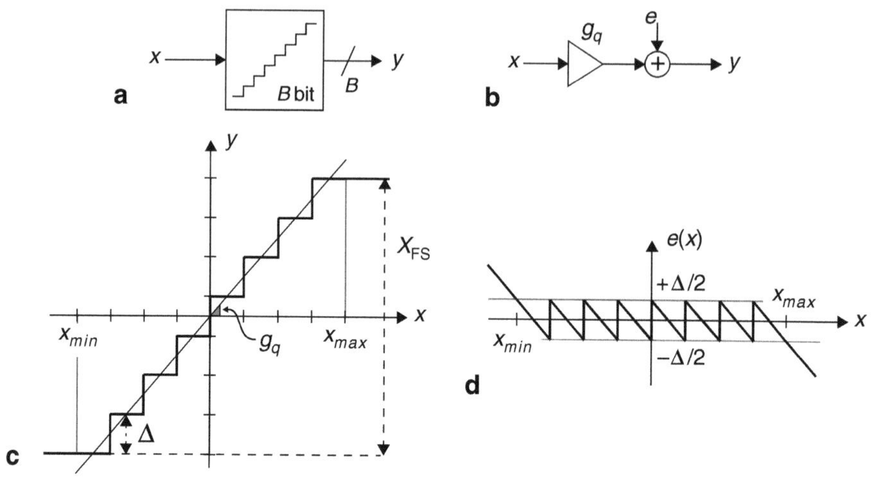

Fig. 2.2 Ideal quantization process: **a** symbolic representation, **b** linear model of an ideal quantizer, **c** ideal transfer characteristic, **d** quantization error of a 3-bit quantizer

function of the input x, as shown in Fig. 2.2d. Note that, if x is confined in the interval $[x_{min}, x_{max}]$, the quantization error is bounded by $\pm\,\Delta/2$. For inputs outside that interval, the absolute value of the quantizer error grows monotonically. This situation is known as quantizer overload.

2.1.3 *White Noise Approximation of Quantization Error*

In order to evaluate the performance of an ideal quantizer, some assumptions are usually made on the statistical properties of the quantization error which are collectively called the 'additive white noise approximation'. As shown in Fig. 2.2d, the quantization error is strongly dependent on the input signal value. Nevertheless, if x is assumed to change randomly from sample to sample in the interval $[x_{min}, x_{max}]$ and the number of levels in the quantizer is large, the quantization error can be assumed uncorrelated from sample to sample. Quantization can therefore be viewed as a random process, with the quantization error exhibiting a uniform probability density function (PDF) in the range $[-\Delta/2, +\Delta/2]$ [Benn48, Srip77]. The power associated to the quantization error is then

$$\sigma^2(e) = \int_{-\infty}^{+\infty} e^2 \mathrm{PDF}(e)de = \int_{-\Delta/2}^{+\Delta/2} e^2 \frac{1}{\Delta} de = \frac{\Delta^2}{12} \tag{2.3}$$

This power will be uniformly distributed in the band $[-f_s/2, +f_s/2]$ as the quantized signal is sampled at rate f_s and the quantization error will therefore exhibit a constant

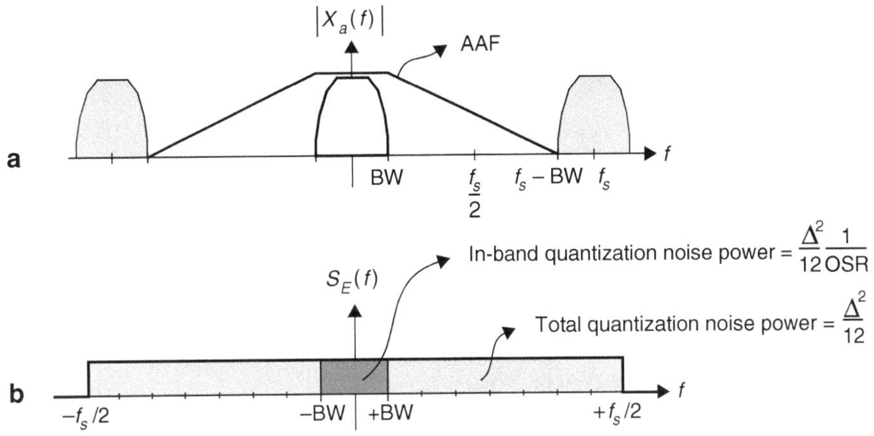

Fig. 2.3 Benefits of oversampling on the: **a** AAF, **b** in-band quantization noise

power spectral density (PSD) in that frequency interval

$$S_E(f) = \frac{\sigma^2(e)}{f_s} = \frac{\Delta^2}{12f_s} \tag{2.4}$$

The degradation introduced by the quantizer in the performance of an ADC can be expressed through the in-band quantization 'noise' power P_Q, calculated as

$$P_Q = \int_{-\mathrm{BW}}^{+\mathrm{BW}} S_E(f)df = \frac{\Delta^2}{12} \tag{2.5}$$

Note that f_s equals 2BW in a Nyquist ADC and all the quantization noise power therefore falls inside the signal band and passes to the ADC output as a part of the signal itself.

2.1.4 *Oversampling*

Oversampling consists in sampling a signal faster than the minimum rate imposed by the Nyquist theorem to avoid aliasing. How much faster than required the signal is sampled is expressed through the oversampling ratio, defined as OSR $= f_s/(2\mathrm{BW})$.

Oversampling has two noticeable effects in an ADC. First, as illustrated in Fig. 2.3a, since f_s is larger than the Nyquist rate, the images of the input created by the sampling process are more separated than in a Nyquist ADC. Spectral components of the input signal in the range $[\mathrm{BW}, f_s - \mathrm{BW}]$ do not alias within the signal band and, consequently, the transition band of the AAF can be smoother in an oversampling ADC, what greatly simplifies its design. Second, as illustrated in Fig. 2.3b, when an oversampled signal is quantized, the quantization noise is uniformly distributed in the range $[-f_s/2, +f_s/2]$ and only a fraction of the total power lays within the signal band. The in-band quantization noise power can therefore be calculated as

$$P_Q = \int_{-\mathrm{BW}}^{+\mathrm{BW}} S_E(f)df = \int_{-\mathrm{BW}}^{+\mathrm{BW}} \frac{\Delta^2}{12f_s}df = \frac{\Delta^2}{12\mathrm{OSR}} \tag{2.6}$$

and decreases with OSR at a rate of 3 dB/octave.

2.1.5 Quantization Noise Shaping

The quantization noise power within the signal band can be further decreased through the processing of the quantization error. Let us consider the quantization of an oversampled signal. If OSR is large, the input signal value will only slightly change from one sample to another and most of the changes in the quantization error will occur at high frequencies—i.e., low-frequency components of consecutive samples of the quantization error $e[n]$ will be similar. Hence, low-frequency in-band components of the quantization error can be attenuated by subtracting the previous sample from the current one

$$e_{\mathrm{HPF}}[n] = e[n] - e[n-1] \tag{2.7}$$

or further reduced by involving a larger number of previous samples in the error processing

$$\begin{array}{ll} e_{\mathrm{HPF},1}[n] = e[n] - e[n-1], & \text{1st} - \text{order error processing} \\ e_{\mathrm{HPF},2}[n] = e[n] - 2e[n-1] + e[n-2], & \text{2nd} - \text{order error processing} \\ e_{\mathrm{HPF},3}[n] = e[n] - 3e[n-1] + 3e[n-2] - e[n-3], & \text{3rd} - \text{order error processing} \\ \ldots & \end{array} \tag{2.8}$$

This procedure can be formulated in an unified manner in Z-domain as

$$E_{\mathrm{HPF},\,L}(z) = (1 - z^{-1})^L \cdot E(z) \tag{2.9}$$

indicating that the processed error is a (high-pass) filtered version of the original. The filtering transfer function on the quantization error due to this processing, called noise transfer function (NTF), is therefore obtained as

$$\mathrm{NTF}(z) = (1 - z^{-1})^L \tag{2.10}$$

where L denotes the order of the filtering. If OSR is large enough, NTF takes small values within the signal band and can be approximated to

$$\left|\mathrm{NTF}\left(e^{j2\pi f/f_s}\right)\right|^2 = \left|1 - e^{-j2\pi f/f_s}\right|^{2L} = 2^{2L}\sin^{2L}\left(\frac{\pi f}{f_s}\right) \approx 2^{2L}\left(\frac{\pi f}{f_s}\right)^{2L} \tag{2.11}$$

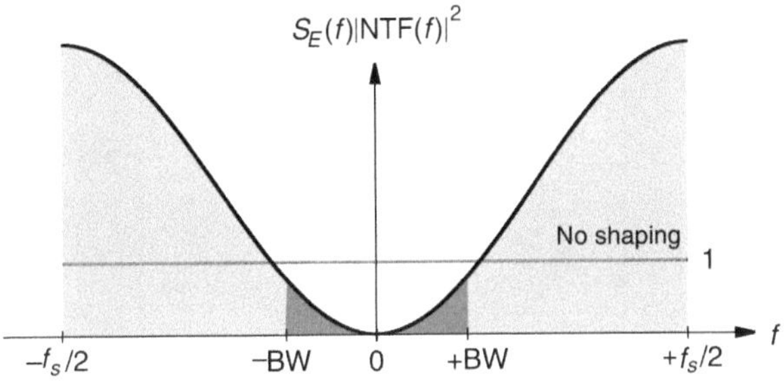

Fig. 2.4 Illustration of the quantization noise shaping

Hence, the in-band power of the filtered quantization error results in

$$P_Q = \int_{-\mathrm{BW}}^{+\mathrm{BW}} S_E(f)|\mathrm{NTF}(f)|^2 df \approx \frac{\Delta^2}{12} \cdot \frac{\pi^{2L}}{(2L+1)\mathrm{OSR}^{(2L+1)}} \tag{2.12}$$

that is much smaller than only applying oversampling. The resulting error reduction is illustrated in Fig. 2.4.

2.2 Basics of ΣΔ A/D Converters

In contrast to Nyquist ADCs, oversampling sigma-delta ADCs—usually referred to as ΣΔ ADCs—make use of oversampling and noise shaping to decrease the quantization noise power within the signal band and increase the accuracy of the A-to-D conversion. Fig. 2.5 illustrates the basic scheme of a ΣΔ ADC, as well as the signal processing involved. As shown, a ΣΔ converter comprises three main blocks:

- **Anti-Aliasing Filter (AAF).** Its function is the same as in Nyquist ADCs; i.e., to band limit the input signal in order to avoid aliasing during sampling. As stated above, oversampling considerably relaxes the attenuation requirements for this analog filter and smooth transition bands are sufficient (see Fig. 2.3a).
- **Sigma-Delta Modulator (ΣΔM).** It simultaneously performs the oversampling and quantization of the band-limited input signal. Quantization error is also high-pass filtered by means of a given noise-shaping technique. This is accomplished by placing an appropriate loop filter $H(z)$ before a low-resolution quantizer and closing a negative feedback loop around them. The in-band quantization noise is therefore greatly decreased in comparison to that of the embedded quantizer. The output of the ΣΔM is a B-bit digital stream at f_s sampling rate.
- **Decimator.** It reduces the rate of the ΣΔM output stream down to the Nyquist rate. Jointly, the word length increases from B to N in order to preserve resolution as the word rate decreases. Although the block scheme of a decimator may differ in practice from that illustrated in Fig. 2.5, it conceptually consists of a high-selectivity digital filter and a downsampler. Frequency components of the stream

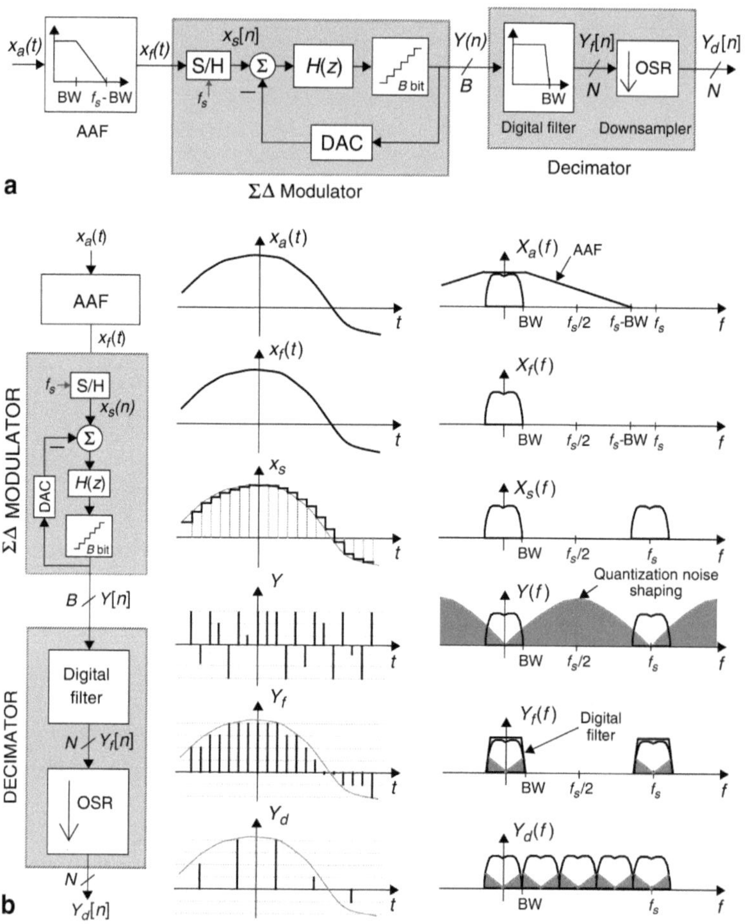

Fig. 2.5 ΣΔ ADC: **a** generic scheme, **b** signal processing involved

above BW are removed[1]—and, therefore, most part of the shaped quantization error—to avoid aliasing during the subsequent downsampling, in which the stream rate is divided by OSR keeping only one out of every OSR samples.

The ΣΔ modulator is the block that has most influence upon the ADC performance, basically because it is the responsible of the sampling and quantization processes and, therefore, ultimately limits the accuracy of the A-to-D conversion.

---

[1] A large steepness in the transition band is demanded usually for the filter in order to avoid degrading the signal band. However, this specification is imposed on a digital filter and is a priori easier to fulfill than for the analog AAF of a Nyquist ADC.

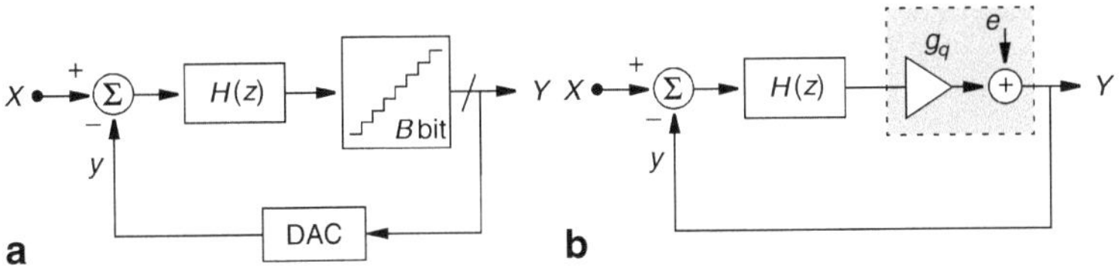

Fig. 2.6 ΣΔM architecture: **a** basic scheme, **b** linear model

2.2.1 *Signal Processing in a ΣΔ Modulator*

Figure 2.6a shows the basic scheme of a ΣΔ modulator. It consists of a feed-forward path formed by a loop filter $H(z)$ and a B-bit quantizer and a negative feedback path around them using a B-bit DAC [Inos62]. The operation of the ΣΔM can be explained as follows. Assume that $H(z)$ exhibits large gain inside the signal band and small gain outside of it. Due to the negative feedback, the error signal $x - y$ will become practically null in the signal band; i.e., the input signal x and the analog version of the output y will practically coincide within this band. Most of the differences between x and y will therefore be placed at higher frequencies, shaping quantization error and pushing it outside the signal band.

Figure 2.6b shows the linear model of a ΣΔM, in which the DAC is assumed to be ideal, the quantizer is replaced by the model in Fig. 2.2b and the additive white noise approximation is considered for the quantization error. This way, the modulator can be viewed as a two-input system whose output is represented in Z-domain as

$$Y(z) = \mathrm{STF}(z)X(z) + \mathrm{NTF}(z)E(z) \tag{2.13}$$

where $X(z)$ and $E(z)$ are the Z-transform of the input signal and the quantization noise, respectively, and STF(z) and NTF(z) are the respective transfer functions, given by

$$\mathrm{STF}(z) = \frac{g_q H(z)}{1 + g_q H(z)} \quad \mathrm{NTF}(z) = \frac{1}{1 + g_q H(z)} \tag{2.14}$$

Since the signal and the noise pass through different transfer functions, $H(z)$ can be chosen so that the noise shaping does not affect the signal. Using a loop filter with large gain within the signal band, the signal and noise transfer functions can be approximated in that range to

$$\mathrm{STF}(z) \approx 1 \quad \mathrm{NTF}(z) \approx \frac{1}{g_q H(z)} \ll 1 \tag{2.15}$$

The noise-shaping functions in (2.10) can be built with proper selection of $H(z)$. The easiest loop filter that exhibits the desired frequency performance is an integrator, whose Z-domain transfer function is

$$H(z) = \frac{z^{-1}}{1 - z^{-1}} \tag{2.16}$$

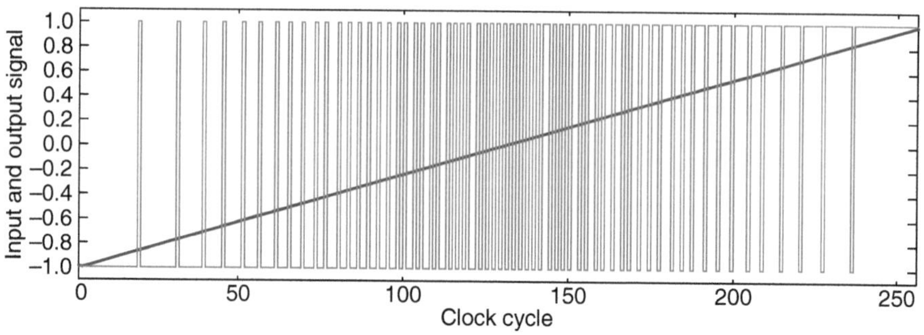

Fig. 2.7 PDM output stream of a 1st-order ΣΔM for an input ramp

Assuming that the quantizer gain g_q equals unity, the ΣΔM output yields

$$Y(z) = z^{-1}X(z) + (1 - z^{-1})E(z) \tag{2.17}$$

and the modulator is called a 1st-order ΣΔM, referring to the order of the noise shaping.

Figure 2.7 shows the output of a 1st-order ΣΔM with a 1-bit embedded quantizer for a ramp input signal. Due to the combined action of oversampling and negative feedback, the ΣΔM output is a pulse-density modulated (PDM) signal that locally tracks the input on average: when the input level is low, the ΣΔM output contains more −1's than +1's; when it is high, the +1's are dominant; and when the input signal is close to zero, the density of +1's and −1's practically coincides. If the quantizer resolution is larger, the output tracks the input much closer, since the separation between the discrete levels decreases.

2.2.2 *Performance Metrics of ΣΔ Modulators*

For the sake of clarity, it is convenient at this time to define the most important parameters commonly used to quantify the performance of ΣΔ modulators; namely:

- **Signal-to-noise ratio, SNR.** It is the ratio of the output power at the frequency of an input sinusoid to the uncorrelated in-band error power. Due to non-idealities of the circuitry that implements the modulator, other (linear and non-linear) errors apart from quantization noise contribute to the in-band error. SNR accounts for the linear performance of the modulator the in-band power associated to harmonics is therefore not included. For an ideal ΣΔ modulator and taking only quantization error into account, the SNR can be approximated to

$$\text{SNR}\Big|_{\text{dB}} = 10\log_{10}\left(\frac{A^2}{2P_Q}\right) \tag{2.18}$$

where A is the amplitude of the output sinusoid.

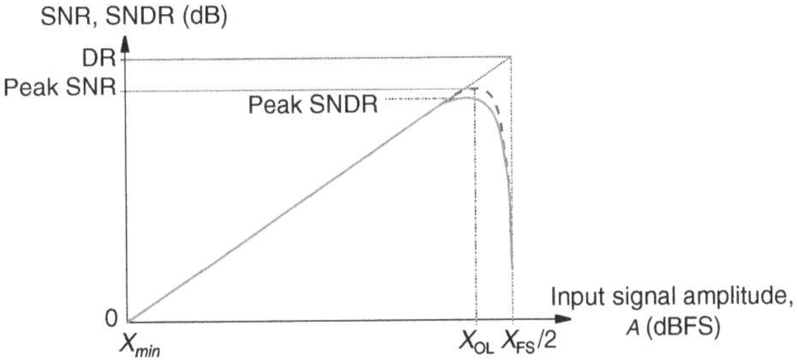

Fig. 2.8 Illustration of the performance parameters of a ΣΔM on a typical SNR/SNDR curve

- **Signal-to-(noise + distortion) ratio, SNDR.** It is defined as the ratio of the output power at the frequency of an input sinusoid to the total in-band error power, taking therefore into account harmonics at the modulator output.
- **Dynamic range, DR.** It is defined as the ratio of the output power at the frequency of an input sinusoid with maximum amplitude to the output power for a small input for which SNR = 0 dB (i.e., so that it cannot be distinguished from the error). Ideally, a sinusoid with maximum amplitude at the modulator input $X_{FS}/2$ will provide an output sinusoid sweeping the full scale of the ΣΔM quantizer and hence

$$\mathrm{DR}\Big|_{\mathrm{dB}} = 10\log_{10}\left[\frac{(X_{\mathrm{FS}}/2)^2}{2P_Q}\right] \tag{2.19}$$

- **Effective number of bits, ENOB.** Since the DR of an ideal N-bit Nyquist ADC equals $6.02N + 1.76$, a similar relationship is established for ΣΔ ADCs for comparison purposes [Bose88]:

$$\mathrm{ENOB} = \frac{\mathrm{DR}\,|_{\mathrm{dB}} - 1.76}{6.02} \tag{2.20}$$

  where the ENOB represents the number of bits needed for an ideal Nyquist ADC to achieve the same DR as the ΣΔ converter.
- **Overload level, X_{OL}.** The SNR of a ΣΔM does not increase monotonously for input amplitudes in the range [0, $X_{FS}/2$], where X_{FS} stands for the quantizer full scale. In practice, the embedded quantizer overloads for large amplitudes close to $X_{FS}/2$, causing an increase in the in-band error and a sharp drop in the SNR. The maximum value of the SNR before that drop defines the peak SNR and the corresponding input level is defined as the overload level X_{OL} of the ΣΔM.

The performance parameters defined above are illustrated in Fig. 2.8 on typical SNR and SNDR curves of a ΣΔM as a function of the amplitude of an input sinusoid signal. Both curves usually coincide up to medium input levels, since distortion is submerged into the modulator noise floor. For large input levels, harmonic distortion becomes more evident and degrades performance, causing the deviation of the SNDR curve.

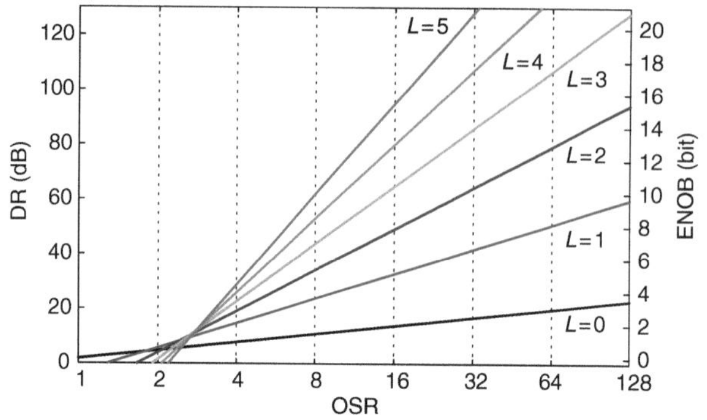

Fig. 2.9 DR and ENOB versus OSR for different orders of an ideal ΣΔ modulator. (A single-bit embedded quantizer is assumed)

2.2.3 *Ideal Performance of* ΣΔ *Modulators*

The dynamic range of an ideal Lth-order ΣΔM with a B-bit embedded quantizer that operates at a given oversampling ratio can be obtained from (2.1), (2.12) and (2.19) as

$$\mathrm{DR} \approx 10\log_{10}\left(\frac{3}{2}(2^B-1)^2 \cdot \frac{(2L+1)\mathrm{OSR}^{(2L+1)}}{\pi^{2L}}\right) \quad (2.21)$$

and can therefore be increased if L, OSR and/or B are augmented. The pros and cons of each possibility are discussed below:

- **Increasing the modulator order** L considerably improves the performance of a ΣΔM, since quantization error will be more attenuated at low frequencies and pushed to high frequencies. For a given OSR, the rise in DR when increasing L in one leads from (2.21) to

$$\Delta\mathrm{DR}\,|_{\mathrm{dB}} \approx 10\log_{10}\left[\frac{2L+3}{2L+1}\cdot\left(\frac{\mathrm{OSR}}{\pi}\right)^2\right] \quad (2.22)$$

  This means, e.g., that the DR of a 4th-order ΣΔM with OSR = 32 is improved in 21.3 dB (3.5 bit) in comparison to a 3rd-order one. However, stability problems arise from using high-order shapings ($L > 3$). These problems can be circumvented using different techniques [Nors97a], but at the price of reducing DR in comparison to the ideal value given by (2.21).
- **Increasing the oversampling ratio OSR** leads, according to (2.21), to an increase in the dynamic range of $3(2L+1)$dB/octave for an ideal Lth-order ΣΔM. This is shown in Fig. 2.9, where DR and ENOB are plotted versus OSR for different modulator orders. Note that for OSR > 4, the combined action of oversampling and noise shaping considerably improves performance. However, for a given signal band, larger OSRs lead to higher sampling frequencies and a faster operation of the modulator internal circuitry. The latter, if achievable, penalizes power dissipation.

- **Increasing the resolution B of the modulator embedded quantizer** leads to an increase in the DR of approximately 6 dB (1 bit) per extra bit in the quantizer [Geer02]. However, ΣΔMs with an internal multi-bit quantizer require a multi-bit DAC in the feedback loop that—contrary to a single-bit one, with only two levels—is not inherently linear. Note from Fig. 2.6 that non-linearities in the multi-bit DAC will be directly added to the modulator input. The linearity required in the DAC equals therefore in practice that wanted for the ΣΔ modulator.

2.3 Classification of ΣΔ Modulators

The above-mentioned strategies can be combined in many different ways giving rise to a pleiad of ΣΔM topologies reported in literature and that can be grouped attending to different classification criteria [Rodr03]:

- The nature of the signals being converted: low-pass versus band-pass ΣΔMs.
- The type of dynamics of the loop filter. Historically, most of ΣΔ modulators employed only discrete-time (DT) loop filters, as has been assumed in previous sections. Per contra, continuous-time (CT) ΣΔ modulators are also actually implemented. They use CT loop filters but DT quantizers. Also hybrid CT-DT modulators have been more recently reported.
- The number of quantizers employed. ΣΔMs employing only one quantizer are called single-loop structures. Those employing several quantizers have different names: cascades, dual-quantizer ΣΔMs, etc.
- The number of bits in the embedded quantizer. Historically, ΣΔ modulators employed mostly single-bit quantizers because they are inherently linear. Today, multi-bit ΣΔMs—i.e., those using multi-bit quantizers—are widely spread.
- Type of circuitry employed, devices available in the fabrication process, voltage supply, etc. Most of the reported DT implementations employ switched-capacitor (SC) circuits with high-quality passive capacitors—mixed-signal technology options—, but others employ capacitors available on standard CMOS technologies, active capacitors built with MOS transistors, switched-current (SI) circuits, etc.

Describing all possible ΣΔM architectures derived from previous classification criteria goes beyond the scope of this book. A detailed study of them can be found in many papers and books [Nors97a, Mede99, Geer02, Schr05b, Ortm06]. Instead, we will hereafter focus on low-pass DT ΣΔ architectures implemented using SC techniques, field to which the main contributions of this book refer to.

2.4 Single-Loop ΣΔ Architectures

The fundamentals and ideal performance of generic DT ΣΔMs have already been introduced. The 1st-order ΣΔM has been also presented in Sect. 2.2.1 for illustrating the operation of the simplest ΣΔ topology. However, its scope of application is very

Fig. 2.10 Second-order ΣΔ modulator

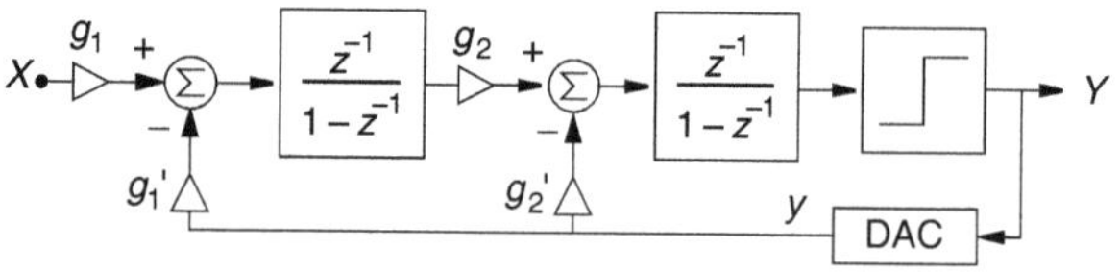

limited in practice due to the high correlation between the quantization error and the input signal, which severely deviates from the white noise approximation and leads to non-linear dynamic phenomena. This section is dedicated to modulator topologies with a larger number of integrators and only one quantizer, called single-loop ΣΔMs. Their linear performance—thanks to a better decorrelation of quantization error—will be discussed together with some aspects that are not covered by the white noise approximation, such as instabilities. A single-bit embedded quantizer will be assumed.

2.4.1 Second-Order ΣΔ Modulator

We will start the review of basic modulator architectures with the single-loop second-order topology [Cand85] shown in Fig. 2.10. Assuming a linear model for the single-bit quantizer (comparator)[2] as in Fig. 2.6b, the modulator output in Z-domain yields

$$Y(z) = \frac{g_1 g_2 g_q z^{-2} X(z) + (1 - z^{-1})^2 E(z)}{1 + (g_2' g_q - 2) z^{-1} + (1 + g_1' g_2 g_q - g_2' g_q) z^{-2}} \qquad (2.23)$$

so that the following conditions must be fulfilled for a pure 2nd-order NTF:

$$\left.\begin{array}{l} g_1' g_2 g_q = 1 \\ g_2' = 2 g_1' g_2 \end{array}\right\} \Rightarrow Y(z) = \frac{g_1}{g_1'} \cdot z^{-2} X(z) + (1 - z^{-1})^2 E(z) \qquad (2.24)$$

Note that the modulator can provide signal gain $G = g_1/g_1'$ if different coefficients are used in the first integrator for the input and the feedback paths. However, usually $g_1 \equiv g_1'$ and it will be so assumed hereinafter.

The conditions shown in (2.24) for pure 2nd-order shaping define relationships between the integrator coefficients, but not the values themselves. In general, selecting the coefficients of a ΣΔM involves solving several trade-offs between architectural, circuit and technological aspects of the practical implementation; namely:

- Keeping the integrator outputs bounded to ensure the modulator stability[3].
- Maximizing the overload level X_{OL} of the ΣΔM to ensure a high peak SNR.

---

[2] Note that a two-level quantizer has no inherent gain and, therefore, there is no obvious value for g_q. The linear model used henceforth for comparators assumes that g_q is such that the product of the loop gain factors is forced to unity by the feedback loop [Will91]. This model for comparators in a ΣΔM is empirical, but its results usually compare well to computer simulations using the true non-linear quantizer function.

[3] A ΣΔM is considered stable if, for bounded inputs and whatever integrators initial conditions, the internal state variables (integrator outputs) remain also bounded over time.

Table 2.1 Some reported coefficients for 2nd-order single-bit ΣΔMs

Coefficients	[Bose88]	[Yin94]	[Mede99]	[Marq98]
g_1, g_1'	0.5, 0.5	0.25, 0.25	0.25, 0.25	1/3, 1/3
g_2, g_2'	0.5, 0.5	0.5, 0.25	1, 0.5	0.6, 0.4
OS/Δ (at − 4 dBFS)	1.50	0.75	1.25	1.00
Unitary capacitors[a]	6 (3 + 3)	10 (4 + 6)	9 (5 + 4)	20 (4 + 16)

[a] Capacitor sharing between coefficients in an integrator is assumed

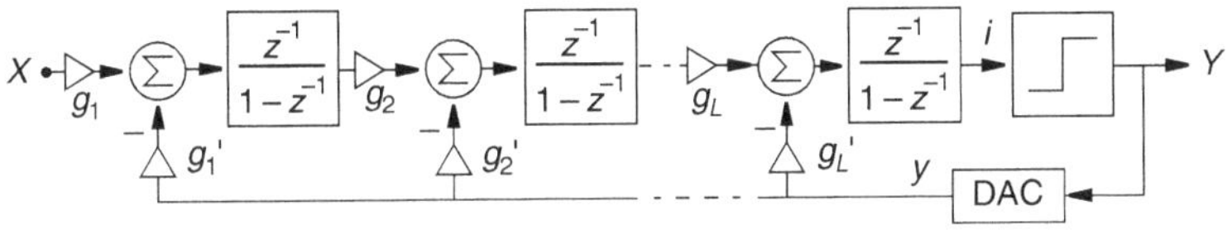

Fig. 2.11 Lth-order single-loop ΣΔM with distributed feedback

- Minimizing the required signal swing at the integrator outputs; i.e., the integrator output swing (OS) specifications should be, first, achievable in practice, and second, as low as possible in order to reduce power consumption and to ease circuit design.
- Simplifying the implementation of the coefficients. In SC ΣΔMs they are implemented as capacitor ratios using unitary elements for improved matching. A set of coefficients involving a small number of unitary capacitors leads to a saving of silicon area.

For comparison purposes, Table 2.1 shows some sets of coefficients for 2nd-order ΣΔ reported in literature. All of them exhibit an overload level $X_{\mathrm{OL}} \approx -4$ dBFS (− 4 dB below the full-scale amplitude, Δ/2). The requirements on integrator OS and the total number of unitary capacitors are also included.

2.4.2 *High-Order ΣΔ Modulators*

The simplest way to extend a ΣΔM to an arbitrary Lth-order shaping consists of including L integrators before the quantizer [Ritc77]. Extending the 2nd-order ΣΔM in Fig. 2.10, the architecture in Fig. 2.11 can be obtained, often called Lth-order single-loop ΣΔM with distributed feedback. Using a linear model, its output would be given by

$$Y(z) = z^{-L}X(z) + (1 - z^{-1})^L E(z) \tag{2.25}$$

if a set of relationships between the coefficients were fulfilled, similar to those in (2.24) for 2nd-order ΣΔMs. The in-band quantization error for this NTF would ideally be given by (2.12), therefore achieving very low P_Q for large L, even for low OSR.

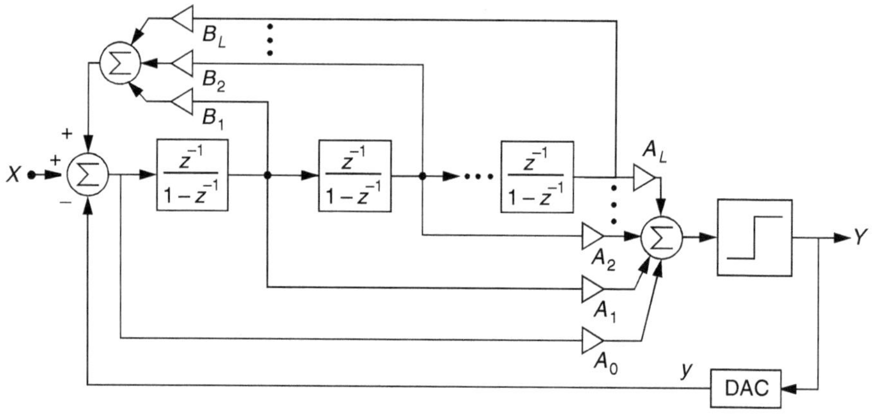

Fig. 2.12 Lee-Sodini Lth-order ΣΔ modulator

Nevertheless, this performance is not achievable in practice because ΣΔMs with pure differentiator NTFs—i.e., FIR filters like $(1 - z^{-1})^L$—are prone to instability[4] if $L > 2$, exhibiting unbounded states and poor SNR in comparison to that predicted by the linear model. This tendency to instability can be qualitatively explained as follows [Adam97]. For a ΣΔM to be stable, the input to the quantizer must not be allowed to become too large. Since the quantizer input in Z-domain is given by

$$I(z) = \mathrm{STF}(z)X(z) + [\mathrm{NTF}(z) - 1]E(z) \tag{2.26}$$

the gain of $\mathrm{NTF}(z) - 1$, or simply $\mathrm{NTF}(z)$, must not be too large. However, the gain of NTFs of the form $(1-z^{-1})^L$ rapidly increases for high frequencies if $L > 2$, having a maximum $\|\mathrm{NTF}\|_\infty = \max[\mathrm{NTF}(z)] = 2^L$ at $z = -1 (f = f_s/2)$.

The determination of exact conditions to guarantee the stability of higher-order ($L \geq 3$) ΣΔMs is still an open question, but, despite the absence of general stability conditions, high-order ΣΔMs have been successfully designed since the late 1980s.

2.4.2.1 Optimization of NTFs

Figure 2.12 illustrates the so-called interpolative architecture introduced by Lee and Sodini [Lee87], which allows to obtain high-order stable designs. Thanks to the large set of coefficients in multiple feed-forward and feedback paths, more complex high-pass NTFs can be built with sufficiently low gain at the high-frequency region. Let us first consider that the feedback coefficients B_i are set to zero. In this case, the following IIR NTF is obtained

$$\mathrm{NTF}(z) = \frac{(z-1)^L}{D(z)} \tag{2.27}$$

[4] Instability appears at the modulator output as a large-amplitude low-frequency oscillation, leading to long strings of alternating +1's and −1's [Adam97].

where all zeros are located at $z = 1$ (DC) and $D(z)$ is a polynomial determined by the feed-forward coefficients A_i. These coefficients can be adjusted to build a high-pass Butterworth or Chebyshev filter for NTF, with cutoff frequency beyond the signal band and almost flat gain in the filter pass-band. For the ΣΔM to remain stable, this gain must be adjusted to satisfy[5]

$$\|\mathrm{NTF}\|_{\infty} = \max[\mathrm{NTF(z)}] \sim 1.5 \tag{2.28}$$

However, with all zeros at DC, NTF rises monotonically in the signal band like an Lth-order function and its gain at the end of the signal band will therefore practically determine the total in-band error power. The feedback coefficients B_i in the Lee-Sodini ΣΔM are in practice used to modify the NTF in (2.27) and place notches in the signal band for further shaping of the quantization error. The values of B_i are set small in comparison to A_i in order to obtain NTF poles mostly controlled by the latter coefficients, preserving therefore the flat out-of-band gain of NTF and the modulator stability.

In [Lee87] the zeros are fixed to obtain an NTF with equal-ripple response over the signal band, but other alternatives are also feasible. The approach followed in [Schr93] leads to minimizing P_Q by optimally placing complex-conjugate zeros of NTF at frequency positions obtained by solving the minimization problem

$$\min\left[\int_0^{\mathrm{BW}} |\mathrm{NTF}(f)|^2\, df\right] \Rightarrow \min\left[\begin{cases} \displaystyle\int_0^1 \prod_{i=1}^{L/2} (f^2 - f_{z_i}^2)^2\, df, & L \text{ even} \\ \displaystyle\int_0^1 f^2 \prod_{i=1}^{(L-1)/2} (f^2 - f_{z_i}^2)^2\, df, & L \text{ odd} \end{cases}\right] \tag{2.29}$$

where f_{z_i} represents the location of the complex zeros normalized to BW. The solutions for (2.29) up to $L = 8$ can be found in [Schr93]. Fig. 2.13 illustrates these alternatives for implementing stable high-order NTFs for a 5th-order ΣΔM with OSR = 64. Note from Fig. 2.13a that the position of the poles has been fixed following a Butterworth configuration, so that the cutoff frequency of NTF is beyond the signal band—vertical line in Fig. 2.13b—and its out-of-band gain is 1.5 (3.5 dB). Two notches can be introduced in the signal band by slightly moving four of the zeros at $z = 1$ along the unit circle. As shown in Fig. 2.13b, this considerably reduces $|\mathrm{NTF}(f)|$ at the upper edge of the signal band and therefore the final P_Q (18-dB reduction for the case considered).

Figure 2.14 shows the maximum SNR achievable by single-loop ΣΔMs of various orders following these two approaches. The ideal performance for pure differentiator

[5] This empirical condition is commonly accepted as a rule of thumb for designing stable high-order NTFs [Adam97]. Other stability criteria also exist, but none of them can ensure the stability of an arbitrary high-order single-bit ΣΔM [Schr93]. Computer simulation is still the most reliable method to verify stability.

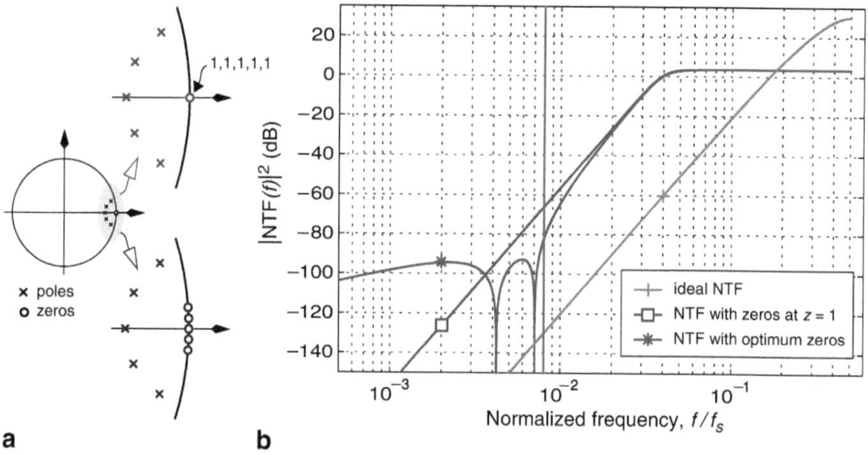

Fig. 2.13 Comparison of different implementations of a 5th-order NTF: **a** illustration of the pole-zero placement in the unit circle, **b** magnitude response of the corresponding NTFs. (OSR = 64 is assumed for the design of the Butterworth poles)

NTFs is included for comparison purposes in this figure as in Fig. 2.13, indicating that the SNRs obtained in practice are much lower. Nevertheless, as shown in Fig. 2.14b, the optimal placement of notches within the signal band can lead to stable high-order ΣΔMs with high SNR at moderate OSR.

2.4.2.2 Common High-Order Topologies

Many ΣΔ modulator topologies have been developed for implementing high-order NTFs with the characteristics of most common IIR filter families. Most of them make use of use multiple weighted feedback or feed-forward paths (or both) [Adam97]. The two architectures of 5th-order ΣΔMs illustrated in Fig. 2.15 can be considered typical examples.

Figure 2.15a illustrates a topology with feed-forward summation of the integrator outputs before the quantizer. If $\gamma_i = 0$, a high-pass NTF is obtained with all zeros at DC. By adding small negative feedback locally, around pairs of integrators in the loop filter, pairs of zeros can be moved along the unit circle to create notches in $|\mathrm{NTF}(f)|$ at frequencies $2\pi f_i/f_s \approx \sqrt{\gamma_i}$. Once NTF is set for the desired noise shaping, the modulator topology fixes the signal transfer function to $\mathrm{STF}(z) = 1 - \mathrm{NTF}(z)$.

The architecture in Fig. 2.15b that includes distributed feedback and feed-forward paths can be used if a certain degree of freedom is desired in designing both NTF and STF. In this topology, the zeros of STF can be fixed with coefficients b_i without affecting the pole placement. Local resonator feedbacks can be also included to set notches in $|\mathrm{NTF}(f)|$.

A common drawback of most high-order topologies implementing IIR filters for NTF is the increased circuit complexity due to the many feedback and feed-forward

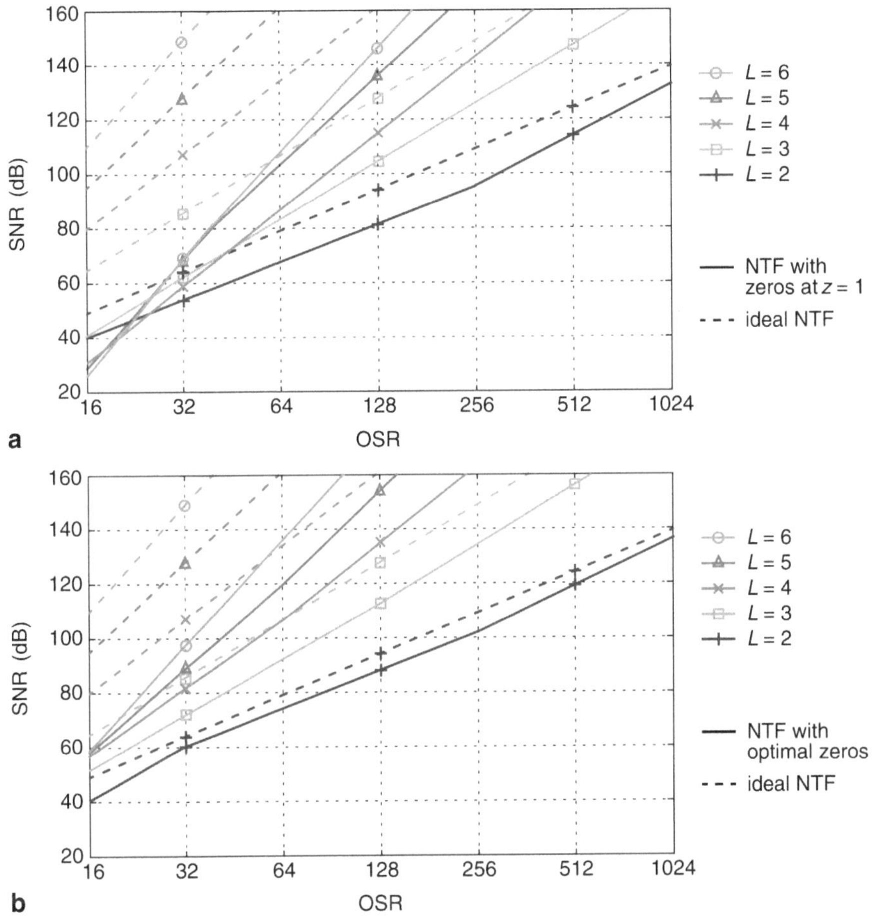

Fig. 2.14 Maximum SNR achievable by Lth-order single-loop ΣΔMs versus OSR: **a** NTF with all zeros at $z = 1$, **b** NTF with zeros optimally spread over the signal band. (Data taken from [Schr93])

coefficients required. Furthermore, some of the coefficients may be considerably small [Chao90, Kuo99], leading to large capacitor ratios in the final SC implementation and, therefore, to great power consumption and area occupation.

Another strategy for designing stable high-order single-loop ΣΔMs basically consists of extensively exploring the design space by means of behavioral simulations searching for sets of coefficients that are easy to implement and maximize the modulator SNR, while achieving a reasonable overload level. This procedure is followed in [OptE90] and [Marq98] for the distributed feedback topology in Fig. 2.11. High-pass IIR NTFs are obtained with zeros at DC, but the obtained placement of poles does not follow any particular filter family configuration.

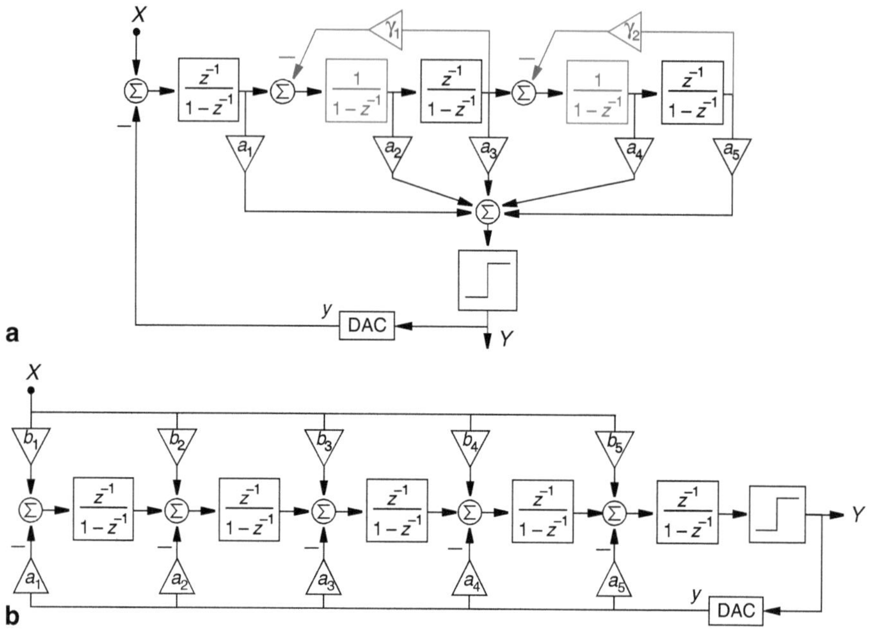

Fig. 2.15 5th-order single-loop ΣΔMs: **a** with feed-forward summation and local resonator feedbacks, **b** with distributed feedback and distributed feed-forward input paths

2.4.2.3 Non-linear Stabilization Techniques

In spite of the procedures described above, the resulting high-order single-loops are only conditionally stable and instabilities may therefore appear for inputs above certain bounds or for certain initial conditions. Non-linear techniques can be used to ensure global stable operation. Limiters can be included at the integrators in order to preclude their outputs from exceeding 'safe' values of the state-variable, previously identified during stable operation of the modulator [OptE90]. Instead of clipping the integrators outputs, they can be reset to zero or some other initial condition at power-up of the system or when unstable operation is detected [Mous94]. The detection of instability can be done at the integrator level, by placing comparators to determine if a state variable has surpassed a certain limit, or by monitoring the length of consecutive $+1$'s and -1's at the modulator output.

2.5 Cascade ΣΔ Architectures

As stated in the previous section and illustrated in Fig. 2.14, precluding instabilities in high-order single-loop ΣΔ modulators leads to attainable values of SNR that are considerably far from ideal. An alternative to circumvent instabilities while obtaining

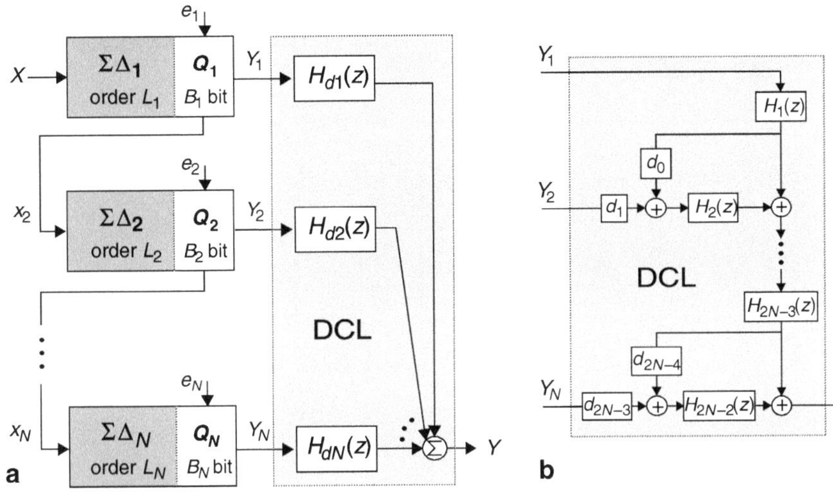

Fig. 2.16 **a** Generic N-stage cascade $\Sigma\Delta$ modulator; **b** structure of the digital cancellation logic (DCL)

high SNR can be found in the so-called MASH (multi-stage noise-shaping) $\Sigma\Delta$Ms, often referred to as cascade $\Sigma\Delta$Ms [Mats87, Long88]. The generic scheme of a cascade $\Sigma\Delta$M is shown in Fig. 2.16. It consists of several low-order $\Sigma\Delta$ modulators or stages, in which each stage remodulates a signal containing the quantization error generated in the previous one. The outputs Y_i of the low-order $\Sigma\Delta$Ms are properly processed and combined in digital domain in order to cancel out in the overall cascade output Y the traces of the quantization errors of all the stages, but the last one. This latter error appears at the overall modulator output, after the digital cancellation logic (DCL block in Fig. 2.16), with a noise shaping of order equal to the summation of the orders of the cascaded stages. Furthermore, since all feedback loops are local and there is no inter-stage feedback, unconditionally-stable high-order shaping can be obtained provided that only $\Sigma\Delta$Ms with order $L \leq 3$ are cascaded. The performance of a MASH $\Sigma\Delta$M is, therefore, similar to that of an ideal high-order loop with no stability issues.

For an N-stage cascade $\Sigma\Delta$M like that in Fig. 2.16, by means of adequately processing in digital domain the stages outputs, only the modulator input signal $X(z)$ and the last-stage quantization error $E_N(z)$ remain in the overall modulator output, yielding

$$Y(z) = \mathrm{STF}(z)X(z) + \mathrm{NTF}_N(z)E_N(z) = z^{-L}X(z) + d_{2N-3}(1 - z^{-1})^L E_N(z) \tag{2.30}$$

where $L = L_1 + L_2 + \cdots + L_N$ and d_{2N-3} is a scaling factor related to the transmission of the input signal from one stage to another (in order to avoid overloading of the stages) that amplifies the last-stage quantization error. The in-band quantization error

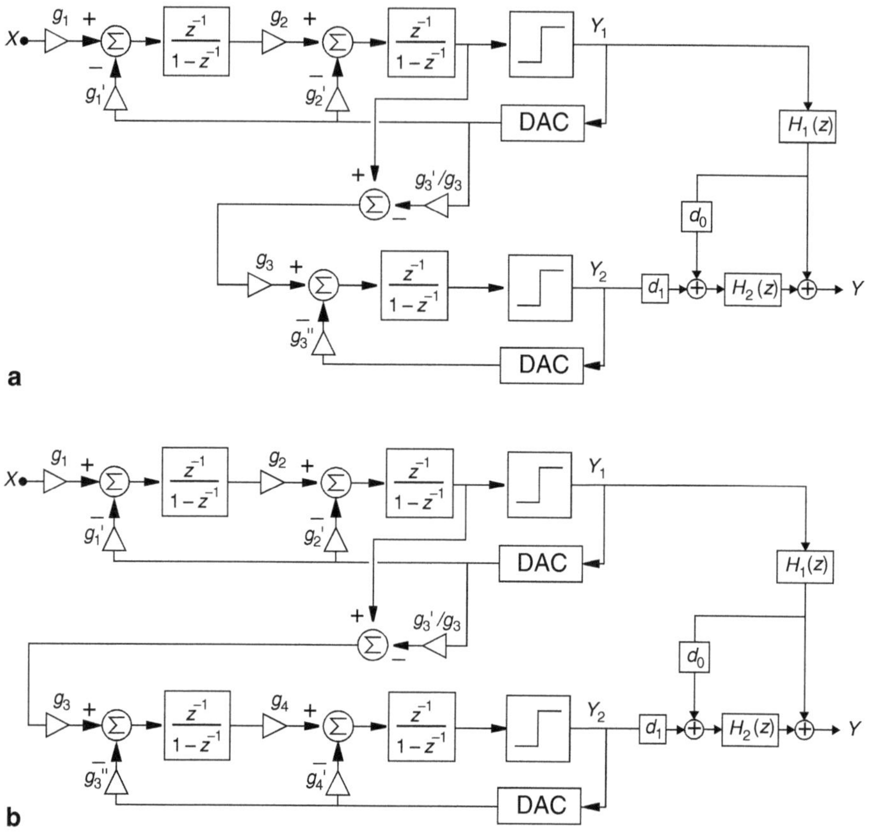

Fig. 2.17 Illustration of 2-stage cascade topologies: **a** 3rd-order 2-1 ΣΔM, **b** 4th-order 2-2 ΣΔM

power of a N-stage cascade is, therefore, given by

$$P_Q \cong d_{2N-3}^2 \cdot \frac{\Delta_N^2}{12} \cdot \frac{\pi^{2L}}{(2L+1)\mathrm{OSR}^{(2L+1)}} \tag{2.31}$$

with Δ_N being the level spacing in the B_N-bit quantizer of the Nth modulator stage. Hence, the performance corresponds to that of an ideal Lth-order B_N-bit ΣΔM, except for the scalar d_{2N-3} that causes a systematic loss of performance. Common values for this amplifying factor are 2 and 4, which lead to a reduction in the attainable SNR of 6 dB (1 bit) and 12 dB (2 bit), respectively. These performance losses that are inherent to cascade ΣΔMs, are however considerably lower than those resulting for optimized high-order single loops (see Fig. 2.14). Moreover, in the case of MASH ΣΔMs, they are independent of OSR.

The aforementioned benefits have favoured the development of a great number of cascade ΣΔ topologies. Fig. 2.17 illustrates two different 2-stage cascades. A 3rd-order ΣΔM that is formed by a 2nd-order stage followed by a 1st-order stage

Table 2.2 Relationships to be fulfilled for digital cancellation in the 2-1 and the 2-2 ΣΔMs

2-1 ΣΔM			2-2 ΣΔM		
Analog	Digital		Analog	Digital	
$g_2' = 2g_1'g_2$	$d_0 = \frac{g_3'}{g_1'g_2g_3} - 1$	$H_1(z) = z^{-1}$	$g_2' = 2g_1'g_2$	$d_0 = \frac{g_3'}{g_1'g_2g_3} - 1$	$H_1(z) = z^{-2}$
	$d_1 = \frac{g_3''}{g_1'g_2g_3}$	$H_2(z) = (1 - z^{-1})^2$	$g_4' = 2g_3''g_4$	$d_1 = \frac{g_3''}{g_1'g_2g_3}$	$H_2(z) = (1 - z^{-1})^2$

[Long88] (so-called 2-1 ΣΔM) is shown in Fig. 2.17a, whereas a 4th-order cascade built up by two 2nd-order stages [Kare90, Bahe92] (2-2 ΣΔM) is shown in Fig. 2.17b. Table 2.2 also summarizes the relationships between integrator coefficients (analog coefficients) and the blocks in the digital cancellation logic (digital coefficients and filters) that must be fulfilled for a proper performance of the cascades in accordance with (2.30).

Many other MASH topologies have been proposed: a 4th-order 3-stage cascade (2-1-1 ΣΔM) [Yin94], 5th-order cascades implemented as a 2-2-1 ΣΔM [Vleu01] or as a 2-1-1-1 ΣΔM [Rio00], a 6th-order 2-2-2 architecture [Dedi94], etc. Although cascade ΣΔMs can a priori be extended to whatever number of stages to increase the order of the noise shaping, non-idealities in the analog circuitry of the final implementation (e.g., mismatching) limit in practice the effectiveness of this approach. They cause deviations in the performance of the analog section of the cascade (integrator coefficients and transfer functions) which impede that the DCL completely cancels the NTFs of the first $N-1$ stages at the overall output. This effect, called noise leakage, causes traces of the lower-order quantization errors to appear at the cascade output, where they can dominate the overall in-band error power[6] [Rebe97].

Note that the relationships in Table 2.2 for the digital section of the cascade assume that the digital cancellation logic is structured according to Fig. 2.16b. For other kind of structures, the equations may be different, but anyhow the digital part will have to fulfil given relationships with the analog one. Once these relationships are fulfilled, the value of the free analog coefficients is fixed for minimum loss of performance and simpler circuit implementation of the cascade ΣΔM. Proper concerns for their selection are:

- Minimizing the loss of performance in comparison to an ideal ΣΔM.
- Maximizing the modulator overload level to ensure high peak SNR.
- Minimizing the integrators output swing, especially in low-voltage scenarios.
- Easing the implementation of the analog coefficients as capacitor ratios with unitary elements by means of enabling capacitor sharing within the integrators, reducing the total number of unitary capacitors to save silicon area, etc.

[6] Obviously, non-idealities in the analog circuitry also affect the practical performance of single-loop ΣΔMs. Nevertheless, their operation does not rely in the cancellation of quantization errors and their sensitivity to noise leakages is therefore much lower than for cascade ΣΔMs.

Table 2.3 Some sets of analog coefficients reported for the 4th-order 2-2 ΣΔM

Coefficients	[Miao98]	[Marq98]-A	[Marq98]-B
g_1, g_1'	0.25, 0.25	0.5, 0.5	0.5, 0.5
g_2, g_2'	0.5, 0.25	0.5, 0.5	0.5, 0.5
g_3, g_3', g_3''	0.5, 0.125, 0.25	1, 0.5, 0.5	0.5, 0.25, 0.5
g_4, g_4'	0.5, 0.25	0.5, 0.5	0.5, 0.5
d_0, d_1	1, 4	1, 2	1, 4
ΔSNR in comparison to ideal	−12 dB (−2 bit)	− 6 dB (−1 bit)	− 12 dB (−2 bit)
$X_{OL}/(\Delta/2)$	−2 dBFS	−5 dBFS	−2 dBFS
Unitary capacitors[a]	28 (4+6+12+6)	13 (3+3+4+3)	16 (3+3+7+3)

[a] Capacitor sharing between coefficients in an integrator is assumed

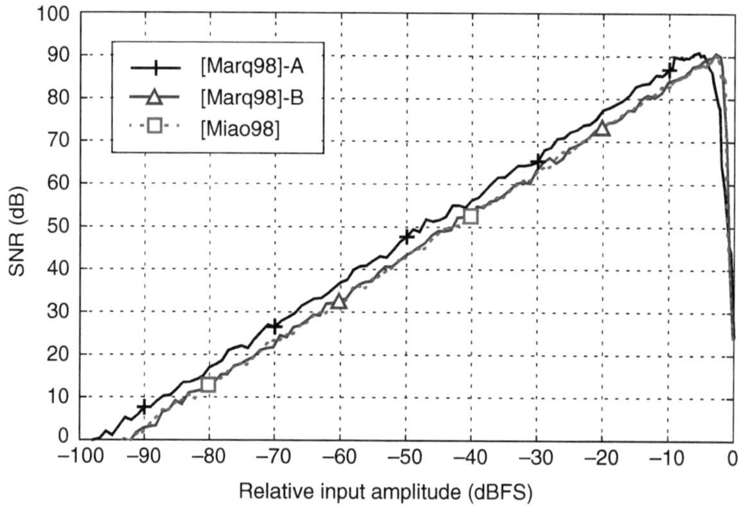

Fig. 2.18 SNR curves of the 2-2 cascade ΣΔMs in Table 2.3 for OSR = 32

- Facilitating the implementation of the digital cancellation logic; e.g., power-of-two coefficients are often preferred in order to use simple shift registers.

For comparison purposes, Table 2.3 shows some sets of analog coefficients reported for the 4th-order 2-2 cascade ΣΔM. Related aspects, like nominal performance loss, overload level and minimum number of unitary capacitors, are also included. Obviously, many trade-offs exist between the aforementioned aspects, but once they are worked out the SNR obtained with a cascade ΣΔM can be considerably larger than with its single-loop counterpart. Fig. 2.18 shows the SNR curves of the three 2-2 ΣΔMs in Table 2.3 for an oversampling ratio of 32, in which peak SNRs of approximately 91 dB are obtained. This value surpasses in 30 dB the peak SNR achieved by a stable 4th-order single-loop ΣΔM with all NTF-zeros at DC (see Fig. 2.14a). The improvement is still around 10 dB in comparison to optimized 4th-order NTFs with two notches within the signal band (see Fig. 2.14b).

2.6 Multi-bit ΣΔ Architectures

As previously stated, the dynamic range of a ΣΔ modulator can be enhanced by means of increasing the order of the quantization noise shaping. Nevertheless, the expected improvement in performance for large modulator orders can vanish due to instabilities in single-loop ΣΔ architectures (see Sect. 2.4.2) or due to noise leakages in cascade ΣΔ topologies (see Sect. 2.5). In accordance with (2.19), an alternative way for further increasing the modulator DR consists of using embedded quantizers with larger resolution. The main advantages of multi-bit ΣΔ modulators are:

- The in-band quantization error power is roughly reduced 6 dB per additional bit in the embedded quantizer thanks to the smaller internal quantization step Δ.
- Internal non-linearities are weaker in multi-bit ΣΔMs than for single-bit counterparts. The quantizer operation better fits the white noise approximation and phenomena caused by non-linear dynamics are less evident.
- For a given order in the loop filter, the stability properties for multi-bit ΣΔMs are better than for single-bit ΣΔ architectures.

The aforementioned benefits suggest that, for a targeted modulator performance, multi-bit quantization can be traded for noise shaping and/or oversampling. Indeed, multi-bit ΣΔMs are often employed in wideband applications, since the oversampling ratio can be lower than in their single-bit counterparts. This helps to reduce the operation frequency and, therefore, the power consumption not only in the ΣΔM itself, but also in the decimation filter. Nevertheless, besides the increase in circuit complexity when moving from single- to multi-bit quantization, other aspects related to linearity requirements have a strong impact on the modulator operation. Contrary to 1-bit quantizers that are intrinsically linear because only two levels are used in the quantization process, multi-bit quantizers exhibit in practice some non-linearities in their transfer characteristic mostly due to device mismatching. As we indicate below, these errors have a significant influence on modulator performance and may represent an important drawback that counters the aforementioned advantages.

2.6.1 *Impact of DAC Non-linearities*

Figure 2.19 illustrates the linear model of a multi-bit ΣΔM, in which errors related to the multi-bit conversion are added to the quantization error e that has been taken into account so far; namely, an error e_{ADC} associated to the A-to-D conversion process and an error e_{DAC} in the subsequent D-to-A conversion required to reconstruct the analog feedback signal. Note that e_{ADC} is injected in the same path as the quantization error e and is also therefore attenuated by the gain of the loop filter within the signal band. However, DAC errors are injected in the feedback path and the corresponding non-linearities are therefore directly added to the modulator input where they are not

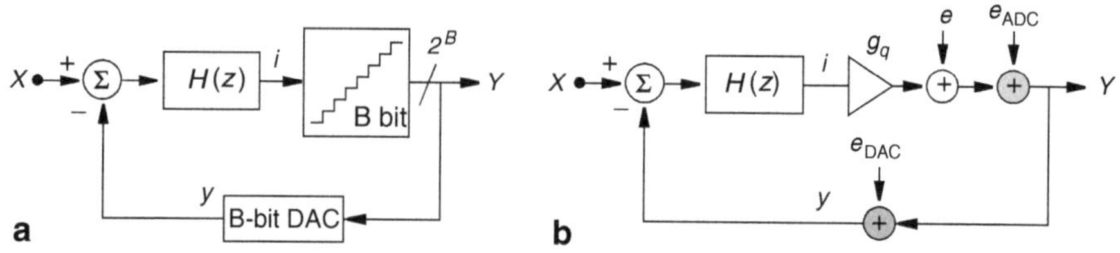

Fig. 2.19 Impact of errors in ΣΔMs with multi-bit embedded quantizers: **a** basic scheme, **b** corresponding linear model

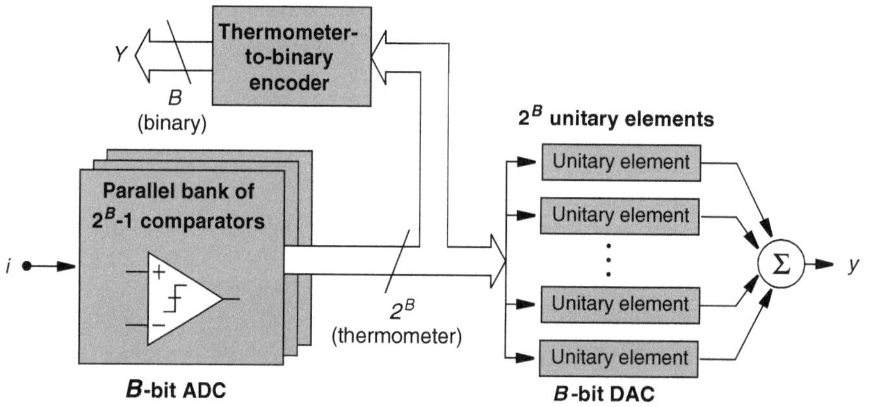

Fig. 2.20 Typical scheme of the multi-bit ADC and DAC embedded in a ΣΔM

be mitigated by noise shaping; i.e., they pass to the modulator output as the input signal does. Consequently, the linearity of a multi-bit ΣΔM will be no better than that of the multi-bit embedded DAC [Carl97] and the latter must be designed to reach the linearity targeted for the whole ΣΔ ADC, what is quite challenging due to the influence of component mismatching.

Figure 2.20 shows the architecture typically employed for the embedded quantizer and the feedback DAC in multi-bit ΣΔ modulators, which is fully parallel since the number of internal bits is normally low ($B \leq 5$). The B-bit ADC consists of a bank of $2^B - 1$ comparators that digitizes the output of the loop filter into thermometer code, which is subsequently coded into binary at the back end. On the other hand, the DAC employs $2^B - 1$ unitary elements (capacitors, resistors, current sources, etc.) to reconstruct the analog feedback signal using 2^B levels. The ith analog output level is generated by activating i unitary elements and adding their outputs (charges or currents). Errors in the DAC are originated by the mismatching between its unitary elements, which causes the deviation of the DAC output levels from their nominal values. Assuming that the actual value of each unitary element follows a Gaussian distribution, the worst-case relative error in the DAC output y can be estimated as [Carl97]

$$\sigma\left(\frac{\Delta y}{y}\right) \approx \frac{1}{2\sqrt{2^B}}\sigma\left(\frac{\Delta U_e}{U_e}\right) \tag{2.32}$$

where $\sigma(\Delta U_e/U_e)$ is the relative error in the value of the unitary element. Obviously, the accuracy in the DAC increases with the number of unitary elements thanks to the parallel topology. However, for a ΣΔM with 4-bit internal quantization to achieve 16-bit linearity, the required matching of the unitary elements in the DAC should be better than 0.01% (13 bits). Unfortunately, the matching of devices that can be achieved in present-day CMOS processes is in the range of 0.1% (10 bits) and the required accuracy in the unitary components could only be obtained through the parallel connection of many more elements. This clearly means that achieving linearities larger than 12 or 13 bits in multi-bit ΣΔMs, while relying only on standard device matching, leads to prohibitive area occupation.

A straight-forward method to improve the standard device matching consists of trimming their values [Carl97], what can be sometimes done at the foundry, but at the expense of additional fabrication and/or measurement steps and increased cost. Analog calibration running periodically or in background can also be used to this purpose and to compensate for drifts or aging, but at the cost of a significant increase in circuit complexity. Another strategy consists of converting DAC errors into digital form and correcting them in digital domain using look-up tables [Carl97].

Among the different alternatives that have been developed through the years for achieving high-linear multi-bit ΣΔMs, two of them clearly prevail because of the modest component matching and circuit complexity involved. These are discussed below.

2.6.2 Dynamic Element Matching

As previously stated, mismatching between the unitary elements of the feedback DAC in a multi-bit ΣΔM causes non-linear errors to be directly added at its input, therefore limiting the overall modulator linearity. For a given DAC with the structure in Fig. 2.20, there is a univocal correspondence between the thermometric input code and the respective DAC error, because the same unitary elements are always used to generate a given DAC output level. The fundamental idea underlying dynamic element matching (DEM) consists of breaking this direct correspondence by varying over time the set of elements that are used to generate a given DAC output level, therefore transforming its fixed error into a time-varying one. As shown in Fig. 2.21, a digital block is added to control the selection of elements at each clock cycle according to an algorithm that tries to drive the average error in each DAC level to zero over time. Part of the DAC error power that laid in the low-frequency range will therefore be moved to higher frequencies, where it will be removed by the decimation filter.

Many of the different DEM techniques that have been developed are detailed in [Geer02]. They can roughly be grouped in the following clusters:

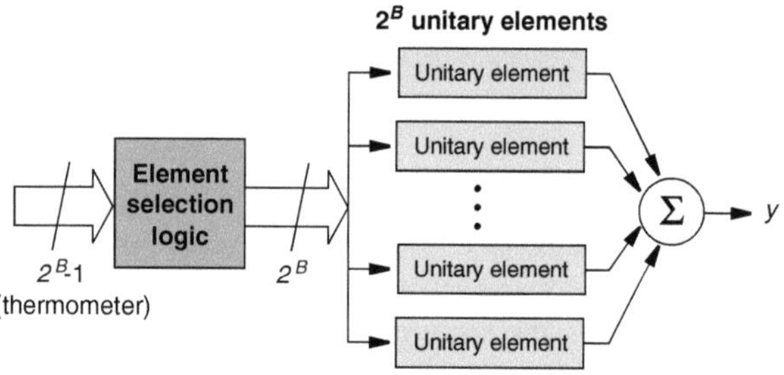

Fig. 2.21 Architectural concept of DEM

- Randomization algorithms, in which the DAC unitary elements are selected according to pseudo-randomly configured networks (e.g., butterfly structures). Harmonic distortion induced by the DAC is transformed into white noise, whose out-of-band power will be removed by decimation. The DAC error power laying within the signal band will however increase the modulator noise floor.
- Rotation algorithms, in which the DAC unitary elements are selected in a periodic manner for shifting harmonic distortion out of the signal band. The modulator noise floor is not increased, but the processing can originate mixed frequency components that fold back into the baseband. Clocked averaging (CLA) is an example of this kind of DEM techniques.
- Mismatch-shaping algorithms, in which the DAC unitary elements are selected according to algorithms that conform the mismatching error in order to push most of its power to higher frequencies. The order of the mismatch shaping is normally limited to one or two. Individual level averaging (ILA), data weighted averaging (DWA and its many modifications) and data directed scrambling (DDS) pertain to these kind of algorithms, whose use is very extended.

2.6.3 Dual-Quantization Techniques

An alternative approach to reduce the impact of DAC non-linearities in multi-bit ΣΔMs can be found in dual-quantization techniques, in which both single- and multi-bit quantizers are used at a time in the same ΣΔM. The underlying idea is to combine their benefits: two-level quantization for its the intrinsic linearity and multi-bit quantization for its reduced error power. Architectural examples of these techniques will be discussed below.

2.6.3.1 Leslie-Singh Architecture

Figure 2.22 shows the general scheme of the dual-quantizer ΣΔM proposed by Leslie and Singh [Lesl90]. Single-bit quantization is used in the ΣΔ feedback loop because of its intrinsic linearity and a path containing only a multi-bit quantizer is connected

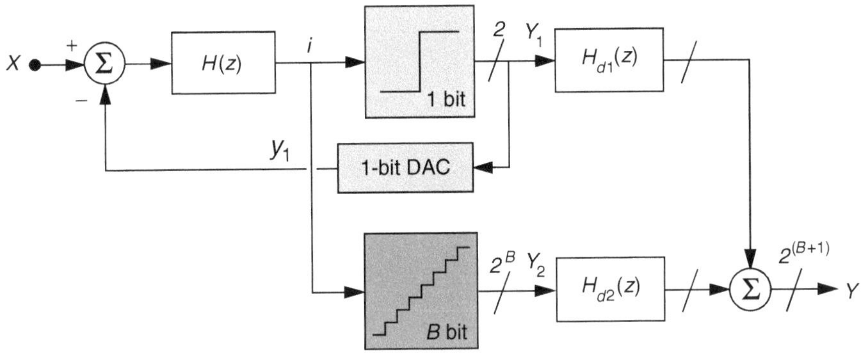

Fig. 2.22 Leslie-Singh ΣΔ modulator

to the loop filter output. The two quantizer outputs are then properly combined in digital domain to reduce the quantization error at the output to that of the multi-bit quantizer. Note that the scheme in Fig. 2.22 can be simplified in practice by removing the comparator and feeding the most-significant bit of the B-bit quantizer back to the ΣΔ loop.

The Leslie-Singh topology can be viewed as a MASH ΣΔM, in which the first stage is an Lth-order single-bit ΣΔM and the second stage is a 0-order B-bit ΣΔM. It therefore requires perfect cancellation of the 1st-stage 1-bit quantization error at the modulator output and also suffers from noise leakage problems. Note also that, although the modulator output ideally contains only the quantization error coming from the multi-bit quantizer, the stability of the architecture does not improve as in standard multi-bit ΣΔMs, since the loop is closed through single-bit feedback. The NTF cannot therefore be optimized to more aggressive high-order shapings without jeopardizing stability.

2.6.3.2 Dual-Quantizer Single Loops

An alternative way of employing dual quantization in high-order ΣΔMs is illustrated in Fig. 2.23. In the 3rd-order ΣΔM shown, the first two integrators are fed by a 1-bit DAC, while the third one is fed by a multi-bit DAC. The modulator linearity is not menaced, since DAC non-linearities are suppressed by the gain of the first two integrators. At the same time, the topology benefits from improved stability thanks to the multi-bit feedback in the last integrator. In practice, the most-significant bit of the B-bit quantizer can be used to close the 1-bit feedback. Under linear analysis, the modulator output is ideally obtained as

$$Y(z) = z^{-1}X(z) + 2(1 - z^{-1})^3 E_2(z) - 2z^{-1}(1 - z^{-1})^2 E_{\mathrm{DAC}}(z) \qquad (2.33)$$

yielding 3rd-order shaping for the B-bit quantization error, $E_2(z)$, and 2nd-order shaping for the DAC errors, $E_{\mathrm{DAC}}(z)$. Note that the topology suffers however from noise leakages.

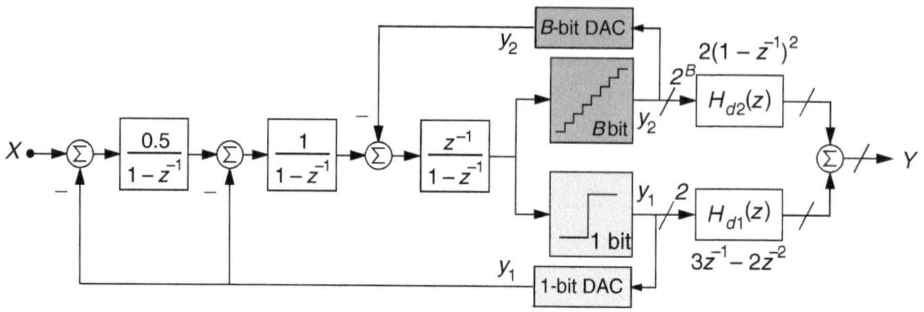

Fig. 2.23 A 3rd-order single-loop ΣΔM with dual quantization. [Hair91]

The concept in Fig. 2.23 can be generalized to higher-order ΣΔMs. As the order of the loop filter increases, the number of back-end integrators with multi-bit feedback can be traded off with aggressive noise shaping (for improved stability) and linearity requirements of the multi-bit DAC.

2.6.3.3 Dual-Quantizer Cascades

Dual-quantization techniques can be easily incorporated to generic MASH ΣΔMs. As shown in (2.30), the output of a cascade ΣΔM ideally contains only the input signal and the last-stage quantization error, whereas the quantization errors from the remaining stages are removed by the DCL. The dynamic range of the ΣΔM can therefore easily be increased by using multi-bit quantization only in the last stage of the cascade. The remaining quantizers are usually single-bit to retain linear feedback in the front-end stages. This way non-linearities in the multi-bit DAC are injected in the back-end stage of the cascade and they will appear at the modulator output with the attenuation provided by the integrators in the preceding stages. This is usually sufficient for achieving a good linearity performance in the MASH ΣΔM, with no need for correction techniques in the multi-bit DAC.

The resulting topology is that illustrated in Fig. 2.16, considering $B_i = 1$ for $i = 1, \ldots, N-1$ and $B_N = B$. Under ideal linear analysis, the output of this generic N-stage cascade ΣΔM with multi-bit quantization at the back end yields

$$\begin{aligned} Y(z) &= \mathrm{STF}(z)X(z) + \mathrm{NTF}_N(z)E_N(z) + \mathrm{NTF}_{\mathrm{DAC}}(z)E_{\mathrm{DAC}}(z) \\ &= z^{-L}X(z) + d_{2N-3}(1-z^{-1})^L E_N(z) + d_{2N-3}(1-z^{-1})^{(L-L_N)}E_{\mathrm{DAC}}(z) \end{aligned} \tag{2.34}$$

where L is the summation of the stage orders, d_{2N-3} is the scaling factor due to inter-stage coupling (usually 2 or 4), $E_N(z)$ is the last-stage quantization error and $E_{\mathrm{DAC}}(z)$ is the error in the multi-bit DAC. Note that the DAC error will be $(L - L_N)$th-order shaped—i.e., the overall cascade order minus that of the back-end stage.

Figure 2.24 illustrates the block diagram of the first reported cascade employing dual quantization [Bran91]. It is a 3rd-order cascade ΣΔM employing a 2nd-order

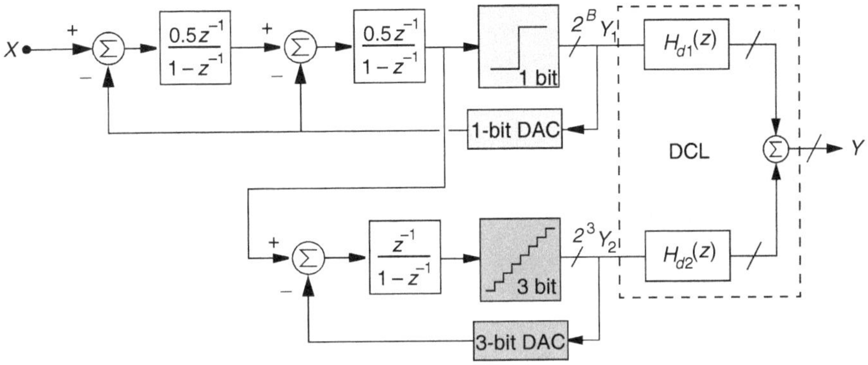

Fig. 2.24 Dual-quantizer 2-1 cascade Σ ΔM. [Bran91]

front-end stage with 1-bit quantization for linear feedback and a 1st-order back-end stage with a 3-bit embedded quantizer for enhanced dynamic range. Many integrated cascade Σ ΔMs using this dual-quantization scheme can be found in the literature [Nors97a, Rodr03].

Cascade Σ ΔMs employing multi-bit quantization in all stages have been also reported [Fuji00, Vleu01]. Note that, under ideal conditions, the quantization errors of the first stages are cancelled out at the modulator output by the DCL. Multi-bit quantization can be used in these stages for the main purpose of reducing the corresponding quantization errors that will in practice leak to the output. Being that the case, DEM techniques can be incorporated to the multi-bit DAC of the front-end stage in order to achieve the required modulator linearity (if required), whereas succeeding multi-bit stages often rely just in the in-band attenuation provided by the preceding integrators [Vleu01]. Using multi-bit quantization in all stages of a cascade Σ ΔM presents an additional appealing feature: the coupling factors between stages can be increased in comparison to a 1-bit approach without overloading the quantizers [Fuji00]. This way, the scaling factor d_{2N-3} that amplifies the last-stage quantization error [see (2.34)] can be smaller than unity, leading to an improvement of the global performance[7].

Cascades using tri-level (1.5-bit) quantizers can also be found in literature [Dedi94] in order to reduce quantization errors in comparison with 1-bit quantization—over 3-dB SNR improvement. Although tri-level coding is not inherently linear, it is often used in fully-differential SC Σ ΔMs, since highly-linear tri-level DACs can be easily implemented using just one extra switch [Reut02].

[7] In [Fuji00], 4-bit quantization is used in all stages of a 2-1-1 cascade Σ ΔM. The resulting scaling factor is $d_3 = 1/32$, which leads to a 30-dB increase in DR in comparison to an ideal 4th-order 4-bit Σ ΔM.

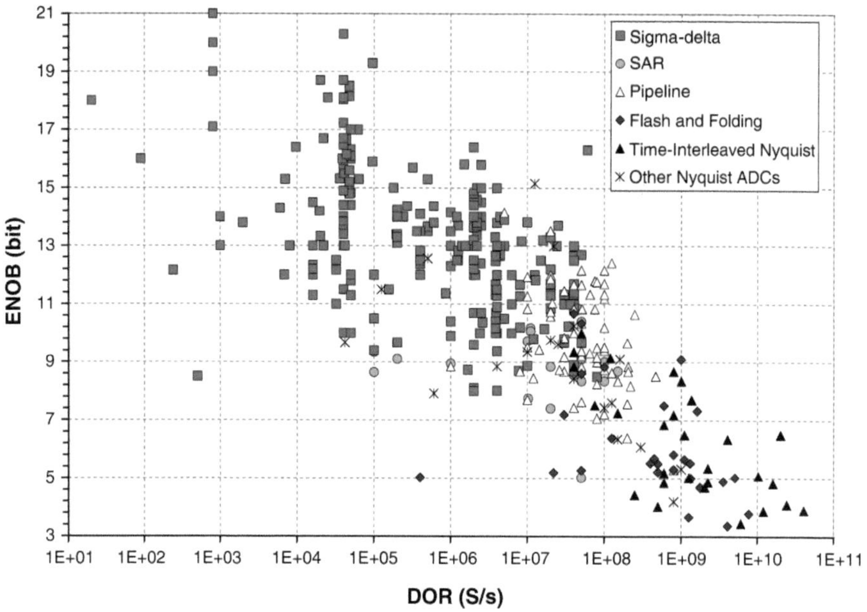

Fig. 2.25 State-of-the-art CMOS ADCs in the ENOB-DOR plane

2.7 State of the Art in ΣΔ ADCs

Although ΣΔ converters were originally conceived for low-frequency, high-resolution applications (like audio and precision measurement) in which they clearly outperform other existing A/D conversion techniques, their use has progressively extended to medium- and high-frequency applications with the development of VLSI technologies. Fig. 2.25 illustrates the state of the art in ADCs implemented in CMOS processes reported up to year 2010 and places them in the ENOB-DOR plane, where DOR = 2 BW stands for the digital output rate of the ADC (i.e., the Nyquist rate). Data in Fig. 2.25 corresponding to CMOS Nyquist-rate ADCs has been taken from [Murm10]. It can be noted that ΣΔ ADCs cover a wide frequency range, ranging from 10 Hz to 50 MHz. Larger conversion bandwidths are still dominated by Nyquist-rate ADCs, especially flash and folding. Oversampling techniques are less efficient in these applications because of the excessive operation speed that is required in the analog blocks. They however coexist with algorithmic, subranging and especially pipeline converters in communication applications.

As discussed in previous sections, many different alternatives exist for the realization of ΣΔ modulators. Fig. 2.26 looks closely at the reported ΣΔMs taking their topologies into account. For comparison purposes, they have been classified into three categories—single- and multi-bit single loops and cascades—distinguishing between discrete-time (DT) and continuous-time (CT) implementations. Note from Fig. 2.26 that, although the majority of reported ICs correspond to DT ΣΔMs,

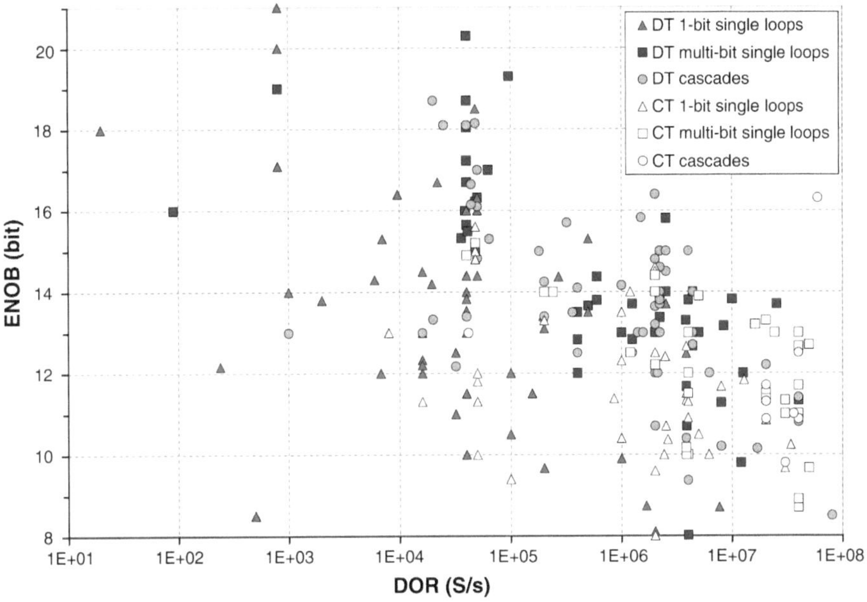

Fig. 2.26 State-of-the-art CMOS ΣΔMs in the ENOB-DOR plane

CT ΣΔMs almost reach up to 35%. DT implementations are clearly dominant for high resolutions (>16 bit), whereas CT ΣΔMs dominate for large bandwidths (DOR > 10 MS/s). Nevertheless, both types of implementations coexist for large-bandwidth, medium-resolution (10–14 bit) applications.

The performance of ΣΔMs is globally quantified in terms of their main specifications (effective resolution, signal bandwidth and power consumption of the circuit) through the following figures of merit (FoMs), respectively proposed by [Good96, Rabi97],

$$\mathrm{FoM}_1|_{\mathrm{pJ/conv}} = \frac{\mathrm{Power(W)}}{2^{\mathrm{ENOB(bit)}}\mathrm{DOR(S/s)}} 10^{12} \tag{2.35}$$

$$\mathrm{FoM}_2 = 2\mathrm{kT}\frac{3 \cdot 2^{2\mathrm{ENOB(bit)}}\mathrm{DOR(S/s)}}{\mathrm{Power(W)}} \tag{2.36}$$

where k is the Boltzmann constant and T is the temperature (measured in Kelvin, K). Note that FoM_1 emphasizes power consumption, whereas FoM_2 stresses effective resolution, represented by the ENOB. Therefore, the smaller the FoM_1 value and the larger the FoM_2 value, the "better" the ΣΔM is. For comparison purposes, the aforementioned information is graphically illustrated in Figs. 2.27 and 2.28.

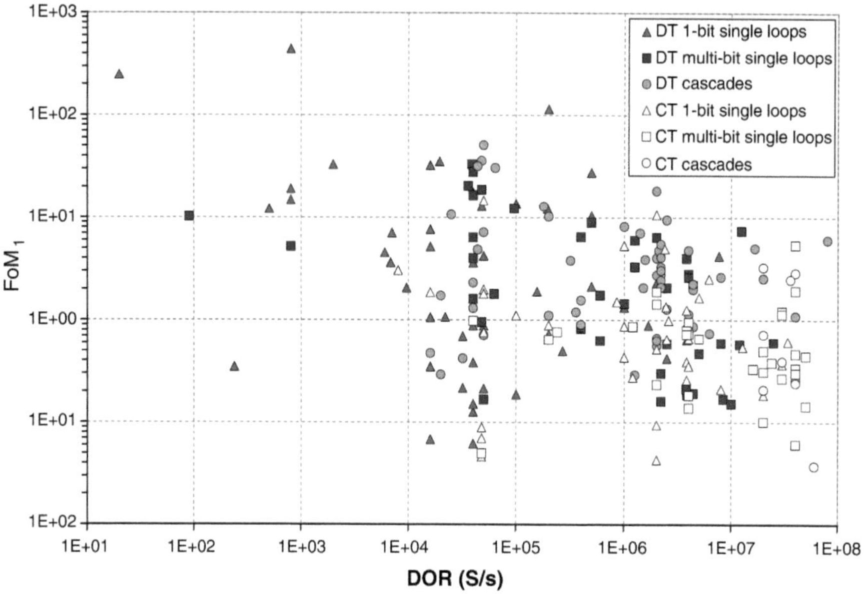

Fig. 2.27 FoM_1 versus DOR of the reported low-pass ΣΔM ICs

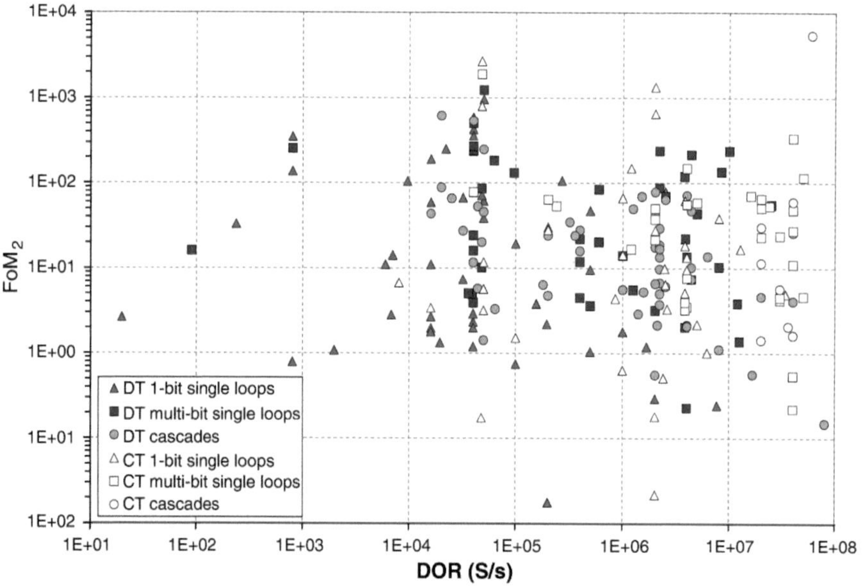

Fig. 2.28 FoM_2 versus DOR of the reported low-pass ΣΔM ICs

2.8 Summary

The basic principles of $\Sigma\Delta$ modulation have been presented in this chapter. The benefits of oversampling and noise shaping on the ADC performance have been discussed and the generic scheme, ideal performance and metrics of $\Sigma\Delta$ converters have been defined and compared with Nyquist ADCs.

Topological alternatives for the practical implementation of $\Sigma\Delta$ modulators have been also presented, addressing the realization of stable high-order $\Sigma\Delta$Ms by means of optimized single loops or by cascading of low-order stages. The use of multi-bit internal quantization has also been presented as an alternative to enhance the modulator effective resolution. Besides reducing quantization error, it provides better stability properties to single-loop architectures, but jeopardizes linearity. Non-linearity error in the multi-bit DAC due to component mismatching has been discussed, together with techniques to palliate its impact on the modulator performance, such as DEM and dual-quantization schemes.

The many design alternatives for the practical implementation of $\Sigma\Delta$ ADCs are finally summarized in the state of the art of reported low-pass ICs.

Chapter 3
New $\Sigma\Delta$ Cascade Modulators for the Next Generation of Wireless Telecom

THE MANDATORY USE OF LOW OSRs IN WIDEBAND STANDARDS has forced $\Sigma\Delta$Ms to achieve the required DRs by resorting to different strategies, namely: high-order shaping—that can be implemented with high-order single-loop $\Sigma\Delta$Ms or, alternatively, by cascading low-order stages—and multi-bit embedded quantizers—that leads to the use of multi-bit DACs and DEM for medium-to-high ENOB specifications. A different approach, fully compatible with the former strategies, is based on the optimal spreading of notches within the signal band to increase ENOB [Schr93].

This chapter presents a number of techniques for building efficient cascade architectures intended for wideband and multi-mode applications. The resulting cascade $\Sigma\Delta$Ms can easily implement both reconfiguration strategies and parallel processing, making them very appropriate for the A/D conversion of transceivers for SDR.

The chapter is divided as follows: Sect. 3.1 shows a number of techniques for A/D conversion in SDR applications; namely, modulator order reconfiguration and concurrency. It then goes on to describe previously reported cascade $\Sigma\Delta$Ms that can apply these techniques—such as a family of expandable cascades, unity-STF cascade $\Sigma\Delta$Ms, the so-called sturdy MASH (SMASH) $\Sigma\Delta$M and resonation-based cascade $\Sigma\Delta$Ms. Sect. 3.2 combines all previous architectural strategies, giving rise to more efficient modulators. The resulting topological advantages are discussed based on a set of case studies for both wideband and multi-mode applications. However, a number of the proposed architectures suffer from timing issues as described in Sect. 3.3. Nevertheless, the same section examines two reliable topologies that combine most of the proposed architectural improvements while removing the corresponding timing problems.

3.1 Strategies for Efficient Cascade $\Sigma\Delta$Ms in Multi-Mode Applications

This section first discusses two concepts; namely: architectural reconfiguration and parallel processing strategies. Both ideas are very appropriate for an SDR scenario and can be easily implemented in several kinds of cascade $\Sigma\Delta$Ms. Then, an example

A. Morgado et al., *Nanometer CMOS Sigma-Delta Modulators for Software Defined Radio*, DOI 10.1007/978-1-4614-0037-0_3, © Springer Science+Business Media, LLC 2011 of a cascade architecture capable of expanding its order is presented. This last topology can implement the aforementioned ideas very efficiently. A number of the previously reported concepts used to increase the performances of cascade ΣΔMs is also shown. These improvements involve increasing the modulator resolution and its operation signal bandwidth or reducing the requirements of the modulator building blocks in order to diminish their power consumption.

3.1.1 Modulator Order Reconfiguration and Concurrency

The most commonly employed reconfiguration strategy at the architectural level is to adapt the order of the loop filter in the modulator to the requirements of the operation modes[1] [Dezz03, Chri07, Morg07b, Bos10, Chri10, Morg10]. In general, the modulator architecture needs to be modified for this purpose. A single-loop modulator would require turning a number of its integrators on or off. The order of the filter would therefore be modified, while the in-loop quantization and feedback operations remain. Several switchable connections are probably needed in this case to connect (or disconnect) the corresponding integrators. A structure with an order potentially easier to reconfigure is the cascade architecture, because of its modular structure. Let us consider, for instance, a 2-2 cascade whose stages outputs can be processed separately. The output of the overall modulator can be obtained after digital cancellation logic processing and provides a fourth-order shaping of the last-stage quantization error, whereas the output of the first stage directly corresponds to a second-order filtering. A cascade ΣΔM is therefore by definition a system with adjustable order. Furthermore, concurrency can easily be implemented in a cascade, as will be described below, while this last feature would require a large amount of additional circuitry in a single-loop ΣΔM.

3.1.1.1 Expandable and Modulator Order Reconfiguration for Cascade ΣΔMs

Expandable cascades can be defined as those cascade ΣΔMs the order of which can be increased just by introducing more stages in the cascade. These kinds of cascades are usually made up of low-order and unconditionally stable stages [Rebe97]. Figure 3.1 shows the block diagram of a generic cascade ΣΔM made up of N stages. The input of each stage is the quantization error of the previous stage, except for the first stage. All stages are independently switchable, so they can be connected or disconnected to the cascade as desired. The digital cancellation logic needs to be adjusted according to the cascade order, L.

[1] Note that the most used reconfiguration strategy consists of changing OSR [Rosa09]. To this purpose, the sampling frequency is adjusted for a given bandwidth. This can therefore be considered a circuit-level rather than an architectural-level reconfiguration strategy.

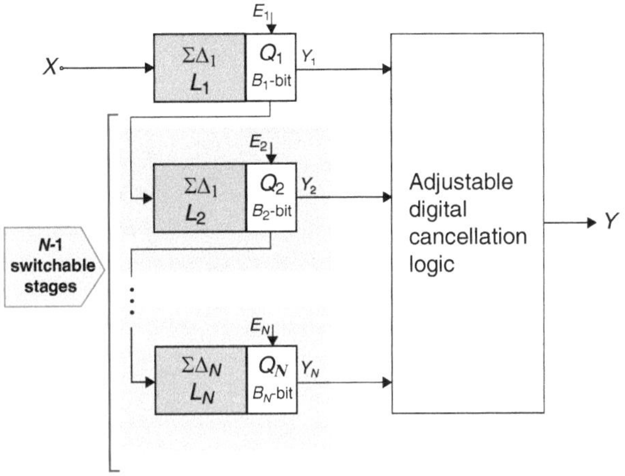

Fig. 3.1 Conceptual diagram of the order reconfiguration in an N-stage cascade $\Sigma\Delta$M

A FIR filtering of the in-band quantization error is obtained for such systems, the order being that of the filter; i.e., L is equal to the number of integrators connected in the cascade $\Sigma\Delta$M. If a stage is turned off, its building blocks can be powered down to save power. Power is therefore traded by performance, as in conventional $\Sigma\Delta$Ms but in a reconfigurable manner with adaptive L. More efficient multi-mode solutions can also be obtained by combining the adjusting modulator order with other reconfiguration strategies, such as the adaptation of the OSR. Chapters 4, 5 and 6 make use of this reconfiguration strategy at the architectural level, together with others.

3.1.1.2 Concurrent $\Sigma\Delta$Ms

Concurrency can be used to expand the functionality of multi-standard $\Sigma\Delta$Ms by allowing the parallel processing of N parallel input signals. In general, N $\Sigma\Delta$Ms in parallel are needed to implement this feature.

To achieve concurrency in single-loop $\Sigma\Delta$Ms extra circuitry may have to be included in the original modulator. In addition, the integrators might need to be made switchable to enable their connection from one $\Sigma\Delta$M to another of the N parallel $\Sigma\Delta$Ms. On the other, N embedded quantizers are required, while only one is used for the original single-loop $\Sigma\Delta$M. Therefore, not only do complex switchable connections need to set up inside the loop but also extra circuitry is required for the implementation of concurrent single-loop $\Sigma\Delta$Ms.

Cascade $\Sigma\Delta$Ms can be adapted to work in parallel more easily. In general, an N-stage MASH $\Sigma\Delta$M can reuse a maximum of N stages as concurrent $\Sigma\Delta$Ms [Rosa09, Silv09]. Figure 3.2 illustrates this concept by showing that an N-stage cascade $\Sigma\Delta$M can thus operate as a combination of cascades[2] (made up of a number of stages in the

[2] Note that the inputs of the stages can be either those coming from the concurrent input network or those from the previous stage, so, eventually, a number of cascade $\Sigma\Delta$Ms could be operating concurrently as well.

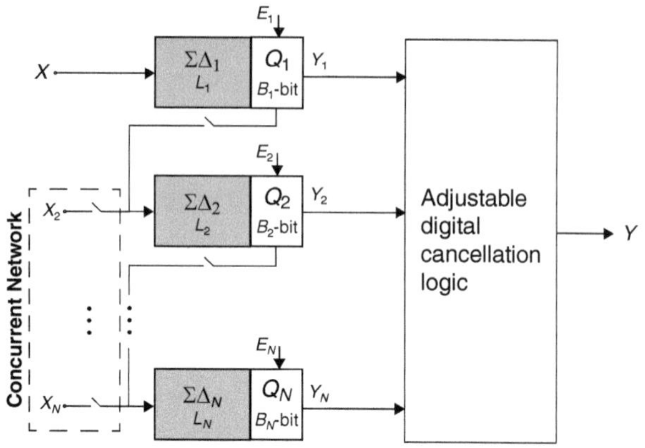

Fig. 3.2 Conceptual diagram of concurrency in an N-stage cascade ΣΔ modulator

figure), or concurrent single-loop stages or both. Virtual switches are needed to select each stage input (except for the first stage); either the inter-stage path coming from previous input—thus enabling a cascade operation— or the corresponding input from the concurrent network. In this last case, this stage operates as an independent single-loop ΣΔM. In practice, the virtual switches might be implemented as SC branches at the input of the front-end integrator of each stage[3]. A number of control signals are needed to enable the switches that correspond to the desired operation mode in each case, either concurrent or cascaded. The digital cancellation logic processing depends on the concurrent configuration and also on the number of stages working as single-loop or part of a cascade [Silv09].

Note that Chap. 6 presents an expandable ΣΔM that includes concurrency and can process up to 3 modes in parallel.

3.1.2 *Expandable Cascade ΣΔMs*

Figure 3.3 shows the generic block diagram of the family of high-order cascades proposed in [Mede03], henceforth called $2-1^{L-2}$ ΣΔM. An Lth-order modulator is formed with a second-order stage followed by $L-2$ identical first-order stages. As in all cascade ΣΔMs, the outputs of the $L-1$ stages are combined and processed in the digital domain through simple operators to cancel out the quantization noise generated in each stage but the last. Linear analysis shows that the output of the $2-1^{L-2}$ ΣΔM can be expressed in Z-domain as

$$Y(z) = z^{-L}X(z) + 2(1-z^{-1})^{L}E_N(z) \tag{3.1}$$

where $X(z)$ stands for the input signal, which is simply delayed, and $E_N(z)$ stands for the last-stage quantization error, which is Lth-order shaped. Note that the cascade

[3] Analog adders will also need these virtual switches for unity-STF ΣΔMs.

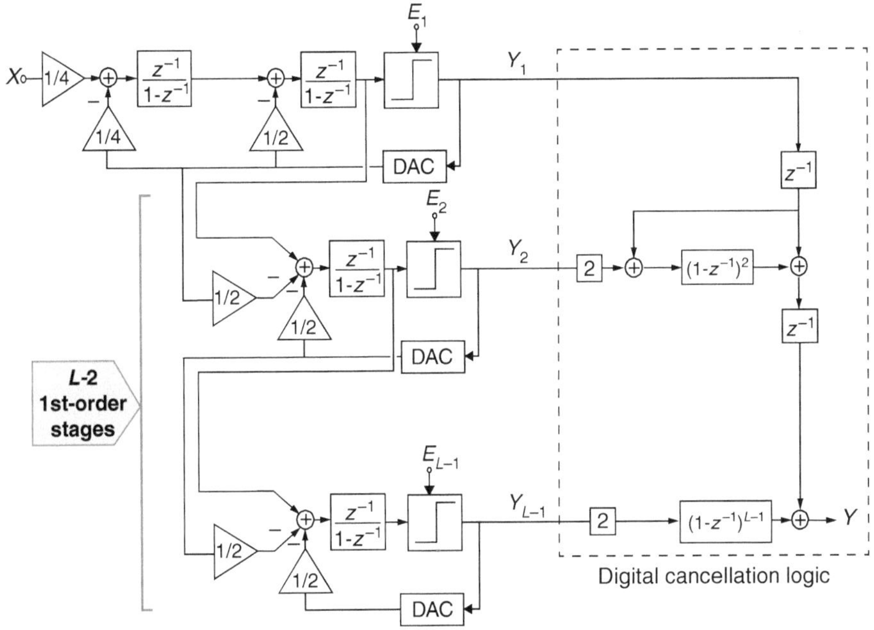

Fig. 3.3 *L*th-order expandable cascade ΣΔM in [Mede03]

response equals that of an ideal *L*th-order ΣΔM, except for the scaling factor 2. This factor derives from the inter-stage scaling of 1/2 required to avoid premature overload when transmitting the signal from one stage to the next.

By integrating the error term in (3.1) over the signal band, the in-band quantization error power is obtained as

$$P_Q = 4\sigma_Q^2 \cdot \frac{\pi^{2L}}{(2L+1)\mathrm{OSR}^{2L+1}} \tag{3.2}$$

where $\sigma_Q^2 = \Delta^2/12$ is the power associated with the quantization error in the last-stage single-bit quantizer ($\Delta = 2V_{ref}$ stands for its full scale). Note that 2 in (3.1) means an amplification factor of 4 in the in-band quantization error power and, thus, leads to a reduction of 6dB (1bit) in the dynamic range of the $2 - 1^{L-2}$ cascade in comparison with an ideal *L*th-order ΣΔM.

The most appealing feature of this architecture—with the set of coefficients proposed—is that, in theory, it can easily be set to any order, just by changing the number of identical first-order stages while maintaining a correct operation[4]. As shown in Fig. 3.4, the overload level of the expandable cascade does not change with the order of the ΣΔM.

[4] However the presence of noise leakages prevents the use of very high orders [Rebe97, Rio06].

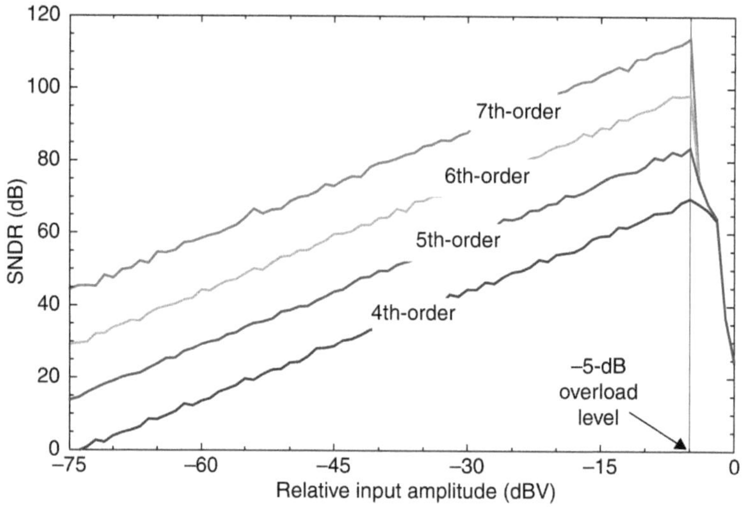

Fig. 3.4 SNDR curves of the $2-1^{L-2}$ cascade for several modulator orders (OSR = 16)

The set of integrator coefficients depicted in Fig. 3.3 also presents the following interesting features:

- The output swing required in all integrators does not exceed the quantizer full scale. This very appealing feature for low-voltage implementations is illustrated in Fig. 3.5 for a fifth-order cascade.
- The largest coefficient of each three-weight integrator can be obtained as the summation of the others, so three-branch SC integrators are not required. By properly sharing the SC input stages, all coefficients can be implemented with just two-branch integrators, minimizing the total number of unit capacitors [Rio06].
- All first-order stages contain the same coefficients, allowing them to be electrically identical. This considerably simplifies the electrical and physical implementation of the modulator.

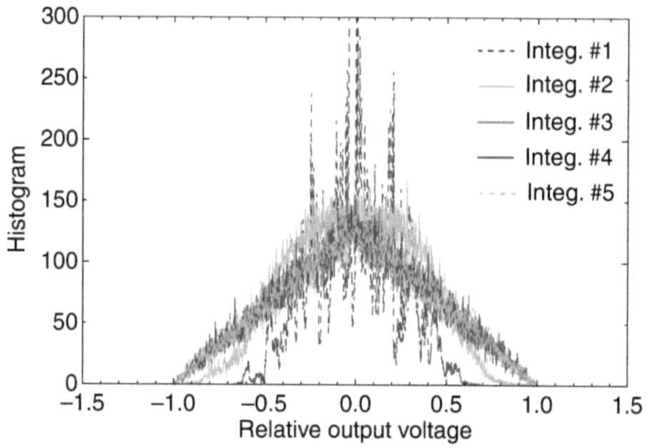

Fig. 3.5 Histogram of the integrator outputs relative to the reference voltage for $L=5$

A dual-quantization operation [Bran91, Dias93, Tan93] can easily be achieved in the $2-1^{L-2}$ cascade ΣΔM by including multi-bit quantization only in the last stage, while the remaining quantizers in the modulator are single-bit. In this case, the coefficients in the last-stage integrator can be multiplied by a factor of 2 in order to have a loop gain of 1. This entails the multi-bit ADC and DAC produce the same full scale, which considerably simplifies their designs.

If errors in the multi-bit DAC are considered in the linear analysis, the output of the dual-quantization $2-1^{L-2}$ ΣΔM can be obtained as

$$Y(z) = z^{-L}X(z) + 2(1-z^{-1})^{L}E_{N}(z) + 2(1-z^{-1})^{(L-1)}E_{DAC}(z) \tag{3.3}$$

where $E_{\mathrm{DAC}}(z)$ stands for the Z-domain transform of the last-stage quantization error, which is $(L-1)$th-order shaped. Thus, provided that errors in the multi-bit DAC are considerably high-pass filtered, its non-linearity can to some extent be tolerated with no need for calibration/correction mechanisms.

The in-band quantization error power can be estimated as [Rio06]

$$P_Q = 4\sigma_Q^2 \cdot \frac{\pi^{2L}}{(2L+1)\mathrm{OSR}^{2L+1}} + 4\sigma_{\mathrm{DAC}}^2 \cdot \frac{\pi^{2(L-1)}}{(2L-1)\mathrm{OSR}^{2L-1}}$$

with

$$\sigma_Q^2 = \frac{1}{12}\cdot\left(\frac{\Delta}{2^B-1}\right)^2, \quad \sigma_{\mathrm{DAC}}^2 \cong \frac{1}{2}\cdot\Delta^2\cdot\left(\frac{\mathrm{INL}}{100}\right)^2 \tag{3.4}$$

where σ_Q^2 is the power of the last-stage quantization error ($\Delta = 2V_{ref}$ stands for the multi-bit quantizer full scale, FS, and B for its resolution) and σ_{DAC}^2 is the power associated with the DAC errors, with INL being the DAC integral non-linearity expressed in %FS.

3.1.3 Unity-STF Cascade Architectures

Over the last few years, unity-STF cascades have proven to be very suitable for wideband and multi-mode applications [Gagg04, Nam05, Chri07, Chri10, Morg10]. This section deals with the architectural implementation of these kinds of topologies.

3.1.3.1 Single-Loop Unity-STF ΣΔMs

Figure 3.6 shows a second-order single-loop ΣΔM that makes use of analog feed-forward paths to implement unity STF [Bena93, Silv01, Silv04a]. Using a linear model for the internal B-bit quantizer, the Z-transform of the modulator output, $Y(z)$, is the following:

$$Y(z) = \mathrm{STF}(z)X(z) + \mathrm{NTF}(z)E(z) \tag{3.5}$$

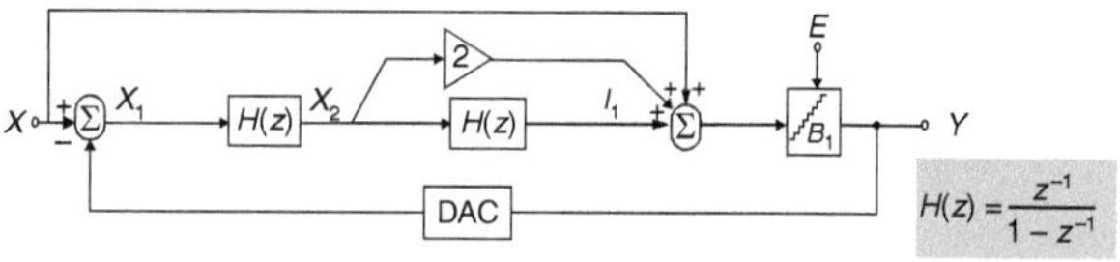

Fig. 3.6 Second-order single-loop ΣΔM with unity STF [Silv01]

where $X(z)$ and $E(z)$ are the Z-transform of the input and the quantization error, respectively, while STF(z) and NTF(z) are given by:

$$\text{STF}(z) = 1$$
$$\text{NTF}(z) = (1 - z^{-1})^2 \qquad (3.6)$$

One of the most remarkable advantages of the modulator in Fig. 3.6 is that, at least ideally, there is no input signal trace processed by the integrators. This is easy to demonstrate by obtaining the Z-transform of the integrator inputs, $X_1(z)$ and $X_2(z)$, given by:

$$X_1(z) = -(1 - z^{-1})^2 \cdot E(z)$$
$$X_2(z) = z^{-1} \cdot (1 - z^{-1})^2 \cdot E(z) \qquad (3.7)$$

The combined used of feed-forward paths—giving rise to unity STF—, together with multi-bit internal quantizers, therefore make the architecture in Fig. 3.6 very suitable for low output swing requirements, reduced sensitivity to amplifier non-linearities and high overload levels [Bena93, Silv01, Silv04a, Yao06]. Moreover, the use of scaling on the integrator coefficients helps to reduce the opamp output swing requirements further on [Gupt02, Schr05b]. This procedure is detailed below step by step.

3.1.3.2 Generic Coefficients for Single-Loop Unity-STF ΣΔMs

Figure 3.6 depicts a second-order ΣΔM with unity STF. Let us assume that a coefficient c_2 is included at the input of the second integrator, as shown in Fig. 3.7a. A feed-forward factor $a_2 = 1/c_2$ is therefore needed to compensate for this architectural modification in order to keep STF and NTF in (3.6) unmodified. In addition to this second-integrator scaling factor, c_1 is also included at the first integrator input as shown in Fig. 3.7b, the resulting feed-forward coefficients thus being the following [Silv04a]:

$$a_1 = \frac{2}{c_1} \quad a_2 = \frac{1}{c_1 \cdot c_2} \qquad (3.8)$$

3.1.3.3 Cascade Unity-STF ΣΔMs

The underlying principles in the block diagram depicted in Fig. 3.6—and Fig. 3.7b when considering integrator scaled coefficients— can be extended to cascade ΣΔMs

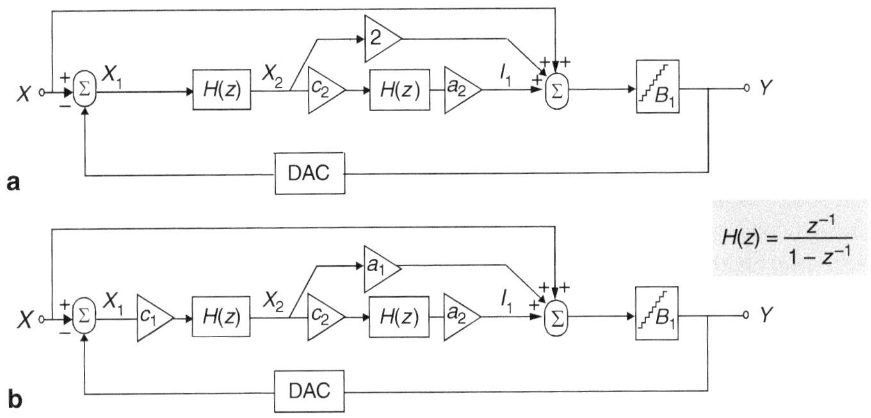

Fig. 3.7 USTF second-order single-loop ΣΔM with scaling on integrator coefficients. **a** Fig. 3.6 with the additional coefficient c_2, **b** block diagram with integrators scaling coefficients c_1 and c_2

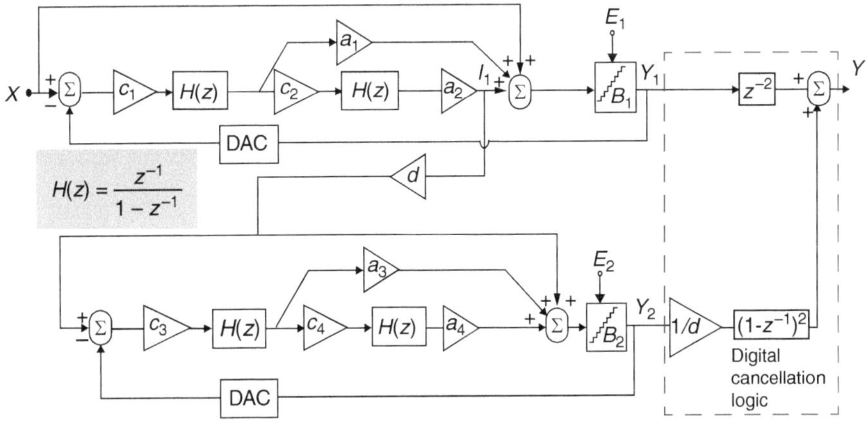

Fig. 3.8 Block diagram of a 2-2 unity-STF cascade ΣΔM architecture [Nam05]

[Silv01, Silv04a]. Ideally, the quantization error can be extracted directly from node $I_1(z)$ as $-z^{-2} \cdot E(z)$ [Silv04a]. Thus, a cascade that uses node $I_1(z)$ as the only inter-stage path will need to include a 2-delay digital block in the cancellation logic (z^{-2}) to compensate for this. Figure 3.8 illustrates a 2-2 cascade based on analog feed-forward paths that implement unity STF [Goth02, Silv04b, Nam05, Chri07, Morg07b, Morg10]. This figure also includes generic coefficients used to reduce the opamp output swing requirements. Following the procedure illustrated above—see the extraction of (3.8), the relation between all the coefficients for each stage is provided by

$$a_1 = \frac{2}{c_1} \quad a_2 = \frac{1}{c_1 \cdot c_2} \quad a_3 = \frac{2}{c_3} \quad a_4 = \frac{1}{c_3 \cdot c_4} \tag{3.9}$$

and the resulting STF and NTF are:

$$\begin{aligned} \mathrm{STF}(z) &= 1 \\ \mathrm{NTF}(z) &= (1-z^{-1})^4/d \end{aligned} \tag{3.10}$$

The inter-stage coefficient d thus modifies the in-band quantization error power. However, excessively increasing its value can lead to an overloading of the next stage. In other words, large values of d mean high output swing requirements for the last-stage opamps.

3.1.3.4 Expandable Unity-STF ΣΔMs with Generic Coefficients

The cascade ΣΔM in Fig. 3.8 can be expanded to increase its order. Fig. 3.9 shows the schematic of an expandable unity-STF ΣΔM architecture. It consists of an extension of the architecture in Fig. 3.8 produced by incorporating second-order stages until the desired order, L, is obtained. There is therefore an Lth-order shaping of the quantization error. In practice, however, the order is limited due to noise leakages coming from the mismatches between the analog circuitry in the front-end stages and the circuitry of the digital cancellation logic (see Sect. 2.5). Nevertheless, the extra robustness of these kinds of topologies against building block non-idealities and non-linearities makes unity-STF topologies better suited for larger values of L

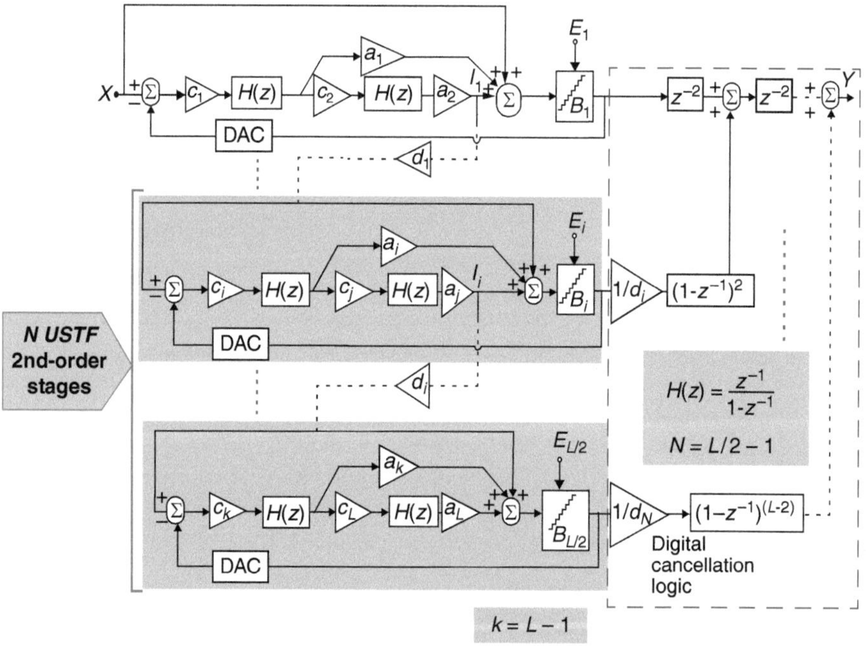

Fig. 3.9 Block diagram of an expandable feed-forward ΣΔM architecture

when compared with conventional expandable cascade ΣΔMs such as, for example, that in Fig. 3.3. Note from Fig. 3.9 that the inter-stage path of each stage is taken from node $I_i\ (z)$ of the previous stage. The unity STF is preserved and the NTF provides a FIR filtering corresponding to the modulator order as follows:

$$\begin{aligned} \mathrm{STF}(z) &= 1 \\ \mathrm{NTF}(z) &= (1-z^{-1})^L/d_N \end{aligned} \tag{3.11}$$

The in-loop coefficients in Fig. 3.9 consider a scaling of the integrator coefficients. Following the above-mentioned procedure to extract the relation between all coefficients in each stage of the cascade, the resulting in-loop coefficients are given as

$$a_{2i-1} = \frac{2}{c_{2i-1}}; \quad a_{2i} = \frac{1}{c_{2i-1}\cdot c_{2i}}; \quad i = 1, 2, \ldots \frac{L}{2} \tag{3.12}$$

3.1.4 Previously Reported Resonation-Based ΣΔMs

Resonation might be a good alternative for increasing modulator resolution when the order cannot be increased mainly due either to stability problems for single-loop ΣΔMs or to the influence of quantization noise leakages coming from the front-end stages in cascade ΣΔMs. Resonation is also a good approach for overcoming the problems of multi-bit quantizers, such as the implementation of complex DEM structures. In addition, resonation is specially beneficial for limited values of OSR because then the aforementioned strategies lose effectiveness.

This section reviews two reported approaches to implement resonation; namely, local and global. The former has been widely used for single-loop ΣΔMs [Schr05b] and can also be implemented in cascade ΣΔMs [Yuan05, Chri07]. Global resonation, on the contrary, has only been conceptually described in [Sanc06].

3.1.4.1 Local Resonation

One efficient way to increase the modulator resolution without penalizing the number of integrators consists of including resonators inside the modulator loop filter. This so-called local resonation technique has been used in ΣΔMs considering either single-loop [Mark04] or cascade topologies [Yuan05]. This section presents a review of previously reported single-loop 2nd-order USTF ΣΔMs that incorporate resonation. If larger orders were considered, stability would have become a major issue. In addition, resonators can be implemented in ΣΔMs with STFs different from 1, but therefore losing the appealing features of USTF ΣΔM.

Resonation allows the zeros of the NTF to be shifted from DC to an arbitrary frequency, thus allowing them to be to distributed in such a manner that the in-band quantization error can be minimized [Schr93]. The conventional second-order

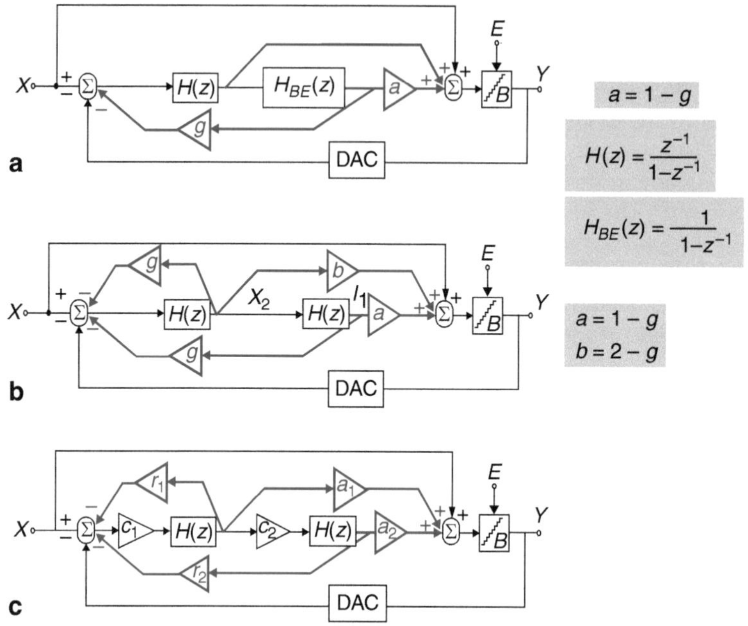

Fig. 3.10 Different alternatives for implementing resonation on 2nd-order ΣΔMs with unity STF [Mark04]: **a** using BE integrators, **b** employing FE integrators only, **c** employing FE integrators only and including a scaling of the integrator coefficients

Table 3.1 Scaling on feed-forward and resonation coefficients

Feed-forward coefficients	Resonation coefficients
$a_1 = \frac{b}{c_1} = \frac{(2-g)}{c_1}$	$r_1 = \frac{g}{c_1}$
$a_2 = \frac{a}{c_1 \cdot c_2} = \frac{(1-g)}{c_1 \cdot c_2}$	$r_2 = \frac{g}{c_1 \cdot c_2}$

high-pass filtering in NTF—see (3.6)—is now replaced by an NTF with notches as follows:

$$\mathrm{NTF}(z) = z^{-2} - (2-g) \cdot z^{-1} + 1 \tag{3.13}$$

Two different architectures are proposed in [Mark04] to implement this NTF. One of them makes use of backward Euler (BE) integrators, while the other does not need them and only uses forward Euler (FE) integrators. The block diagrams of both approaches are depicted in Fig. 3.10a and b, respectively. However, the solution given in Fig. 3.10b needs more coefficients to implement resonation as shown in the figure. The resonation coefficient relations are also included in Fig. 3.10. Figure 3.10c illustrates the use of scaling factors over the integrators of the architecture in Fig. 3.10b. The resulting in-loop coefficients, after this architectural modification, are shown in Table 3.1

Let us decompose the NTF given by (3.13) as

$$\mathrm{NTF}(z) = z^{-2} - (2-g) \cdot z^{-1} + 1 = (1 - z^{-1})^2 + g \cdot z^{-1} \tag{3.14}$$

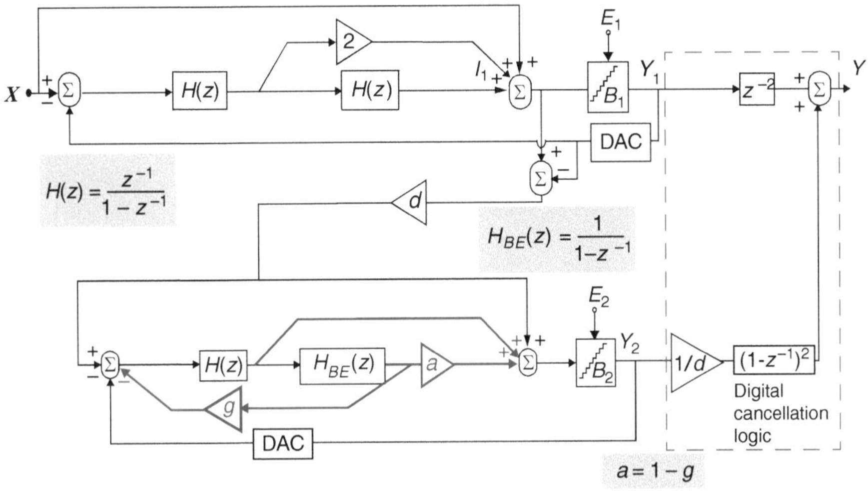

Fig. 3.11 Fourth-order cascade ΣΔM architecture with local resonation and USTF [Yuan05]

where the term $(g \cdot z^{-1})$ is incorporated to a second-order FIR filtering. This is equivalent to add $g \cdot z^{-1} \cdot E(z)$ directly to the modulator output. The same paths in which the input signal is provided (first-integrator input and analog adder) are used to give the above-mentioned term, exploiting thus the unity-STF feature of the architecture in Fig. 3.10b. To this purpose, the delayed version of the quantization error, $E(z)$, is extracted in the scheme in Fig. 3.10b from the sum $X_2(z) + I_1(z)$ and later provided to the first-integrator input and analog adder.

In the case of cascade architectures, only the last stage normally uses resonation in order to reduce the complexity of the digital cancellation logic [Sanc06]. As an illustration, Fig. 3.11 shows a cascade with local resonation in the back-end stage. This topology, originally presented in [Yuan05], takes advantage of the unity STF in the two stages, thus obtaining relaxed output swing requirements.

3.1.4.2 Global or Inter-Stage Resonation

In [Sanc06], a new kind of resonation strategy, named global or inter-stage resonation, is applied to cascade ΣΔMs. This approach, illustrated in Fig. 3.12 for a fourth-order cascade architecture, is obtained by feeding back a scaled version of the 2nd-stage quantization error to the input of the 1st-stage quantizer. In other words, global resonation is based on the partial processing of the quantization error of one stage in a previous stage, so there is an additional resonation term in the NTF. However, this technique cannot be applied to single-loop ΣΔMs because, in this case, there is only one loop and therefore only one quantization error.

In both cases, either using local resonation in the last stage of the modulator (see Fig. 3.11) or global resonation in the overall cascade, the NTF becomes:

$$\mathrm{NTF}(z) = \frac{-(1 - z^{-1})^2 \cdot [1 - (2 - g) \cdot z^{-1} + z^{-2}]}{d} \tag{3.15}$$

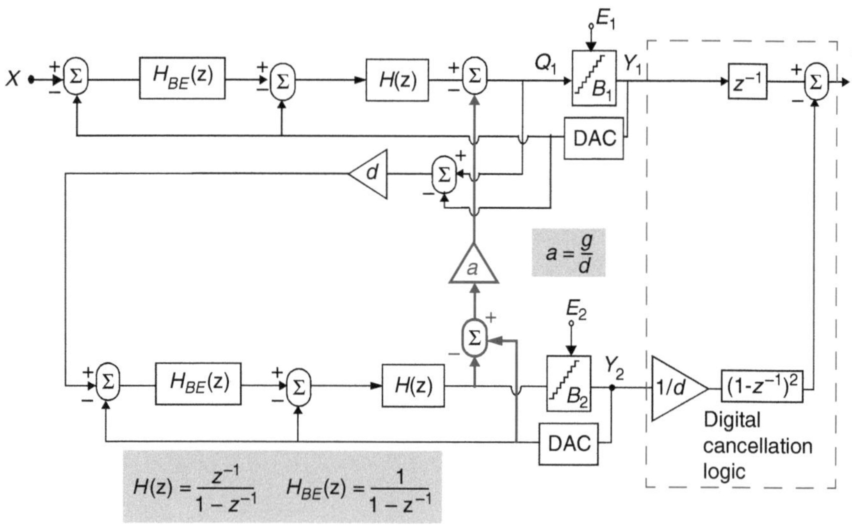

Fig. 3.12 Fourth-order cascade ΣΔM with global resonator [Sanc06]

where d stands for the inter-stage gain. Note that the zeros of NTF are a function of g, which can be optimally chosen to maximize the SNR. Indeed, practical cases may lead to a resolution increase of up to 11dB (see Appendix A) [Hamo06]. Another conclusion deriving from (3.15) is that increasing d—to reduce the quantization noise—yields a reduction of the feedback coefficient a (see Fig. 3.12). In practice, this results in a smaller capacitor ratio, which makes the electrical implementation more difficult and prone to circuit non-idealities.

Most of previously reported resonation-based ΣΔMs, like those shown in Fig. 3.11 and 3.12, make use of BE or non-delayed integrators, which increase the speed requirements of the amplifiers used [Schr05a]. By contrast, Sect. 3.2.1 proposes architectures that use both local and global resonation strategies and include loop-filters based only on FE integrators, which are more suitable for implementing wideband ADCs than previous reported topologies.

3.1.5 SMASH Architectures

Since OSR must be restricted to low values in wideband applications, a usual design choice to achieve the required DR is to employ MASH architectures with multi-bit quantization. In general, these ΣΔ topologies circumvent stability problems related to high-order loops, but are sensitive to quantization noise leakages caused by mismatches between the analog and digital signal processing in the cascade ΣΔM (see Sect. 2.5). MASH modulators therefore usually require integrators with higher accuracy than their single-loop counterparts for limiting noise leakage effects and

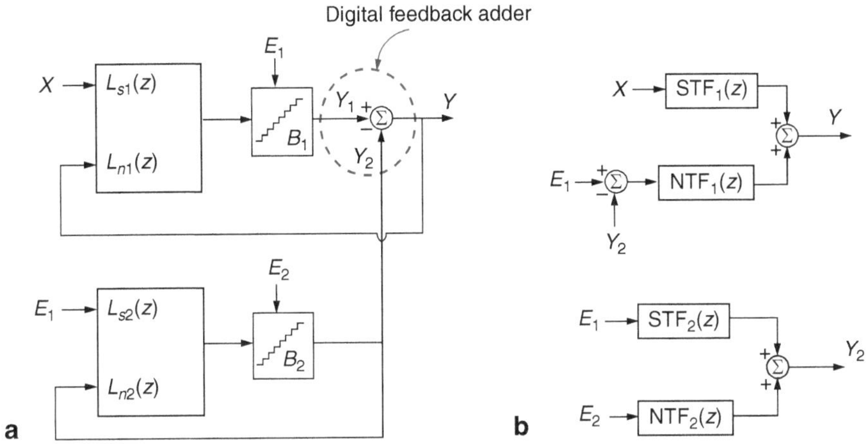

Fig. 3.13 Analysis of a generic SMASH architecture: **a** conceptual block diagram [Magh06], **b** equivalent linear scheme

this increases power consumption in the amplifiers and, consequently, also in the overall ΣΔM.

An alternative ΣΔ architecture that reduces sensitivity to noise leakages of conventional MASH structures is the so-called SMASH modulator, presented in [Magh06]. The main idea behind this family of novel cascades consists of replacing the digital cancellation logic by the own analog processing provided by the ΣΔM integrators together with an additional inter-stage digital feedback path. Figure 3.13a illustrates the underlying concept behind SMASH ΣΔMs [Bena95, Magh06]. Two processing paths are considered for each stage; namely, one for the STF (L_{si}) and another for the NTF (L_{ni}) of each stage, i. The figure shows how a digital feedback adder is used to obtain the model illustrated in Fig. 3.13b. The ΣΔM output is therefore given by [Magh06]:

$$Y(z) = \text{STF}_1 \cdot X(z) + \text{NTF}_1 \cdot (1 - \text{STF}_2) \cdot E_1(z) - \text{NTF}_1 \cdot \text{NTF}_2 \cdot E_2(z) \quad (3.16)$$

where NTF_i and STF_i stand for the NTF and STF of the stage i in the cascade ΣΔM, respectively.

Note that the resulting STF is given by that of the first stage. The NTF corresponding to $E_2(z)$ is similar to that given by a conventional MASH architecture. However, there is an extra term in the NTF multiplying $E_1(z)$. In fact, the first-stage quantization error is filtered not only by NTF_1 but also by $(1 - \text{STF}_2)$. Fig. 3.14 illustrates the case of a 2-2 cascade with $\text{STF}_2 = 2 \cdot z^{-1} - z^{-2}$ as proposed by [Magh06]. Here, a 4th-order NTF high-pass filtering is obtained for both $E_1(z)$ and $E_2(z)$. The digital cancellation logic, required in traditional MASH modulators to properly combine the stage outputs, is replaced by direct feedback paths from the output of the second stage to the input of the first. The former inter-stage feedback paths are marked with ⊗ in Fig. 3.14 for clarity. The ΣΔM output can thus be obtained from the direct

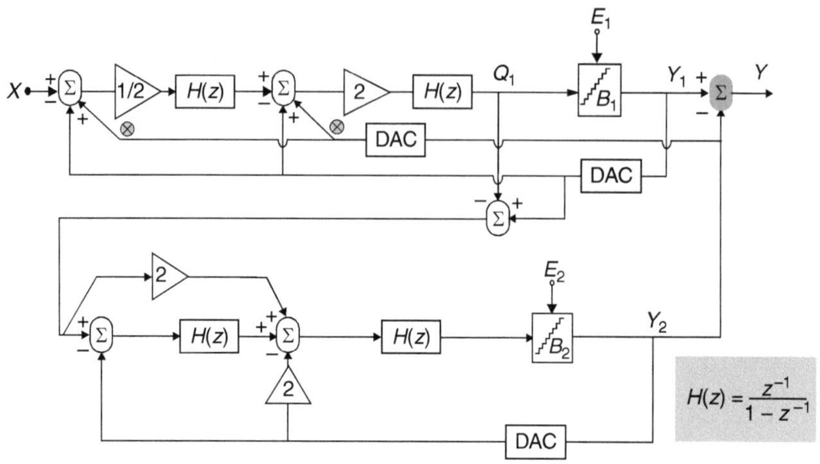

Fig. 3.14 2-2 SMASH modulator presented in [Magh06]

digital subtraction of the two stages outputs, resulting in:

$$Y(z) = z^{-2}X(z) + (1 - z^{-1})^4 E_1(z) - (1 - z^{-1})^4 E_2(z) \qquad (3.17)$$

where $X(z)$ represents the input signal while $E_1(z)$ and $E_2(z)$ stand for the quantization errors in the first and second stage, respectively. Note that, as stated before, both quantization errors are fourth-order shaped thanks exclusively to analog filtering, with no need for digital filtering at the stage outputs. Like that in [Bena96], the topology in [Magh06] eliminates the matching between analog and digital filtering required in traditional MASH ΣΔMs. Nevertheless, two additional DACs are needed, marked with ⊗ in Fig. 3.14.

3.2 Novel Cascade Architectures Based on Previous Strategies

This section provides a number of combinations of previous strategies to obtain efficient cascade ΣΔMs. Unity-STF, resonation-based and SMASH structures are therefore combined. The objective is to obtain architectures capable of drawing the advantages of the former concepts. These topologies will be presented together with some case studies and architectural comparisons in terms of performance. On the one hand, these case studies will help to show how suitable these combined architectures are for wideband and multi-standard applications. On the other, the performance comparisons—also denoted as case studies—will cover the presented architectures and previously shown cascades, such as those in Fig. 3.8[5],

[5] This architecture will employ in-loop coefficients $\{c_1, c_2, a_1, a_2, c_3, c_4, a_3, a_4\} = \{1, 1, 2, 1, 1, 1, 2, 1\}$ in the modulator. Therefore, there will not be any scaling of the integrators coefficients in the 2-2 cascade made of USTF stages for the architectural comparisons under consideration.

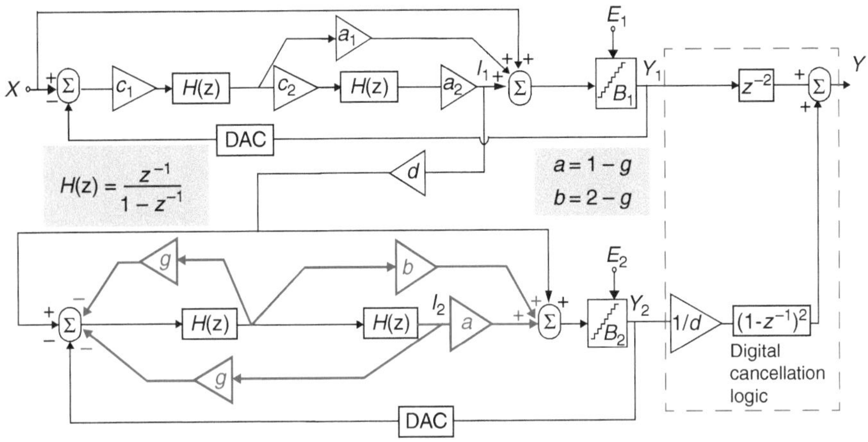

Fig. 3.15 Proposed (I) cascade ΣΔM with local resonator

3.12, 3.14 or a conventional cascade ΣΔM. All the presented case studies and architectural comparisons are based on behavioral simulations. The simulation tool employed is SIMSIDES [Ruiz05], a SIMULINK-based time-domain simulator for ΣΔMs that includes accurate behavioral models for thermal noise, integrator defective settling, distortion, etc. This tool makes extensive use of compiled C-coded functions to speed simulations up. Novel models have been developed when needed.

3.2.1 Proposed Resonation-Based Architectures

This section shows the adaptation of previously reported resonation-based architectures (see Sect. 3.1.4) to a ΣΔM architecture with unity-STF stages (see Sect. 3.1.3). The resulting topologies inherit the architectural advantages of the previous ones; namely, larger SNR due to resonation and low requirements for the opamp output swings due to the unity-STF stages. However, their disadvantages (such as implementation of analog adders, demanding timing requirements and use of extra in-loop coefficients to perform resonation) also extend to the resulting topologies.

3.2.1.1 Local Resonation in Cascade ΣΔMs

Figure 3.15 shows the block diagram of a fourth-order 2-2 USTF cascade ΣΔM with local resonation in the last stage. This topology does not need non-delayed or BE integrators to perform the resonation. The drawback is that additional analog coefficients and an extra feedback path are needed in the last stage, unlike resonators based on BE integrators. Note that this ΣΔM is based on the one depicted in Fig. 3.11

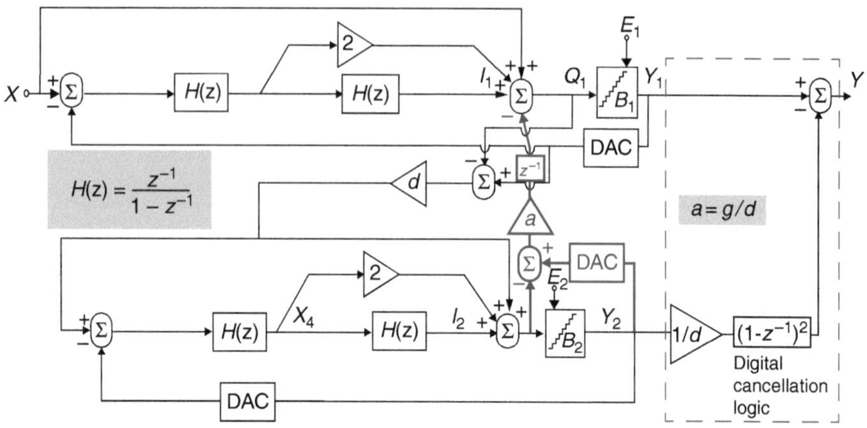

Fig. 3.16 Proposed (II) cascade with global resonation

but with two improvements. First, the resonator of this architecture is based on that depicted in Fig. 3.10b [Mark04], so only FE integrators are required. Second, only one inter-stage path—rather than two, as in Fig. 3.11—is needed to feed the first-stage quantization error to the input of the second stage.

3.2.1.2 ΣΔM Architectures Based on Global Resonation

Figure 3.16 shows a 2-2 cascade ΣΔM that uses unity-STF stages, while resonating through two feedback paths from the last stage to the previous stage; i.e., using global resonation. Since FE integrators are used to implement this kind of resonation, an additional analog delay—that can be implemented by a proper clock-phase scheme and by adding two extra SC branches as described in [Koh05] and [Kana06]—is required, when compared to the architecture in Fig. 3.12. To sum up, the resonation operation needs the quantization error of the second stage $E_2(z)$ to be multiplied by a resonation factor a = g/d and then delayed and subtracted in the first-stage analog adder.

Let us study the contribution of $E_2(z)$ to the overall NTF of the modulator, assuming that $E_1(z)$ and $E_2(z)$ are linear errors. The node where $E_2(z)$, delayed and multiplied by the resonation factor a, is introduced ($Q_1(z)$ in Fig. 3.16) is the same in which $E_1(z)$ is provided. Its transfer function to the ΣΔM output, Y,—i.e., the NTF of the first stage in the cascade ΣΔM—is therefore equal to $(1 - z^{-1})^2$. The global resonation operation contributes to Y as

$$g/d \cdot z^{-1} \cdot (1 - z^{-1})^2 \cdot E_2(z) \tag{3.18}$$

On the contrary, the trace of $E_2(z)$ coming from the output of the second stage, Y_2, after being processed by the digital cancellation logic, is

$$1/d \cdot (1 - z^{-1})^4 \cdot E_2(z) \tag{3.19}$$

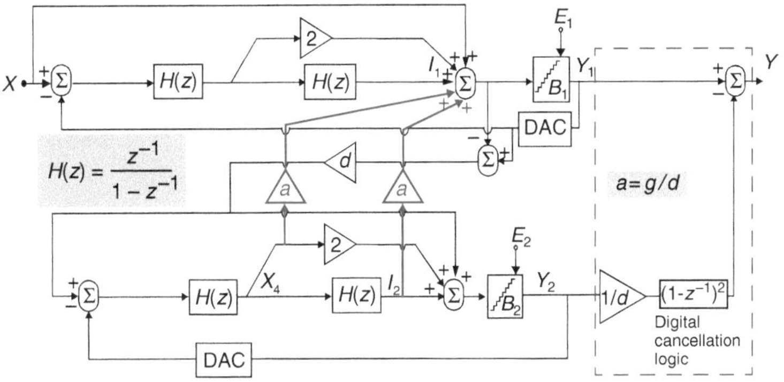

Fig. 3.17 Proposed (III) cascade with global resonation topology

The addition of both terms (3.18) and (3.19) gives the following contribution of $E_2(z)$ to the overall NTF of the architecture in Fig. 3.16:

$$\begin{aligned} &\left[1/d \cdot (1-z^{-1})^4 + g/d \cdot z^{-1} \cdot (1-z^{-1})^2\right] \cdot E_2(z) \\ &\quad = 1/d \cdot (1-z^{-1})^2 \cdot \left[(1-z^{-1})^2 + g \cdot z^{-1}\right] \cdot E_2(z) \end{aligned} \tag{3.20}$$

This is the desired NTF of the $\Sigma\Delta$M with two zeros in optimum locations in the unit circle. Note that there is no trace of the first-stage quantization error because $E_1(z)$ is removed in the digital cancellation logic of the cascade. The idea of processing a delayed version of the quantization error is similar to that employed in the local resonation in Fig. 3.10b, the only difference being the second-order analog filtering performed by the first-stage loop.

There is a more efficient way to perform the global resonation with unity-STF stages, that removes the need for the extra analog delay. An analysis of the scheme in Fig. 3.16 in Z-domain shows that the input and the output of the second integrator in the last-stage are respectively given by:

$$\begin{aligned} X_4(z) &= -z^{-1} \cdot (1-z^{-1}) \cdot E_2(z) \\ I_2(z) &= -z^{-2} \cdot E_2(z) \end{aligned} \tag{3.21}$$

The delayed quantization error of the last stage can therefore be directly obtained as:

$$X_4(z) + I_2(z) = -z^{-1} \cdot E_2(z) \tag{3.22}$$

This modification has been incorporated to Fig. 3.16, resulting in the architecture depicted in Fig. 3.17. Note that the sum $X_4(z) + I_2(z)$ is already done at the input of the first-stage quantizer. Contrary to the modulator in Fig. 3.12, there is therefore no need for an extra addition. Besides avoiding the use of an additional delay, the $\Sigma\Delta$M in Fig. 3.17 requires only one analog coefficient (g) to implement the resonation, instead of three coefficients [g, $(1-g)$ and $(2-g)$] as in Fig. 3.15. Also, only FE integrators are used, which simplifies the electrical implementation and reduces sensitivity to circuit non-idealities.

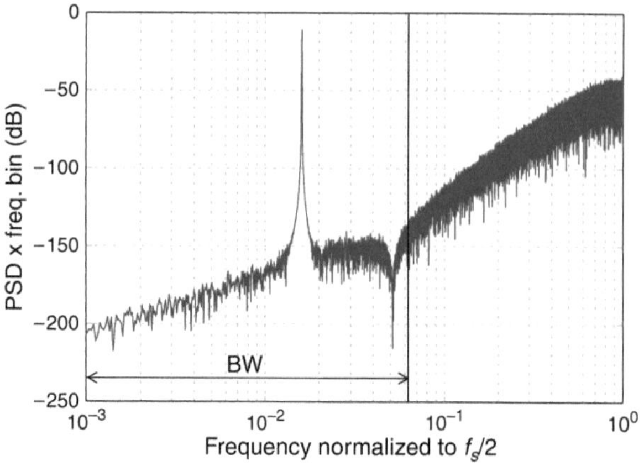

Fig. 3.18 Modulator spectrum with resonation

Note that the architectures in Fig. 3.16 and 3.17 employ two paths to extract $E_1(z)$ and give it to the back-end stage. It would have been more efficient to use one path only for this inter-stage connection as done, for example, in the USTF cascade that implements local resonation in Fig. 3.15. However, this is not possible for the architectures in Fig. 3.16 and 3.17, because the inter-stage feedback operation modifies $I_1(z)$. The signal at this node can be obtained as $-z^{-2} \cdot E_1(z)$ for the architectures in Fig. 3.8 or 3.15, while it becomes $I_1(z) = -z^{-2} \cdot [E_1(z) + g \cdot z^{-1} \cdot E_2(z)]$ when global resonation is implemented in a cascade ΣΔM together with USTF stages (see Fig. 3.16 and 3.17). There is therefore an extra trace of $E_2(z)$ that prevents the use of $I_1(z)$ as the input of the second stage in USTF architectures with inter-stage feedback paths.

In order to compare the performance of the proposed ΣΔM architectures in Fig. 3.15 and 3.17—proposed (I) and (III), several behavioral simulations have been performed using SIMSIDES [Ruiz05]. First, the two architectures are studied in a wideband scenario, considering oversampling ratios of 16 and of 4. Second, the architecture in Fig. 3.17 that uses global resonation is studied in a multi-mode scenario.

3.2.1.3 Case Study 1: Application of the Proposed Architectures to a Wideband Scenario (OSR = 16)

In this case, all topologies under study operate with an OSR of 16, 4-bit internal quantizers, an inter-stage gain (d) of 1 and a 1-V reference voltage. As an illustration, Fig. 3.18 shows the effect of resonation on the output spectrum of the topologies proposed in Fig. 3.15 and 3.17[6]. Note that the resulting spectrum is the same for both architectures, because the shift of the NTF zeros is identical with both local and global resonation techniques. The optimal feedback coefficient that causes the

[6] Note that thermal noise is not included in the simulation model for this case.

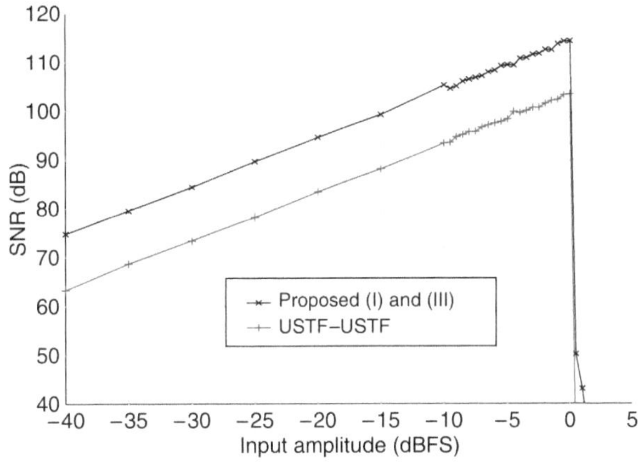

Fig. 3.19 SNDR versus input amplitude

resonation in the topology depicted in Fig. 3.17 is $g = 0.027$, resulting in a shift of two zeros of the NTF from 0 to $0.9865 \pm j \cdot 0.1638$ and a minimization of the in-band quantization noise. This is better illustrated in Fig. 3.19 where the SNR is represented versus the input amplitude. In this example, the peak SNR of the proposed architectures is 11dB larger within the whole input range than that obtained by the 2-2 cascade made up of USTF stages in Fig. 3.8 (USTF-USTF). This increase of the SNDR is due to the use of resonation.

Another advantage of the proposed architectures comes from the use of unity STFs with the subsequent reduction of the amplifiers output swings. This is illustrated in Fig. 3.20 by plotting the histograms of the integrator outputs in the proposed architectures and in previously reported topologies. A conventional cascade 2-2 architecture is also included in the comparison for completeness. Note that the integrator output swings of the proposed $\Sigma\Delta$Ms (lower than 0.15 V for all integrators) are similar to those of the scheme in Fig. 3.8 (USTF-USTF). However, they are lower than the ones in the $\Sigma\Delta$M in Fig. 3.12 (global resonation architecture by [Sanc06]). This improves the linearity of the proposed architectures, as compared to the one in Fig. 3.12. The linearity of the aforementioned architectures will be compared below.

SIMSIDES allows the inclusion of opamp gain non-linearities with a fourth-order polynomial equation defined as

$$Avnl = A_{DC} \cdot (1 - Avnl_1 \cdot v_o - Avnl_2 \cdot v_o^2 - Avnl_3 \cdot v_o^3 - Avnl_4 \cdot v_o^4) \qquad (3.23)$$

where $Avnl$ is the opamp gain, A_{DC} is the DC gain in the quiescent point, v_o is the opamp output voltage and $Avnl_i$ are the non-linear gain coefficients [Ruiz05]. For all the behavioral simulations in this chapter, only a second-order non-linearity will be considered, so the opamp gain can be re-defined as follows

$$Avnl = A_{DC} \cdot \left(1 - Avnl_2 \cdot v_o^2\right) \qquad (3.24)$$

assuming the use of fully differential circuitry in the practical implementation. Fig. 3.21 illustrates the variation of the gain with the opamp output voltage for

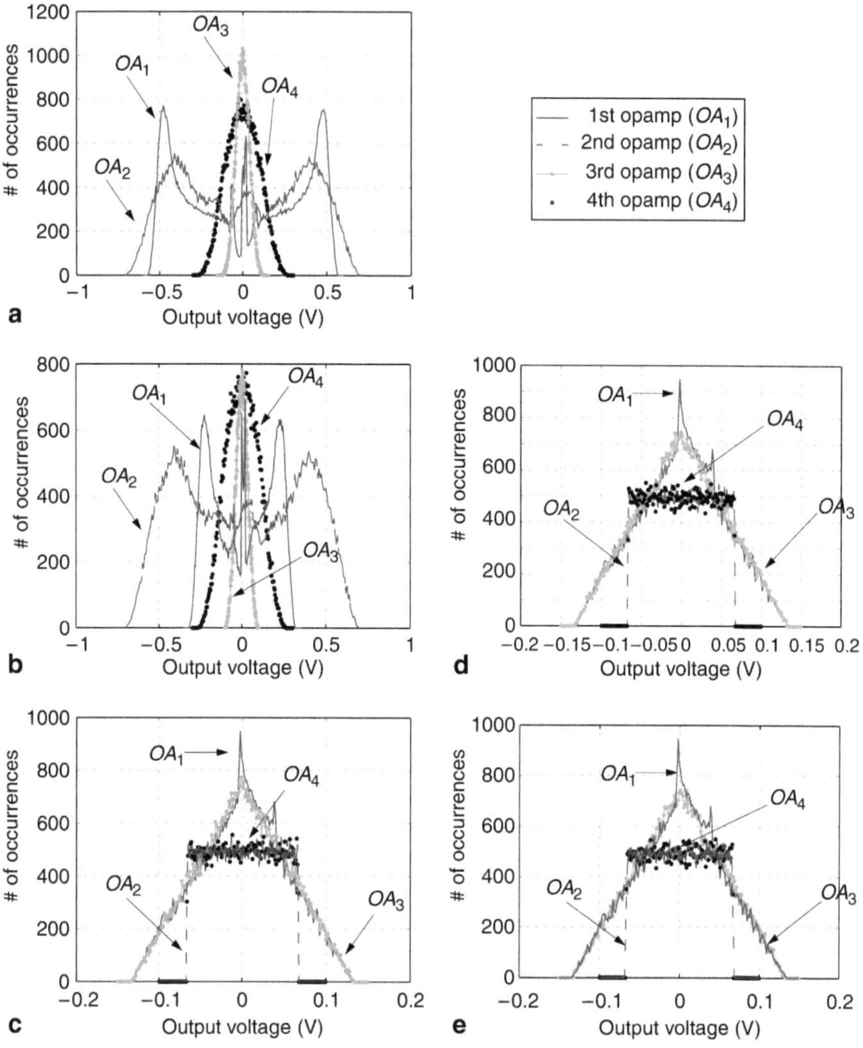

Fig. 3.20 Output swing requirements for the amplifiers of: **a** conventional 2-2 MASH architecture, **b** global resonation structure in Fig. 3.12 [Sanc06], **c** USTF-USTF in Fig. 3.8 [Nam05], **d** proposed topology (I) in Fig. 3.15, **e** proposed topology (III) in Fig. 3.17

different values of the above-mentioned non-linearity factor. Note that the larger the opamp output swing, the greater the degradation of the opamp gain for a given $Avnl_2$.

Fig. 3.22 plots the effect of opamp gain non-linearity on the modulators SNDR, based on the approximation given in (3.24). In this simulation, a finite DC gain of 55dB is considered for all the amplifiers, while the gain second-order non-linearity for the first-stage amplifiers is varied. The results of these simulations are summarized in Table 3.2, which shows the value of $Avnl_2$ that causes the SNDR to fall by 3dB. Note that the robustness of the proposed architectures to non-linearities is similar

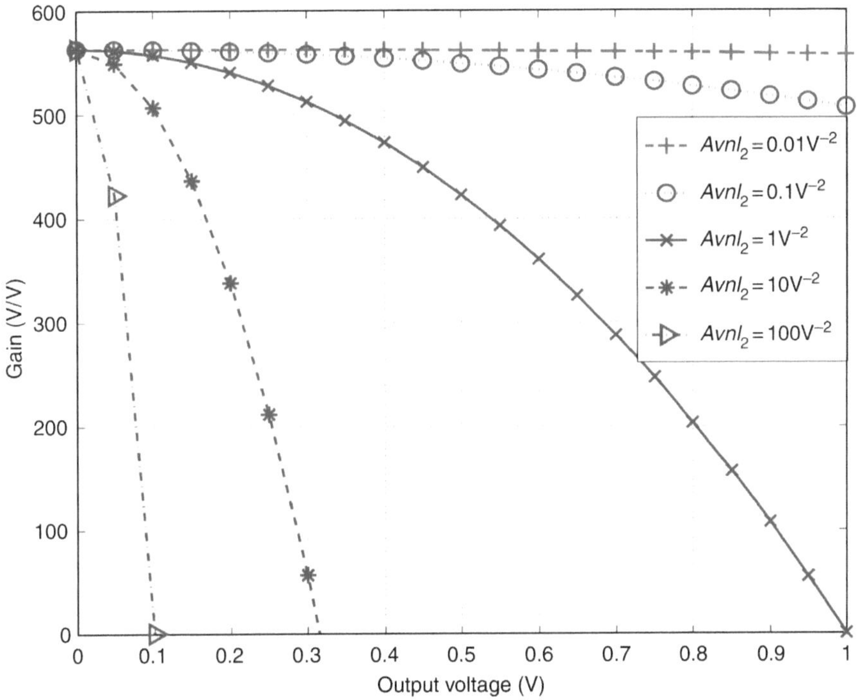

Fig. 3.21 Illustration of the opamp non-linear gain as defined in (3.24)

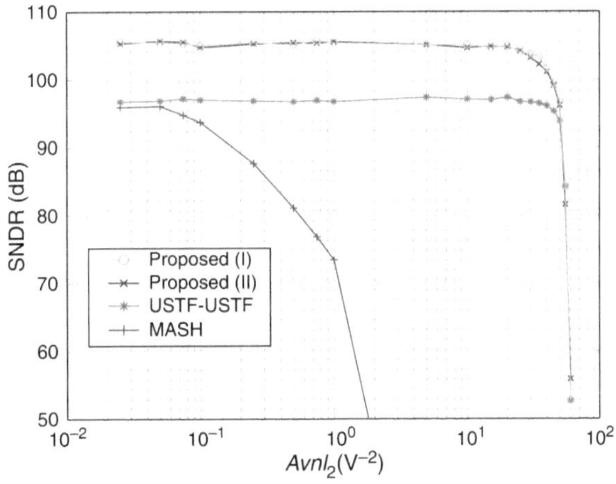

Fig. 3.22 Impact of amplifier gain non-linearities on the modulator SNDR (input level of –6dBFS)

to that of the cascade USTF-USTF shown in Fig. 3.8. Furthermore, the proposed $\Sigma\Delta$Ms achieve larger effective resolution through the action of resonation. Note that cascades based on USTF have better linearity performance when compared with conventional cascade architectures.

Table 3.2 Maximum 2nd-order non-linearity for a 3-dB drop in SNDR

Topology	$Avnl_2(V^{-2})$
Proposed (I)	40.0
Proposed (III)	35.0
USTF-USTF	50.0
Conventional MASH	0.1

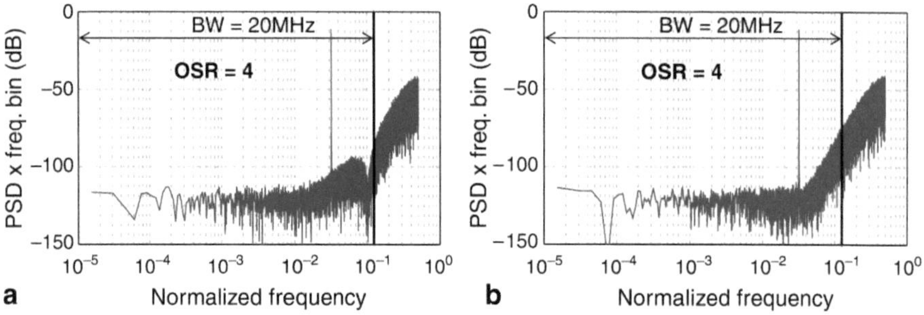

Fig. 3.23 Output modulator spectrum: **a** with resonation [proposed cascades (I) and (III)], **b** without resonation (conventional 2-2 MASH).

3.2.1.4 Case Study 2: Application of the Proposed Architectures to a Wideband Scenario (OSR = 4)

All topologies in this section operate with 4-bit internal quantizers, an inter-stage gain (d) of 1, a 1-V reference voltage and a very much reduced OSR of 4. All the following simulations also include kT/C noise sampled at a clock rate of 160 MHz by 0.25-pF capacitors at the front-end integrator. The embedded DACs are assumed to be linear.

As an illustration, Fig. 3.23 compares the effect of resonation on the output spectrum of the proposed topologies (Fig. 3.15 and 3.17) to that of a conventional 2-2 cascade, for an input level of -6dBFS. Note that, thanks to resonation, a notch—clearly visible in Fig. 3.23a—is produced in the proposed topologies close to the upper limit of the signal band.

The optimal resonation coefficient for the topologies in Fig. 3.15 and 3.17 (implementing local and global resonation, respectively) is, after rounding, $g = 0.4$ for an OSR of 4, resulting in a shift of two zeros of NTF from 0 to $0.8 \pm j \cdot 0.6$. Because of this distribution of the NTF zeros, in-band quantization error is considerably reduced. This is illustrated in Fig. 3.24, where the modulator SNR is represented versus the input amplitude. In this case, resonation increases the resolution of the proposed architectures by roughly 10.5 dB within the whole input range as compared to a conventional cascade.

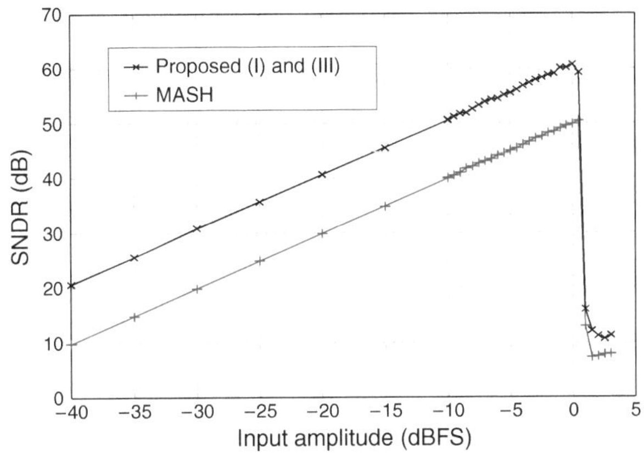

Fig. 3.24 SNDR versus input amplitude

Table 3.3 Modulator specifications in case study 3

Standard	GSM	UMTS	WiMAX	WLAN
SNDR (bit)	14	12	8	10
BW (MHz)	0.2	4	20	20

Table 3.4 Architectural parameters of the proposed modulator

Standard	GSM	UMTS	WiMAX	WLAN
Modulator order (L)	2		4	
Inter-stage gain (d)	—		4	
Resonation coefficient (g)		0 (No resonation)		0.4
Resonation coefficient (g/d)		0 (No resonation)		0.1
OSR	100	10	4	
Sampling frequency (f_s)	40 MHz	80 MHz	160 MHz	

3.2.1.5 Case Study 3: Application of the Proposed Architectures to a Multi-Mode Scenario

This section presents a case study that considers the specifications and standards listed in Table 3.3 in order to show the reconfiguration architectural capabilities of the 2-2 cascade $\Sigma\Delta$M in Fig. 3.17 combining USTF with inter-stage resonation. Four standards are considered (GMS, UMTS, WiMAX and WLAN), covering resolutions that range from 8bits to 14bits and bandwidths from 200 kHz to 20 MHz.

Table 3.4 shows the values of the modulator design parameters reconfigured for the different operation modes, namely: NTF order (L), coefficient for global resonation (g/d) and OSR. In all operation modes, 4-bit internal quantizers and a 1-V reference voltage are considered. Also, kT/C noise, sampled by 0.25-pF sampling capacitors, is contemplated for the front-end integrator in all simulations. The embedded DACs are assumed to be linear, although in practice DEM might be required, at least in the first stage of the cascade. Note that different OSRs and sampling frequencies (f_s)

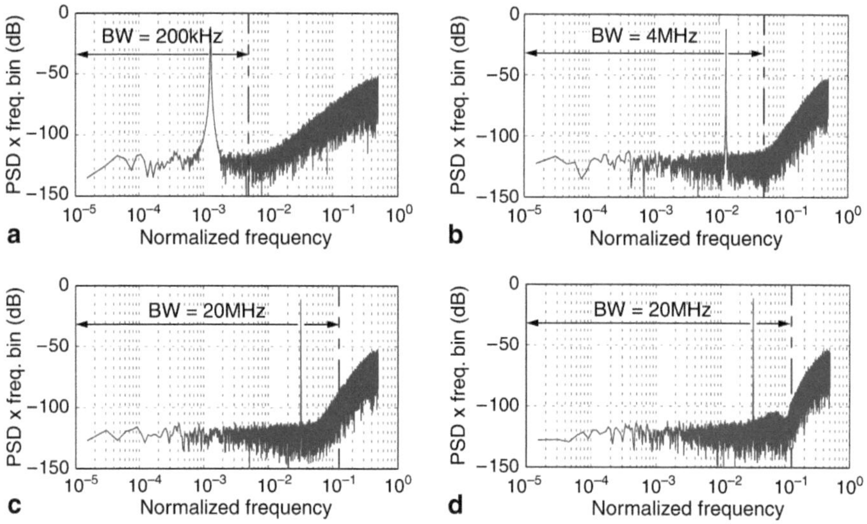

Fig. 3.25 Modulator output spectra for the different modes: **a** GSM, **b** UMTS, **c** WiMAX, **d** WLAN (−6-dBFS input level, 64k-point FFTs)

are employed to cope with the required resolution of each signal bandwidth in the different standards. The feedback paths in Fig. 3.17 that provide the last-stage error to the first stage are disconnected in those operation modes in which no resonation is used (GSM, UMTS and WiMAX). Also, only the first stage is working in GSM mode, whereas all the integrators and comparators of the second stage could be in practice turned off. Global resonation is used only in WLAN mode. In this case, the resonation coefficient *g/d* is 0.1, resulting in a shift of two zeros of the NTF from 0 to $0.8 \pm j \cdot 0.6$. Thanks to this distribution of the NTF zeros and to the employment of an inter-stage scaling $d = 4$, in-band quantization error is considerably reduced, as stated above.

Figure 3.25 shows the modulator output spectra for all the operation modes. In GSM [see Fig. 3.25a], the in-band error is dominated by thermal noise, whereas in UMTS [see Fig. 3.25b] is not well defined between thermal and quantization noise. The effect of resonation is clearly visible by comparing Fig. 3.25c and d, corresponding to WiMAX and WLAN, respectively. This is illustrated in Fig. 3.26 where the SNR is represented versus the input signal amplitude. Note that the effective resolution for WLAN is approximately 9dB larger than that for WiMAX, thanks to resonation.

Sensitivity to noise leakages due to capacitor mismatch has been studied on the basis of Monte Carlo simulations. Figure 3.27 shows the modulator ENOB for a –6-dBFS input signal obtained for a 50-run Monte Carlo simulation considering a standard deviation of 0.1% in all capacitor values. The average [m (bits)] and standard deviation (σ) of the resulting ENOBs are also shown for each operation mode. Note that the use of resonation in the WLAN mode leads to no appreciable performance degradation in comparison with other operation modes.

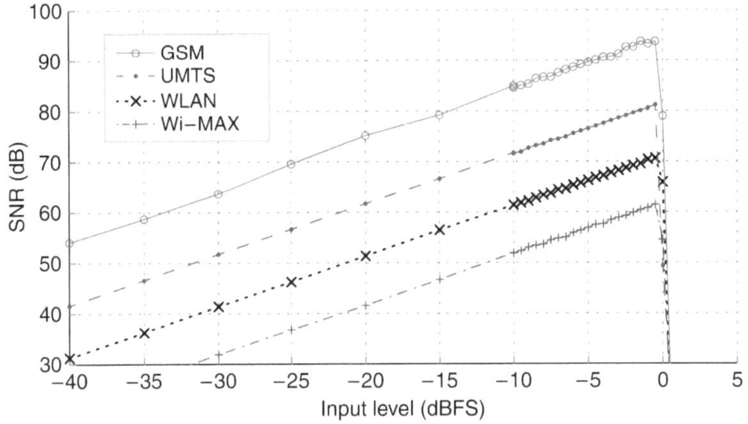

Fig. 3.26 SNR curves of the proposed architecture with parameters in Table 3.4

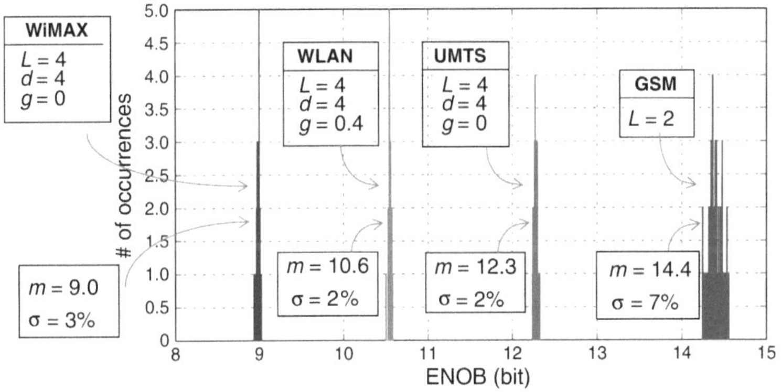

Fig. 3.27 Variation of the modulator ENOB considering a 0.1% capacitor mismatch (−6-dBFS input level, 50-run Monte Carlo simulation)

Figure 3.28 shows the effect of the amplifiers finite DC gain on the modulator SNR for all operation modes. Note that the impact of the amplifier gain on the $\Sigma\Delta$M performance is considerably lower in GSM than in other standards. This is due to the topology used for the $\Sigma\Delta$M in the former case (single-loop), which does not present the sensitivity to noise leakages that is inherent to cascade topologies (used in the remaining standards). In the case of UMTS and WiMAX, there is no appreciable degradation for amplifier DC gains above 40dB, thanks to the robustness of cascade $\Sigma\Delta$Ms made up of USTF. This value rises to 50dB for WLAN due to the SNR increase provided by global resonation. Note that in practice this value of opamp gain is still not a demanding requirement.

One prominent advantage that comes from the use of unity STFs is the subsequent reduction of the amplifier output swing. This is illustrated in Fig. 3.29 by plotting the histogram of the integrator outputs obtained from the operation of the proposed

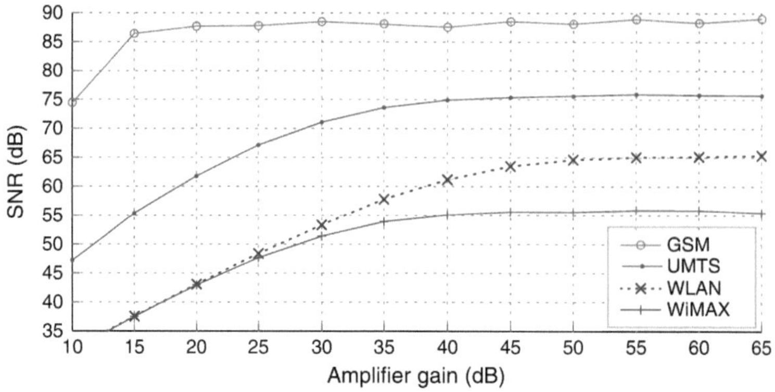

Fig. 3.28 SNR versus amplifier gain (−6-dBFS input level)

Fig. 3.29 Output swing requirements of the amplifiers

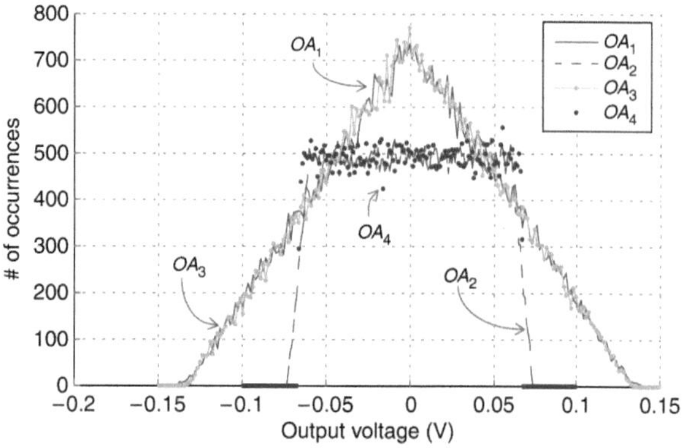

modulator (see Fig. 3.17) in all operation modes. Note that the required output swing is only ±0.15 V for the first and third operational amplifiers (OA_1 and OA_3), whereas it drops to ±0.10 V for the second and fourth (OA_2 and OA_4), producing an excellent linearity behavior in the proposed architecture.

This is illustrated in Fig. 3.30 by plotting the effect of the opamp gain non-linearity on the modulator SNDR. In this simulation, a finite gain of 55dB is considered for all amplifiers, while varying the second-order non-linearity of the amplifier gain in the first-stage amplifiers. Note that the robustness of the proposed ΣΔM to non-linearities is high, even if resonation is used to increase the effective modulator resolution.

3.2.2 *Novel Unity-STF SMASH-Based Architecture*

This section presents a novel topology of ΣΔ cascade intended for high-speed and low-voltage applications, extending the underlying principle of SMASH structures

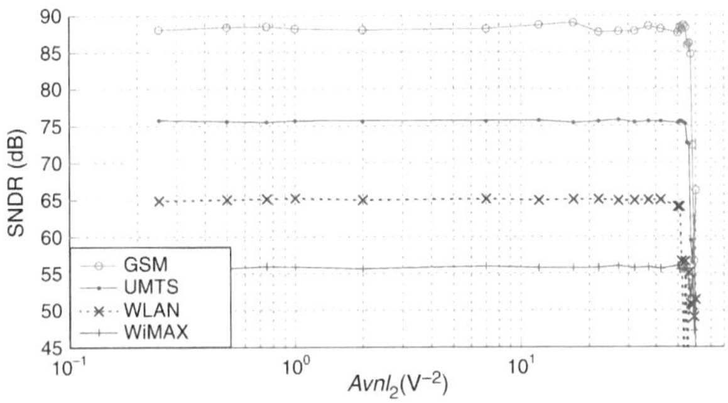

Fig. 3.30 SNDR versus non-linearity of the amplifier gain (−6-dBFS input level, 55-dB amplifier gain)

to the implementation of unity STFs. In recent years, the implementation of $\Sigma\Delta$Ms with unity STF has demonstrated to be an excellent way of reducing the impact of circuit non-linearities on the modulator performance, especially when considering low OSRs in a low-voltage scenario. Several $\Sigma\Delta$ topologies have been reported that apply this technique to either second-order $\Sigma\Delta$Ms [Bena96, Silv01, Silv04a, Kwon04, Gagg04] or to every stage of cascades $\Sigma\Delta$Ms [Nam05, Chri07, Morg10]. Unfortunately, the disadvantages associated with cascades that apply the former concepts (SMASH and USTF) are also inherited by this topology. For example, SMASH forces the use of either extra DACs or a digital adder, and analog adders are required for the USTF implementation.

3.2.2.1 Modulator Architecture

The performance of SMASH architectures, generically shown in Fig. 3.13, can be improved by using USTF in the cascade stages. Note from (3.16) that the contribution of the first-stage quantization error $E_1(z)$ to the overall modulator output can ideally be cancelled if $\mathrm{STF}_2 = 1$. This strategy therefore provides similar quantization noise shaping to that of a conventional MASH $\Sigma\Delta$M. Furthermore, the robustness to noise leakages of SMASH topologies is still present, as will be verified later by a set of behavioral simulations and the implementation of unity STFs allows the reduction of output swing and gain of the amplifiers, what constitutes an appealing feature for low-voltage implementations.

Figure 3.31 illustrates an extension of the 2-2 SMASH $\Sigma\Delta$M in Fig. 3.14 to unity STFs [Silv01]. If multi-bit quantization is used in the stages, a scaling factor $d > 1$ can be implemented to accommodate the inter-stage gain. This factor is compensated in the analog feedback paths from the second stage to the first; i.e., before the digital subtraction of the stage outputs. The output of the modulator in Fig. 3.31 is thus

$$Y(z) = X(z) - 1/d(1 - z^{-1})^4 E_2(z) \tag{3.25}$$

where the overall STF equals unity and $E_1(z)$ is cancelled—contrary to (3.17), while avoiding any filtering in digital domain. Note also from (3.25) that using a scaling

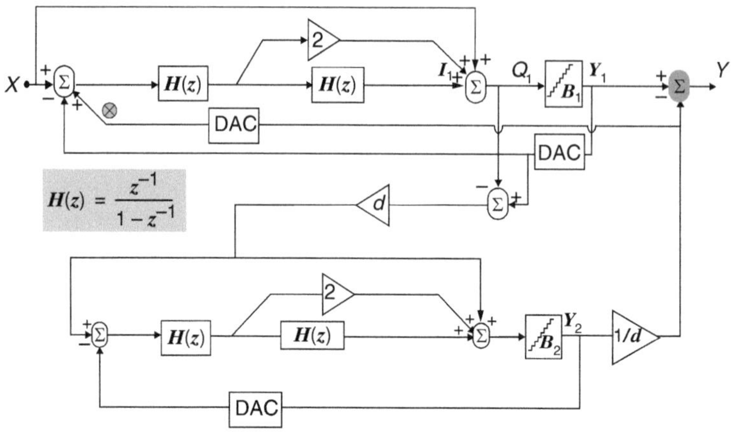

Fig. 3.31 Extensions of the 2-2 SMASH modulator in [Magh06] to unity STFs and inter-stage gain scaling

factor d will help to reduce the power of the second-stage quantization error at the modulator output. If the value of the scaling factor is a power of 2, it will only require a shift register before the in-loop digital subtraction that implements part of the SMASH operation. This scaling strategy cannot be directly applied to the SMASH ΣΔM in Fig. 3.14 [Magh06], since in this case (3.17) transforms into

$$Y(z) = z^{-2}X(z) + (1 - z^{-1})^4 E_1(z) - 1/d(1 - z^{-1})^4 E_2(z) \tag{3.26}$$

in which a larger d therefore reduces the power of the second-stage quantization error at the output but not that of the first stage.

Although improving the performance of the 2-2 cascade in Fig. 3.14, the ΣΔM illustrated in Fig. 3.31 still suffers from several drawbacks associated with the direct feedback path from the second-stage output to the first stage, namely:

- It requires, at least, an extra highly linear DAC in the added feedback path to the first-stage input.
- It is very sensitive to mismatch effects in the added feedback paths with respect to the first-stage analog coefficients, which cause low-order noise leakages [Morg07a].

These setbacks can be circumvented by replacing the feedback paths from the second-stage output to the first stage input, and directly feeding the modulator output to the first stage as shown in Fig. 3.32. The digital subtraction of the quantizer outputs is thus performed inside the first-stage loop of the cascade. This strategy—generally presented in [Magh06] (see Fig. 3.13) but not specifically applied to a SMASH cascade—results in no modification of the overall output compared to that in (3.25), while eliminates the need for extra feedback paths, so the number of linear DACs required is not increased. However, feeding the output of the ΣΔM back to the input requires a DAC with double full scale and one more bit than the largest of the resolutions of the ADCs in the stages. This is due to the digital summation of

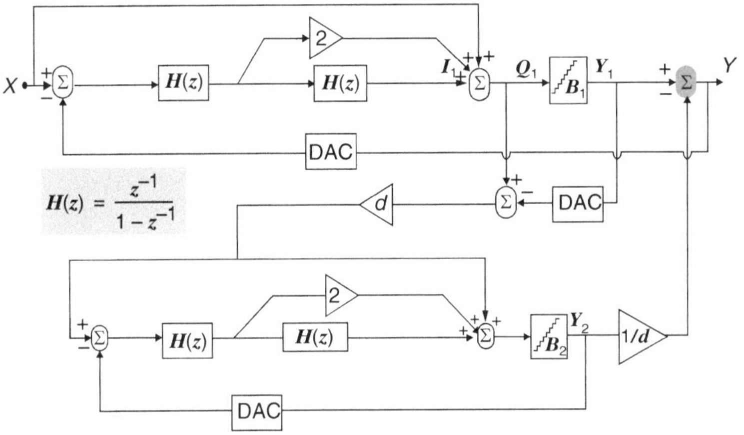

Fig. 3.32 Proposed (IV) topology for the robust implementation of a 2-2 SMASH $\Sigma\Delta$M with unity STFs

the stages outputs (see Appendix B); although, as will be shown in the following case study, the in-loop location of the digital adder helps to considerably increase the robustness of the proposed cascades to mismatches compared to the 2-2 SMASH $\Sigma\Delta$M in [Magh06].

Note from Fig. 3.32 that two paths are used to extract the quantization error of the first stage and give it to the next. Unfortunately, $I_1(z)$ can be no longer used to that purpose, since it yields

$$I_1(z) = -z^{-2} \cdot [E_1(z) - Y_2(z)/d] \tag{3.27}$$

where $Y_2(z)$ stands for the digital output of the second stage.

3.2.2.2 Case Study

The performance of the SMASH architecture proposed in Fig. 3.32 is compared below to that of a 2-2 conventional cascade $\Sigma\Delta$M and of the original SMASH $\Sigma\Delta$M proposed in [Magh06] shown in Fig. 3.14. Behavioral simulations in SIMSIDES [Ruiz05] are used for this purpose. All topologies operate with an OSR of 16, 4-bit internal quantizers and a 1-V multi-bit reference voltage for comparison purposes with data reported in [Magh06]. Multi-bit DACs are assumed linear.

Figure 3.33 depicts the SNR achieved by the different modulators versus the input level when considering quantization errors only. Curves corresponding to the proposed and conventional cascades with an inter-stage gain $d = 1$ can be directly compared with the performance of the SMASH modulator. Note that the overload level of the proposed cascade is considerably larger compared to the SMASH structure and also improves that of conventional cascades. As shown in Fig. 3.33, the attainable peak SNR can be increased by operating the topology with $d > 1$.

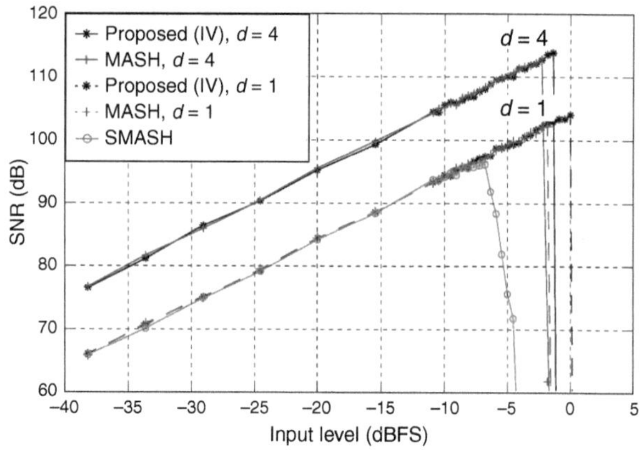

Fig. 3.33 SNR curves of the different modulator architectures

Table 3.5 Overload level and output swing requirements of the different architectures

Architecture	Overload level (dBFS)	Output swing (V)			
		OA_1	OA_2	OA_3	OA_4
MASH, $d = 4$	−2.5	0.60	0.70	0.35	0.50
SMASH	−7.0	0.75	1.24	1.49	2.69
Proposed (IV), $d = 4$	−1.5	0.15	0.10	0.15	0.10

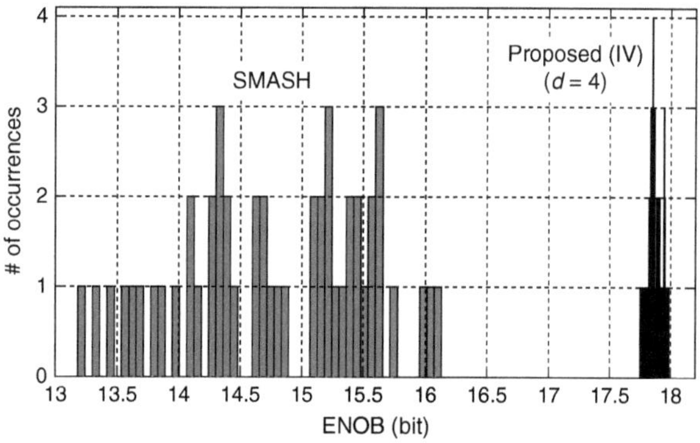

Fig. 3.34 ENOB variation of the SMASH and the topology proposed in Fig. 3.32, considering a 0.1% capacitor mismatch (–6-dBFS input level, 50-run Monte Carlo)

The overload levels of the diverse architectures are shown in Table 3.5, together with the output swing requirements of the amplifiers in the cascades. Note that the combined use of unity STFs and multi-bit quantization leads to a remarkable relaxation in the output swing for the proposed cascade compared to MASH and SMASH topologies, simplifying their low-voltage implementation.

The sensitivity of the different architectures to noise leakages due to mismatches has been studied on the basis of Monte Carlo simulations. Figure 3.34 shows the ENOB for a −6-dBFS input signal obtained for the SMASH modulator in Fig. 3.14 and the cascade in Fig. 3.32 for a 50-run Monte Carlo simulation, considering a

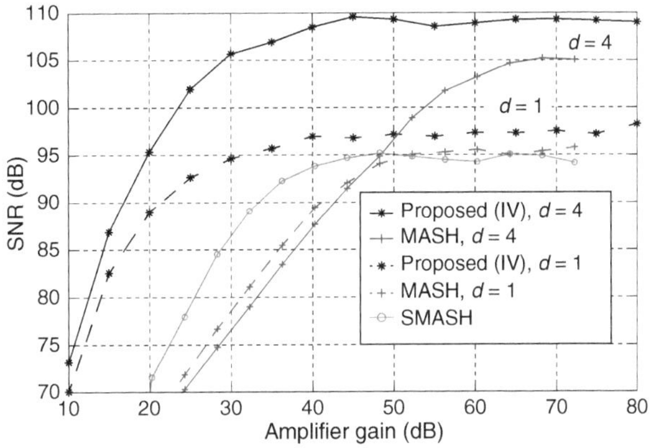

Fig. 3.35 Variation of the SNDR with the amplifier gain for the different architectures (−6-dBFS input level)

standard deviation of 0.1% in all capacitors. Note that mismatches at the extra feedback paths—marked with ⊗ in Fig. 3.14—are responsible for great variation in the resolution in the SMASH topology, which would eventually result in an unreliable practical implementation. However, the location of the digital summation of the stage outputs inside the first-stage loop results in additional filtering and provides the cascade proposed in Fig. 3.32 with great immunity to mismatches. Figure 3.34 shows how the low sensitivity to mismatches is still maintained despite the use of a scaling inter-stage factor $d > 1$ to increase the modulator resolution.

Figure 3.35 compares the SNR obtained for the different $\Sigma\Delta$ structures against the amplifier gain in the integrators for an input level of –6dBFS. Note that the required DC gains in the conventional MASH with $d = 1$ and the SMASH to achieve an SNR of 95dB are 50dB and 40dB, respectively. These values drop to 30–35 dB for the cascade proposed in Fig. 3.32 operating with $d = 1$. Figure 3.35 also shows how SNDRs of 105dB can be addressed with similar gain values if the topology operates with $d = 4$.

Thanks to the implementation of unity STFs, the cascade in Fig. 3.32 also proves to have considerably higher tolerance to non-linearities in the amplifier gain. Figure 3.36 shows the SNDR of the different topologies against the gain non-linearity for an input level of −6dBFS. The amplifiers gain is assumed to be 55dB and non-linearities are contemplated only in the front-end stage. Note that non-linearity requirements are greatly reduced for the architecture in Fig. 3.32.

The above-mentioned results are summarized in Table 3.6. The first column shows the amplifier gain required in all considered $\Sigma\Delta$ topologies to obtain an SNDR that is only 3dB lower than that attainable by making $Avnl_i = 0$ for all i [see (3.23)]. The second column shows the linearity required for the same SNDR drop while assuming a DC gain of 55 dB in all amplifiers. Note from the data in Table 3.6 that the performance of the architecture proposed in Fig. 3.32 outstands in comparison to MASH and SMASH cascades.

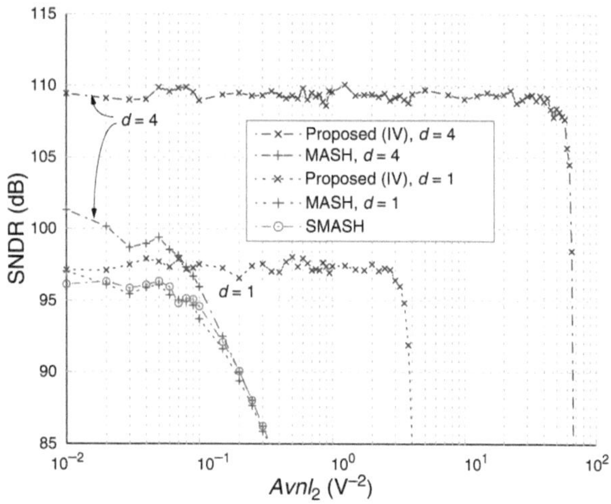

Fig. 3.36 SNDR variation with the non-linearity of the opamp gain (−6-dBFS input level, 55-dB opamp gain)

Table 3.6 Amplifier gain and non-linearity requirements for a 3-dB drop on SNDR

Architecture	Amplifier gain (dB)	$Avnl_2$ (V^{-2})
MASH, $d=4$	63.71	0.06
SMASH	36.57	0.15
Proposed (IV), $d=4$	33.57	60.00

3.2.3 Novel SMASH ΣΔMs with Resonation

As with conventional cascades, two different approaches can be used to implement resonation in SMASH ΣΔMs, namely: by using local resonators or via an inter-stage feedback factor to perform global resonation. The implementation of both resonation strategies, together with USTF, in these ΣΔMs is described in this section. The subsequent architectural advantages are also addressed. However, the implementation of USTF, SMASH and resonation entails the use of analog adders, at least one digital adder and extra SC branches and coefficients for resonation.

3.2.3.1 Modulator Architecture with Local Resonation

Resonation can be included in the front-end or in the back-end stage of the 2-2 SMASH ΣΔM, so either NTF_1 or NTF_2 will be given by (3.13). Equation (3.16) indicates that a resonator in the first stage $[NTF_1(z) = 1-(2-g_1)\cdot z^{-1}+z^{-2}]$ shifts 2 zeros for both quantization errors $E_1(z)$ and $E_2(z)$. On the contrary, including a resonator in the last stage will not modify the DC zeros of $E_1(z)$. However, the use of unity STF in the last stage will remove this term from NTF in (3.16).

Besides, all NTF zeros can be shifted from DC to an optimum location by using two resonators, one for the first stage and another for the second when the last stage implements unity STF. Fortunately, this does not require modifications to the

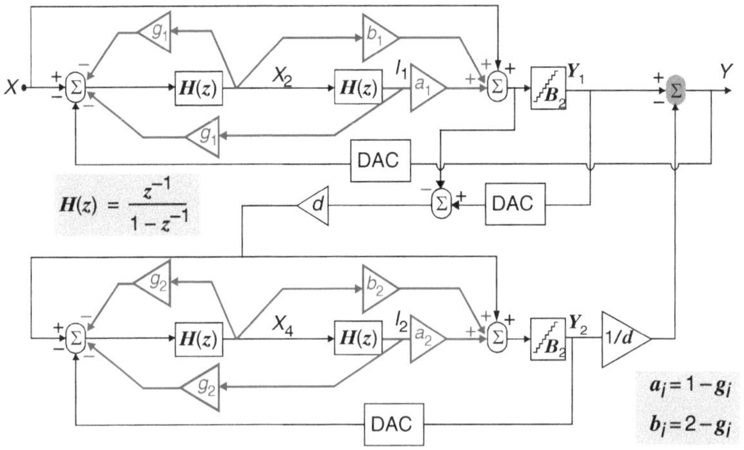

Fig. 3.37 Proposed SMASH $\Sigma\Delta$M with two local resonators

digital cancellation logic, unlike conventional cascades which need to implement the resonation coefficient in the digital cancellation logic[7].

Figure 3.37 shows the block diagram of a 2-2 cascade $\Sigma\Delta$M architecture combining local resonation, SMASH and USTF. Two resonators are employed with the corresponding coefficients g_1 and g_2 for NTF_1 and NTF_2, respectively. The NTF of this $\Sigma\Delta$M is thus given by

$$\mathrm{NTF}(z) = \frac{-[1-(2-g_1)\cdot z^{-1}+z^{-2}]\cdot[1-(2-g_2)\cdot z^{-1}+z^{-2}]}{d} \tag{3.28}$$

where d stands for the inter-stage scaling factor.

3.2.3.2 Modulator Architecture with Global Resonation

Figure 3.38 depicts the scheme of a 2-2 SMASH $\Sigma\Delta$M, which includes unity STF and multi-bit quantization in both stages and global resonation in the cascade. Table 3.7 encloses the feed-forward and resonation coefficients of the $\Sigma\Delta$M topology, which can be easily derived from a procedure similar to that described in Sect. 3.1.3. This SMASH architecture removes the matching requirements between analog and digital filtering by performing a digital subtraction of the quantizer outputs inside the first-stage loop [Morg07b]. This strategy eliminates the need for additional feedback paths as originally proposed in [Magh06], so the number of linear DACs required is not increased in comparison with conventional MASH $\Sigma\Delta$M. The location of the digital adder also helps to considerably increase the robustness of the modulator to capacitor mismatches.

[7] These coefficients are most probably not a number power of two. Their implementation in the digital domain can therefore become costly and complex [Sanc06].

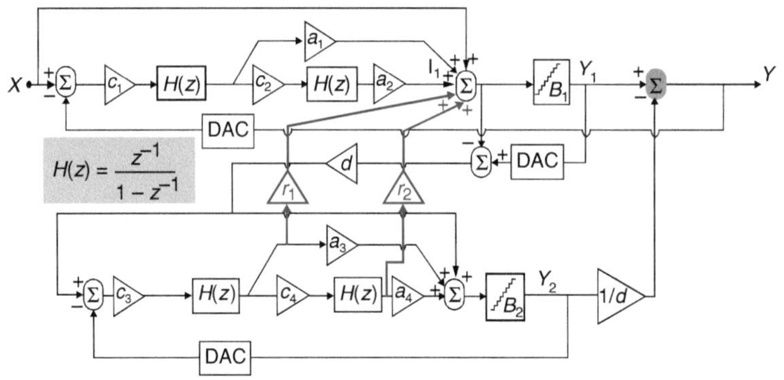

Fig. 3.38 SMASH ΣΔM with global resonation and unity STFs

Table 3.7 Scaling on feed-forward and resonation coefficients

Feed-forward coefficients	Global resonation coefficients
$a_1 = \dfrac{2}{c_1}$	$r_1 = \dfrac{g}{d \cdot c_3}$
$a_2 = \dfrac{1}{c_1 \cdot c_2}$	$r_2 = \dfrac{g \cdot a_4}{d}$
$a_3 = \dfrac{2}{c_3}$	
$a_4 = \dfrac{1}{c_3 \cdot c_4}$	

The digital adder in Fig. 3.38 performs the subtraction of the first- and second-stage outputs (Y_1 and Y_2, respectively), the latter signal being digitally scaled by $1/d$ in this operation ($Y = Y_1 - Y_2/d$). The output of this subtraction is given directly to the multi-level DAC embedded in the front-end loop. As a consequence of the digital addition, the feedback signal in the first-stage loop is implicitly scaled by a factor that depends on the quantizers' full scale and on the inter-stage scaling factor d. Assuming that the full scale of both quantizers is the same and equals the reference voltage, the scaling factor is 2/3 for d equal to 2 (see Appendix B). However, this is equivalent to a reduction of the ΣΔM voltage references by this factor, so the ΣΔM full scale decreases accordingly.

Global resonation is implemented in the modulator in Fig. 3.38 by means of two inter-stage paths with coefficients r_1 and r_2 that feed a delayed version of the last-stage quantization error back to the first-stage quantizer. One advantage of this scheme is that, unlike that reported in [Sanc06], only FE integrators are used, and this reduces the modulator sensitivity to circuit non-idealities.

Assuming a linear model for the embedded quantizers, thanks to global resonation, the NTF of the modulator in Fig. 3.38 is given by

$$\mathrm{NTF}(z) = \frac{-(1-z^{-1})^2 \cdot [1-(2-g)\cdot z^{-1} + z^{-2}]}{d} \tag{3.29}$$

where two of the NTF zeros are a function of g, the value of which can be optimally chosen to minimize the in-band quantization error. If the OSR is low enough, the

Table 3.8 Standards specifications and ΣΔM parameters

Standard	GSM	BT	UMTS	DVB-H	WiMAX	WLANa	WLANn
SNDR (dB)	80	75	65	55	60	65	50
BW (MHz)	0.2	0.5	1.96	3.8	10	10	20
L	2	2	2	2	4	4	4
OSR	200	80	40	20	12	16	8
fs (MHz)	40	80	160	160	240	320	320

value of g/d can easily be implemented, because a reduction in OSR means an increase in g as described in Appendix A. The resonation provided by this ΣΔM is thus specially suited to wideband applications at low OSR.

3.2.3.3 Case study

In order to show the benefits of the ΣΔM architecture in Fig. 3.38, a case study is presented considering all the standards listed in Table 3.8. This table also shows the values of the modulator design parameters varied for the distinctive operation modes, namely: modulator order (L), oversampling ratio (OSR) and sampling frequency (f_s) in order to cope with the required resolution for each signal bandwidth. The in-loop coefficients for the modulator in Fig. 3.38 are $\{c_1, c_2, a_1, a_2, c_3, c_4, a_3, a_4\} = \{0.25, 1, 8, 4, 0.25, 1, 8, 4\}$, while the resolution of the first- and second-stage quantizers are 3 and 5 levels, respectively. Global resonation is used only in WLANn mode with a feedback coefficient of $g/d = 0.05$, resulting in a shift of two zeros of the NTF from 0 to $0.95 \pm j0.32$. This distribution of zeros reduces the IBE power by approximately 9.5 dB within a bandwidth of 20 MHz. A sampling capacitor of 0.25 pF is considered in all simulations. The thermal noise introduced by the sampling operation of this capacitor is also taken into account in the following simulations.

Several behavioral simulations have been performed using SIMSIDES [Ruiz05]. The first-stage feedback DAC is assumed linear here, whereas the alternatives for the practical implementation of the required DEM technique are described in Appendix B.

Since there are no noise leakages due to mismatching between the analog and digital processing in the cascade in Fig. 3.38, the DC gain amplifier requirements are clearly relaxed compared to conventional cascades, as illustrated in Fig. 3.39 for the WiMAX mode. The figure shows that amplifier DC gains of 35 dB are enough to fulfill specifications. Similar results are obtained for the remaining modes in Table 3.8. These low requirements for the opamp gain are specially beneficial for a low-voltage scenario, in which technology scaling degrades both the transistor gain and, subsequently, that of the amplifier [Sans06]. Note from Fig. 3.39 that, except for the third opamp, the amplifier DC gain can be lowered much more in the WiMAX mode. Unlike conventional cascades, in which the most demanding requirements are associated with the front-end amplifiers, this is not the case in the proposed modulator topology thanks to the elimination of the digital processing.

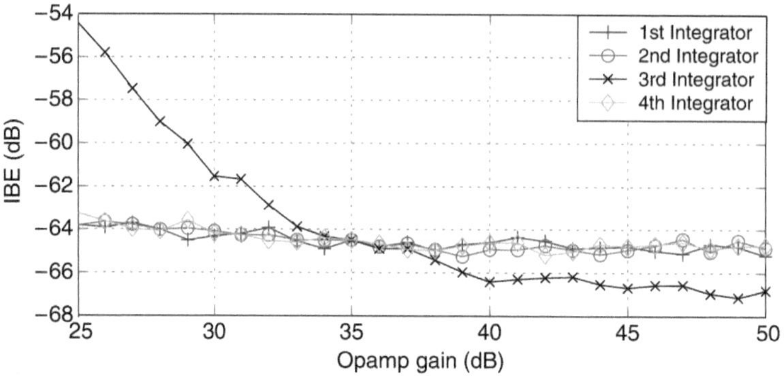

Fig. 3.39 In-band error versus amplifier DC gain for WiMAX (−6-dBFS input level)

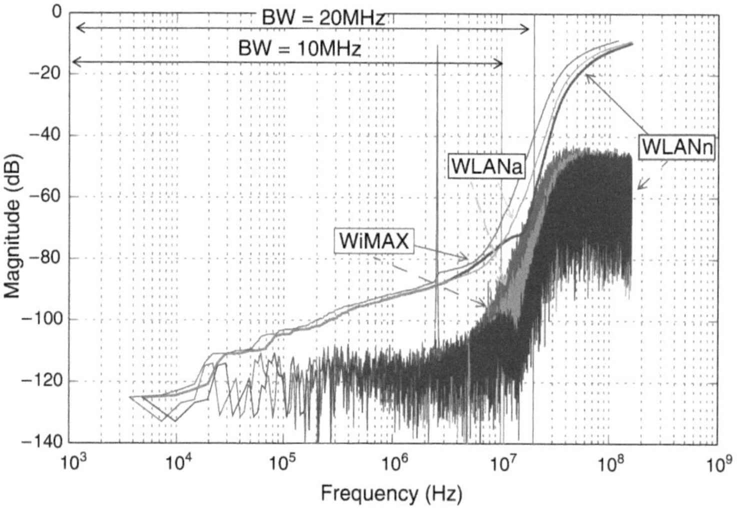

Fig. 3.40 Modulator output spectra for WiMAX and WLAN modes (−6-dBFS input level)

Figure 3.40 shows the output spectra and the integrated IBE powers over frequency for WiMAX, WLANa and WLANn. The IBE power within the 10-MHz band is reduced for WLANa when compared to WiMAX by increasing the sampling frequency. For WLANn, resonation introduces a notch close to the upper limit of the 20-MHz signal band, which helps to further reduce IBE.

Figure 3.41 depicts the obtained modulator SNDR curves for all operation modes, which fulfil the requirements in Table 3.8. Note also that the $\Sigma\Delta$M overloads close to the reference voltage for all modes[8] as a consequence of the great robustness of the modulator topology to amplifier non-linearities. This results in very low requirements

[8] Note that the scaling factor of $2/3$ on the references resulting from the in-loop digital addition makes the SMASH modulator overload 3.5 dB earlier than in a conventional cascade.

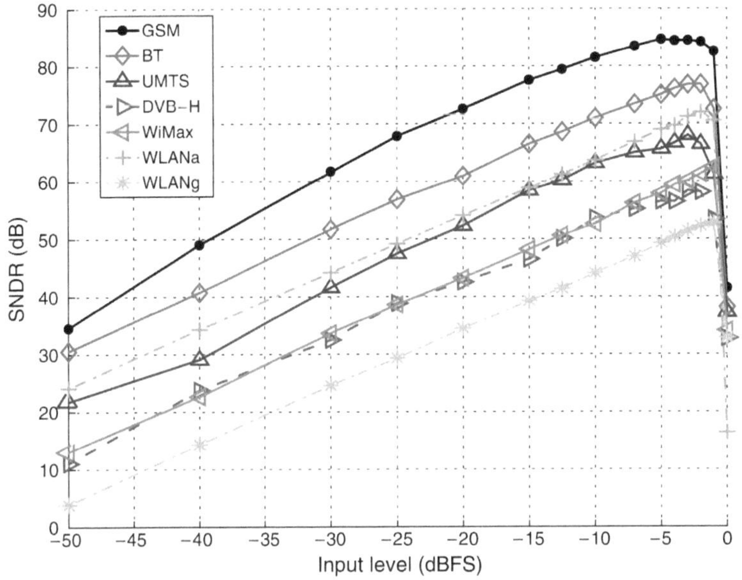

Fig. 3.41 Modulator SNDR versus input signal level

for the amplifier output swings, which are always lower than 0.3 V (25% of the reference voltage).

This case study confirms that the cascade $\Sigma\Delta$M in Fig. 3.38 presents many appealing features, inherited from the combination of USTF, resonation and SMASH concepts in its modulator architecture. These features include reduced opamp output swings (inherited from the USTF implementation in the stages), increased resolution (thanks to global resonation) and robustness against noise leakages such as those that correspond to low DC gain in the first-stage opamps (drawn from the SMASH structure).

3.3 Practical Timing Issues of the Novel Architectures

The previously reported topologies and their proposed combinations shown in Sect. 3.1 and 3.2, respectively, have so far been studied either as linear models in Z-domain or on the basis of behavioral simulations. An appropriate timing analysis is required further down in the design hierarchy. In other words, the SC implementation of these architectures needs to be considered in order to check the overall timing operation of the modulator building blocks. After this analysis (shown below), this section will illustrate why a number of the architectures in Sect. 3.2 present timing issues that make them unreliable in practice. Two novel topologies combining almost all the concepts of the architectures proposed in Sect. 3.1 and 3.2 (USTF, local and global resonation, and SMASH) and circumventing these timing issues are proposed.

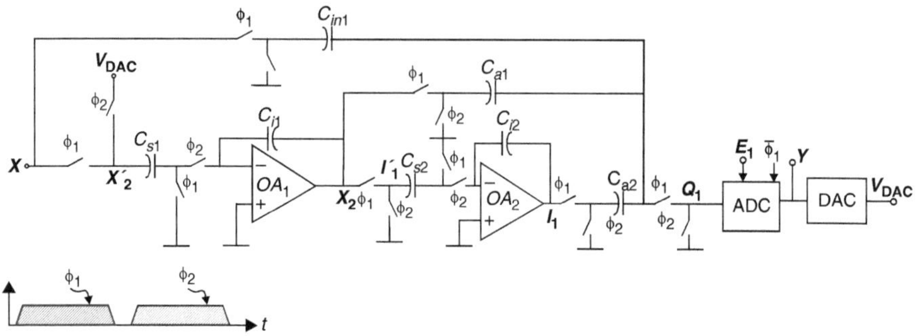

Fig. 3.42 Single-ended SC schematic of the second-order USTF ΣΔM in Fig. 3.6

Most of the architectures presented in previous sections are cascades based on USTF stages, which need to be carefully studied in terms of timing. In fact, USTF implementations in ΣΔMs are known for demanding timing requirements in the loop [Ghar06]. However, several ICs have efficiently implemented this concept in both single-loop [Gagg04, Yao06, Chri10] and cascade ΣΔMs [Nam05, Chri07, Morg10].

3.3.1 Implementation and Timing of USTF ΣΔMs

Let us first study the 2nd-order USTF ΣΔM shown in Fig. 3.6. To this purpose, Fig. 3.42 illustrates its SC implementation. Its timing can be summarized as follows:

- During the sampling phase (ϕ_1), integrators sample their inputs and the analog adder operates. The comparators of the quantizer are triggered at the end of this phase; i.e., when the added value is well settled.
- During the integration phase (ϕ_2), the output voltage of the DAC, V_{DAC}, is fed back to the first integrator. Both integrators integrate in this clock phase. The first integrator processes the subtraction of V_{DAC} to the voltage sampled in the previous phase.

Figure 3.43 shows the timing diagram corresponding to the SC schematic in Fig. 3.42, also illustrating the evolution of all the highlighted nodes. These nodes are:

- The inputs and outputs of the integrators. The inputs of the first integrator are X and V_{DAC}. Note that the former signal stands for the modulator input, while the latter indicates the output of the in-loop DAC. The second integrator samples the output of the first, X_2, with the output of the second being I_1.
- Intermediate nodes used to illustrate the sampling operation of the integrators. These nodes are located at the bottom plate of the integrators sampling capacitors. The first integrator uses X_2' and the second I_1'.
- The output of the analog adder, Q_1.
- The quantizer output, Y, which is also the output of the USTF ΣΔM or stage.

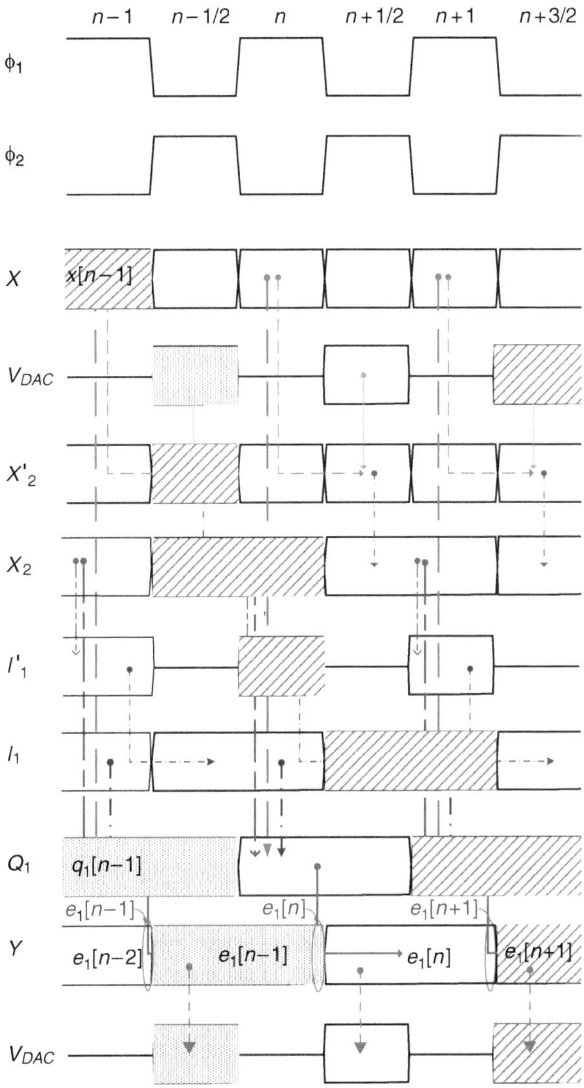

Fig. 3.43 Timing diagram of the SC schematic in Fig. 3.42 (V_{DAC} is duplicated for the sake of clarity)

The top of Fig. 3.43 provides a time reference normalized with respect to the clock period. The numbers of cycles take integer values ($n-1, n, n+1$, etc) that correspond to those points just before the end of the clock phase ϕ_1. On the contrary, the end of ϕ_2 is associated with 1/2 sample of the integer values[9] ($n-1/2, n+1/2, n+3/2$, etc).

[9] Later, the delays (z^{-1}), or half of them ($z^{-1/2}$), can be extracted from a Z-transform. For instance, the value of the node X during cycle n can be defined as $x[n]$ in the time domain. The Z-transform is denoted as $x[n] \leftrightarrow X(z)$, so a delay in this signal will be $x[n-1] \leftrightarrow z^{-1}X(z)$ and half a delay $x[n-1/2] \leftrightarrow z^{-1/2}X(z)$.

Figure 3.43 highlights two paths. The first path (with ruled time slots) stands for the feedback and in-loop operation common to all conventional second-order ΣΔMs; i.e., integrator operation, quantization and D/A conversion. The second path (with filled time slots) is devoted to the feed-forward paths and the associated in-loop operation that is necessary to implement USTF. Note that this is the most critical path and this analysis therefore focuses on it. Special attention has been paid to quantization error generation and its corresponding timing with regard to the remaining nodes, including the preceding analog addition (Q_1) and the in-loop DAC (V_{DAC}). The generation of the quantization error and its time relation with the input signal and the analog addition are highlighted in Fig. 3.43 with ovals. The sequential operation of the analog addition, the in-loop quantization and feedback DAC operation will be critical for the timing of USTF ΣΔMs and, thus, for the overall analysis discussed in this section.

From Fig. 3.43, the output of the analog adder can be defined as

$$q_1[n] = x[n] + x_2[n] + i_1[n] \tag{3.30}$$

The digital output of the modulator—considering a linear quantization error, $e_1[n]$—is

$$y[n + 1/2] = q_1[n] + e_1[n] \tag{3.31}$$

since the quantization is performed just before the end of clock phase ϕ_1. There is therefore neither delay nor half-delay in the quantization operation and the analog addition. Note from (3.30) and (3.31) that the addition of the input voltage, $x[n]$, and the generation of $e_1[n]$ both take place in the same cycle n.

The voltage at the output of the DAC is

$$v_{DAC}[n + 1/2] = \{y[n + 1/2]\} = \{q_1[n] + e_1[n]\} \tag{3.32}$$

where the function $\{.\}$ stands for the D/A conversion. The error generated by this operation can be added as a linear source, as done for the quantization error $e_1[n]$. However, a fully linear DAC is considered. This will have no impact on the timing analysis, but will simplify it. As a consequence, (3.32) transforms into

$$v_{DAC}[n + 1/2] = y[n + 1/2] = q_1[n] + e_1[n] = x[n] + x_2[n] + i_1[n] + e_1[n] \tag{3.33}$$

indicating that there is only $z^{-1/2}$ between the in-loop DAC and the analog addition. In this case, this $z^{-1/2}$ is already half of the z^{-1} in the numerator of the first integrator transfer function[10]. However, it will be demonstrated below that this $z^{-1/2}$ may lead to a timing issue in cascade ΣΔMs made up of USTF stages.

Let us suppose, for example, that V_{DAC} is sampled by the analog adder of an upcoming stage in a cascade. This last node is sampled to extract the quantization

[10] FE integrators are employed, with transfer function $H(z) = z^{-1}/(1 - z^{-1})$.

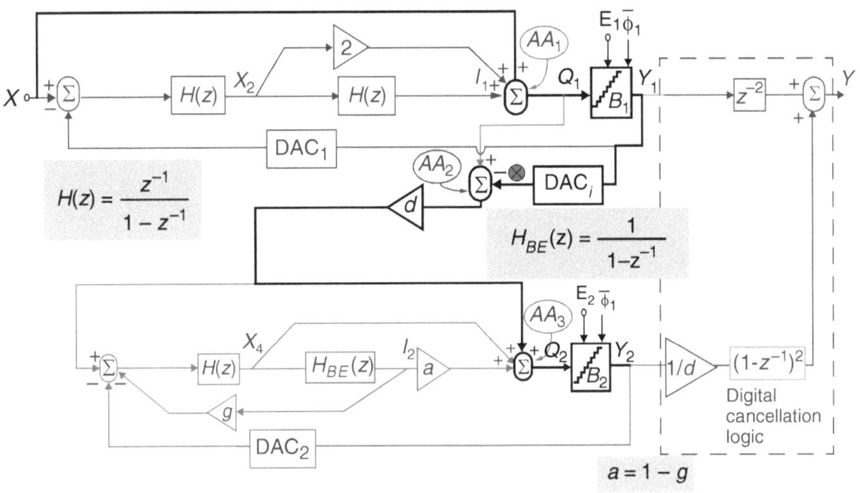

Fig. 3.44 Illustration of timing issues in unity STF-based cascade $\Sigma\Delta$M architectures

error, E_1, as a subtraction of Q_1 and V_{DAC} and used it to feed the latter stage in a cascade $\Sigma\Delta$M. Then, the analog adder both samples its inputs (also V_{DAC} under this presumption) and operates in the clock phase ϕ_1, as depicted in Fig. 3.43 and shown in (3.30). There is thus an extra half a clock cycle ($z^{-1/2}$) derived from the sampling of V_{DAC} in the adder, because the addition and its corresponding sampling are performed in phase ϕ_1. Note from (3.33) that there is a delay (z^{-1}) in the inter-stage addition in question with respect to Q_I and E_1. This is the scenario of the timing issues for most of the architectures presented in Sect. 3.2.

3.3.2 Timing Problems in Unity-STF Cascade $\Sigma\Delta$Ms

In order to analyze potential timing issues in USTF-based cascade $\Sigma\Delta$Ms, let us consider the $\Sigma\Delta$M in Fig. 3.11, in which the 1st-stage quantization error is fed to the second USTF stage by subtracting the 2nd-integrator output and the output of the 1st-stage DAC.

Figure 3.44 shows the critical timing path in Fig. 3.11 highlighting it with ⊗ in the figure. Note from Fig. 3.44 that E_1 is extracted as a subtraction of two paths (performed in the virtual adder AA_2[11]); namely, one from Q_1 and another after the quantization and the inter-stage DAC (labelled as DAC$_i$).

[11] Note that AA_2 will not be implemented in a practical realization, but its operation will be merged with that of AA_3.

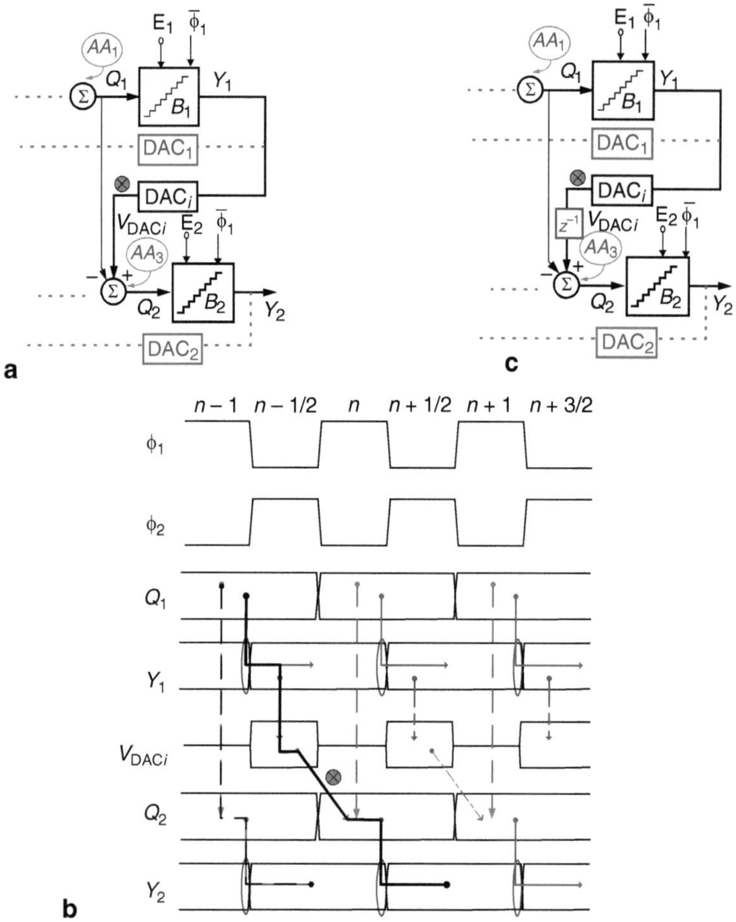

Fig. 3.45 Detailed view of the critical paths in Fig. 3.44: **a** scheme, **b** corresponding timing diagram, **c** actual scheme that includes the effective delay

Practical timing problems arise when the first-stage quantization error, $e_1[n]$, is given to the second-stage adder (AA_3) and to the subsequent quantizer. The analog adders $AA_1$and AA_3 operate in ϕ_1, the embedded quantizers compare their inputs at the end of this phase and the in-loop DAC operates in the following phase, ϕ_2. The main issue is given by the fact that the inter-stage DAC (DAC_i) works at ϕ_2. In practice, therefore, there is an extra delay (z^{-1}) at this point.

Figure 3.45 addresses this in more detail. Figure 3.45a focuses on the region of the scheme in which this timing problem occurs, and then, Fig. 3.45b illustrates the corresponding timing diagram. It can be observed that there is an effective delay between the output of the first-stage analog adder and that of the second-stage adder;

i.e., between nodes Q_1 and Q_2[12]. This path is highlighted with ⊗ in Fig. 3.45b. The signal at node Q_2 can be obtained as

$$q_2[n] = x_4[n] + ai_2[n] + V_{DACi}[n-1/2] - q_1[n] \tag{3.34}$$

Applying (3.33) to V_{DACi}, (3.34) transforms into

$$q_2[n] = x_4[n] + ai_2[n] + q_1[n-1] + e_1[n-1] - q_1[n] \tag{3.35}$$

so not only there is a delay in the delivery of the first-stage quantization error e_1 at the output of AA_3but this delay also affects one of the traces of the first-stage addition q_1 at this node, impeding the cancellation of the two terms in q_1 and the correct extraction of e_1.

Figure 3.45c modifies the scheme in Fig. 3.45a by including the actual extra delay of Q_1 and E_1in the scheme. This delay was not considered for the NTF computation in Z domain and substantially degrades the ΣΔM performance. Cascade ΣΔMs with unity-STF stages cannot thus be implemented in practice by extracting the 1st-stage quantization error through two inter-stage paths due to timing issues. The topologies depicted in previous sections that suffer these timing issues are those depicted in Figs. 3.11, 3.16, 3.17, 3.31, 3.32, 3.37 and 3.38.

This problem[13] is eliminated when only one inter-stage path is used as originally proposed in [Silv01] and [Silv04a], so that unity-STF cascades as those shown in Figs. 3.8, 3.9 and 3.15 do not exhibit this timing issue. This strategy has been applied to reported USTF-based cascade ICs, such as those presented in [Nam05], [Chri07] or [Morg10].

3.3.3 Alternative SMASH- and Resonation-Based Architectures

As stated above, USTF cascade ΣΔMs that need two inter-stage paths to feed the next stage present timing issues. Two topologies that take these issues into account are proposed. One of them makes use of USTF and a resonator in the first stage, and SMASH and global resonation in the overall cascade. This topology solves the timing issues of the proposed cascades that combined USTF stages with either SMASH or global resonation. An alternative architecture, which is also free of the timing issues discussed in this section, is presented. This cascade SMASH ΣΔM makes use of USTF and resonation in the first stage and also global resonation.

---

[12] There is an extra half delay ($z^{-1/2}$) when reaching node Y_2 but it does not affect the NTF and STF of the cascade ΣΔM because it is already part of the delay at the numerator of the third-integrator FE transfer function.

[13] In theory, this problem could be solved if the comparators of the first-stage quantizer are triggered before the end of phase ϕ_1. After a rising edge of this new phase, AA_3must operate and the second-stage quantizer will work at the end of phase ϕ_1. Note that very fast operations are assumed for the analog adders $AA_1$and AA_3. This solution is not robust, specially for high sampling frequencies, because time slots smaller than half-delays are considered.

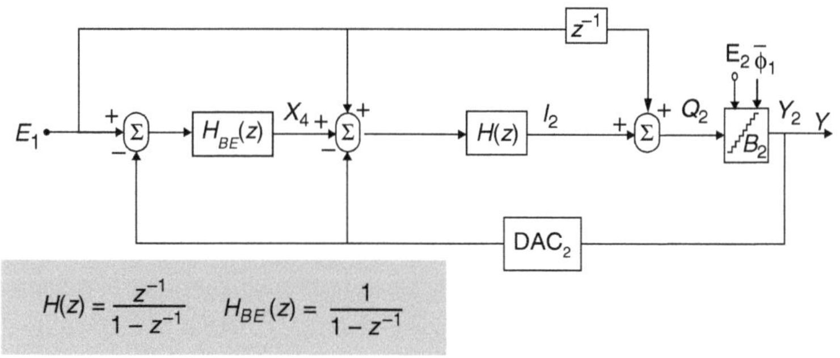

Fig. 3.46 Second stage based on that SMASH ΣΔM in [Magh07]

The first architecture is based on a SMASH ΣΔM proposed by [Magh07]. The second stage of this SMASH, depicted in Fig. 3.46, presents the following STF and NTF:

$$\begin{aligned} \mathrm{STF}_2(z) &= 3 \cdot z^{-1} - 3 \cdot z^{-2} + z^{-3} \\ \mathrm{NTF}_2(z) &= (1 - z^{-1})^2 \end{aligned} \tag{3.36}$$

Let us assume that the first stage is a unity-STF second-order ΣΔM. Then, (3.16) predicts that the output of this SMASH ΣΔM is equal to:

$$Y(z) = X(z) + (1 - z^{-1})^5 E_1(z) - 1/d \cdot (1 - z^{-1})^4 E_2(z) \tag{3.37}$$

where an extra order for the filtering of the first-stage quantization error appears, thanks to the use of the STF_2 in (3.36). If a resonation coefficient g is included in the first stage, the quantization error of both stages making up the ΣΔM will be therefore affected by this first-stage resonator (see Sect. 3.2.3). The resulting output of the proposed modifications for a SMASH ΣΔM is given by:

$$\begin{aligned} Y(z) = X(z) &+ [1 - (2 - g_1) \cdot z^{-1} + z^{-2}] \cdot (1 - z^{-1})^3 \cdot E_1(z) \\ &- 1/d \cdot [1 - (2 - g_1) \cdot z^{-1} + z^{-2}] \cdot (1 - z^{-1})^2 \cdot E_2(z) \end{aligned} \tag{3.38}$$

where a fifth- and fourth-order high-pass filtering, with two complex conjugates zeros, are obtained for both the first and the second quantization errors, respectively.

In general, an FE integrator can operate with inputs sampled in both the sampling and the integration phases with no modification of its transfer function, $H(z) = z^{-1}/(1-z^{-1})$. However, BE integrators need to perform both operations in only one phase because there is no delay in their input transfer function, $H_{BE}(z) = 1/(1-z^{-1})$. In the scheme in Fig. 3.46, $E_1(z)$ is processed as an input by the third and fourth integrators (see the sampling of X and V_{DAC} in the first integrator with the auxiliary signal X_2' in Fig. 3.43), and by the second-stage analog adder, with a delay in this last case. Given that the third integrator is a BE integrator and that its inputs need to extract $E_1(z)$ by operating in different phases—both ϕ_1 and ϕ_2, a cascade with

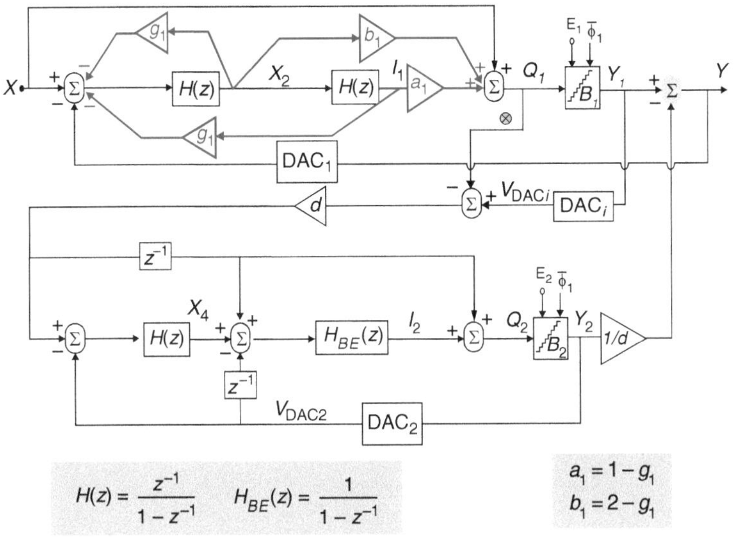

Fig. 3.47 Proposed SMASH ΣΔM free from timing issues and with one local resonator

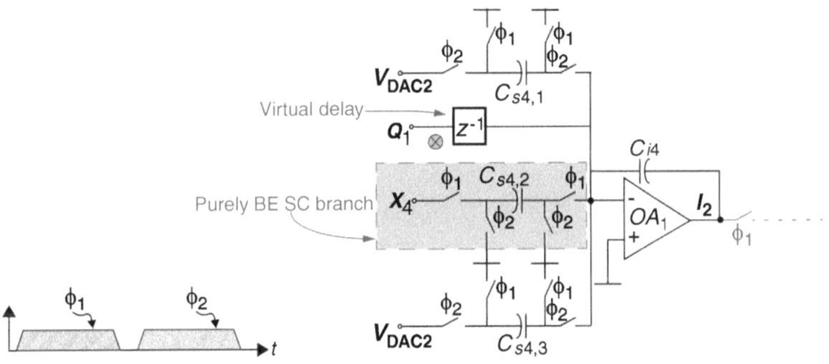

Fig. 3.48 SC schematic of the fourth integrator in Fig. 3.47

USTF first-stage and a second stage similar to that depicted in Fig. 3.46 becomes unfeasible.

Figure 3.47 shows a proposed architecture based on the previous one but solving the problem described above. The signal delay in the numerator of the fourth-integrator transfer function is transferred to its three previous nodes, so the third integrator becomes an FE integrator. This integrator can process $E_1(z)$ taken from the subtraction in different phases of the two inter-stage nodes Q_1 and V_{DACi}. However, the fourth integrator transforms into a BE integrator but only operates in this mode for the output of the third integrator, X_4. Figure 3.48 shows the corresponding SC schematic, where inputs V_{DACi}, Q_1 and V_{DAC2} are processed in an FE mode; i.e., these inputs can be sampled in either ϕ_1 or ϕ_2. Notwithstanding, there is a minor drawback in this architecture from a timing standpoint; the addition in the second

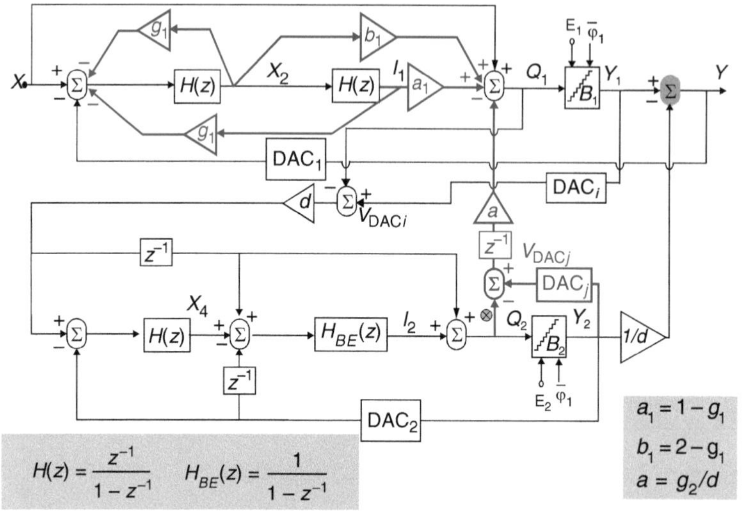

Fig. 3.49 Proposed SMASH ΣΔM with one local resonator and global resonation

stage requires a delay in the signal path of Q_1. To include this required delay, three sample-and-hold SC branches are needed with an appropriate phase scheme as proposed in [Koh05] and [Kana06]. This circuitry has not been included in Fig. 3.48 for the sake of clarity.

Figure 3.49 shows an additional improvement on the topology presented in Fig. 3.47. Global resonation is included as an extra resonation strategy that, unfortunately, only affects the NTF term corresponding to $E_2(z)$. The output of this ΣΔM topology is thus the following:

$$\begin{aligned} Y(z) &= X(z) + \left[1 - (2 - g_1)\cdot z^{-1} + z^{-2}\right]\cdot (1 - z^{-1})^3 \cdot E_1(z) \\ &\quad - 1/d \cdot \left[1 - (2 - g_1)\cdot z^{-1} + z^{-2}\right]\cdot\left[1 - (2 - g_2)\cdot z^{-1} + z^{-2}\right]\cdot E_2(z) \end{aligned} \tag{3.39}$$

Given that the second-stage STF is not equal to 1, there is a component of $E_1(z)$ in the overall modulator NTF. Furthermore, $E_2(z)$ needs to be extracted and then delayed before being added in the first-stage analog adder. As in the extraction of $E_1(z)$, there is an actual delay for only one of the two paths used to feed $E_2(z)$ back from the second to the first stage; namely, the path devoted to the feedback in the inter-stage DAC, DAC_j. Figure 3.50 illustrates this. Figure 3.50a shows the portion of the scheme in Fig. 3.49 that corresponds to the global resonation operation, while Fig. 3.50b illustrates the associated timing diagram. The path that goes from node Q_2 to Q_1 through Y_2 already includes this delay due to the timing requirements of the sequential operations of AA_3, DAC_j and AA_1 during ϕ_1, ϕ_2 and ϕ_1, respectively. However, the path that goes from node Q_2 directly to node Q_1 is not delayed since both adders (AA_1 and AA_3) operate in the same phase. This is the same timing analysis presented in Sect. 3.3.2, the only difference being that the quantization error is given

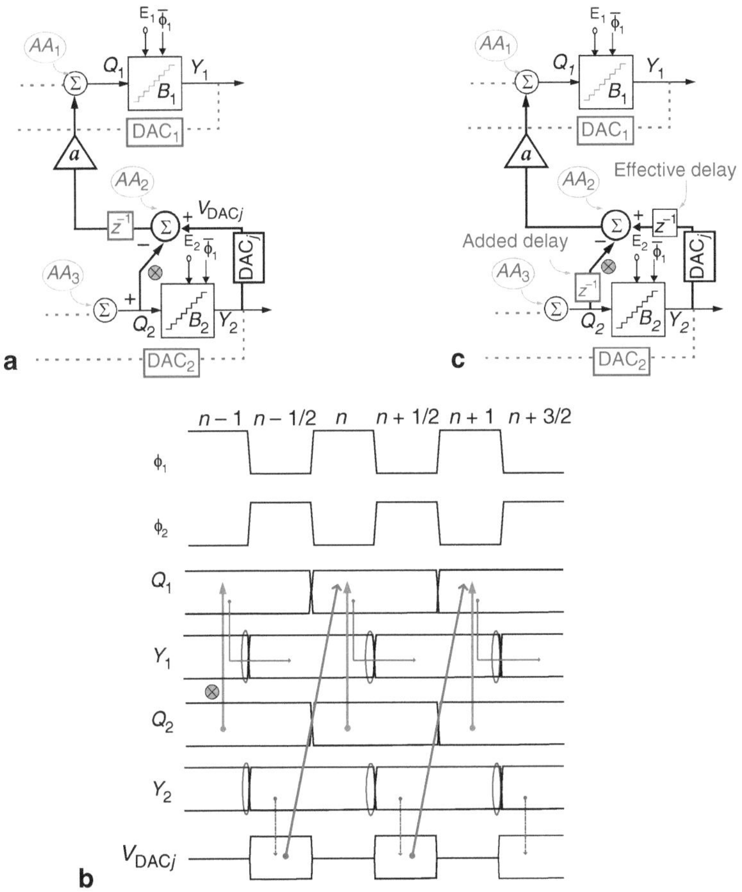

Fig. 3.50 Illustration of the global resonation timing in the schematic in Fig. 3.49: **a** portion of interest of the scheme, **b** timing details, **c** modified schematic for correct timing

from the second stage to the first one. As in the extraction of (3.35), the output of the first-stage adder becomes

$$q_1[n] = x[n] + b_1 x_2[n] + a_1 i_1[n] + aq_2[n-1] + ae_2[n-1] - aq_2[n] \quad (3.40)$$

In practice, an extra delay is thus required to correctly extract the second-stage quantization error as shown in Fig. 3.50c. The inter-stage global resonator path marked with ⊗ in Fig. 3.49 therefore needs to incorporate a delay, following the procedure described in [Koh05] and [Kana06].

The modulator in Fig. 3.51 is also proposed. This modulator combines a SMASH structure with $\mathrm{STF}_2(z) = 2 \cdot z^{-1} - z^{-2}$ as proposed in [Magh06] with a USTF first stage, local resonation in the same stage and global resonation in the overall cascade. Note from Fig. 3.51 that an extra delay—that can be implemented with the procedure

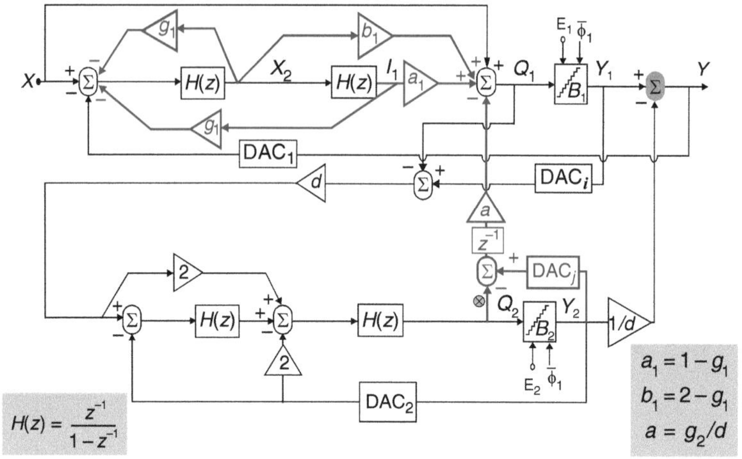

Fig. 3.51 Alternative proposed SMASH ΣΔM with one local resonator and global resonation

described in [Koh05] and [Kana06]—is required for the path with ⊗. The output of the modulator therefore becomes

$$Y(z) = X(z) + \left[1 - (2 - g_1) \cdot z^{-1} + z^{-2}\right] \cdot (1 - z^{-1})^2 \cdot E_1(z)$$
$$- 1/d \cdot \left[1 - (2 - g_1) \cdot z^{-1} + z^{-2}\right] \cdot \left[1 - (2 - g_2) \cdot z^{-1} + z^{-2}\right] \cdot E_2(z) \tag{3.41}$$

where the order of the shaping of the first stage quantization error is reduced to 4. Note from (3.41) that the efficiency of global resonation might be limited because the IBE power may eventually be dominated by the first-stage quantization error.

3.4 Summary

This chapter has shown different techniques—such as concurrency and reconfigurable architectural strategies—for obtaining ΣΔMs suited to SDR. The implementation of these techniques is easier in cascade ΣΔMs that in single-loop topologies. This chapter has focused on the implementation of efficient cascade ΣΔMs capable of incorporating the aforementioned techniques (concurrency and expansion of the modulator order) easily. Several novel cascade ΣΔM topologies have been presented. First, an expandable family of cascade ΣΔMs is shown and previously reported unity-STF and resonation-based cascade ΣΔ topologies are examined, and then other recently reported cascades such as SMASH and global resonation ΣΔMs are also addressed.

All the earlier topologies were combined, resulting in the novel ΣΔMs presented throughout this chapter. These architectures show high robustness against their building block non-linearities and non-idealities, these being specially remarkable in those topologies based on USTF and SMASH $\Sigma\Delta$Ms, respectively. Furthermore, low opamp output swings can easily be obtained with the right selection of in-loop coefficients and/or the resolution of the embedded quantizers in these architectures. All these characteristics make the proposed $\Sigma\Delta$Ms very suitable for the implementation of both low-voltage wideband and multi-mode A/D conversion. In fact, case studies based on behavioral simulations are used as a verification vehicle to show the architectural advantages of the architectures.

Unfortunately, the combined use of cascades made up of unity-STF stages and either SMASH or global resonation involves timing issues. These issues are discussed and solutions are proposed, including two alternative architectures for the practical implementation of most of the ideas presented in the proposed topologies.

A number of the architectures and ideas presented in this chapter will be used for the electrical design and fabrication of three ICs, shown in Chaps. 4, 5 and 6. Some of these ideas, such as the use of expandable cascade topologies, unity-STF cascade $\Sigma\Delta$Ms, resonation and concurrency, together with specific reconfigurable strategies lead to $\Sigma\Delta$M ICs intended for the efficient A/D conversion of multi-standard wireless applications.

Chapter 4
A 130-nm CMOS Reconfigurable 2-1-1 Cascade SC $\Sigma\Delta$M for GSM/BT/UMTS

BEYOND-3G WIRELESS TELECOM SYSTEMS WILL REQUIRE low-power multi-standard chipsets that are capable to operate over a number of different co-existing communication protocols, signal conditions, battery status, etc. [Bran05]. An efficient implementation of these chipsets demands for reconfigurable transceiver blocks that can adjust their circuit parameters to the diverse specifications with adaptive power consumption and at the lowest cost. One of the most challenging building blocks in multi-standard receivers is the ADC, because of the assorted signal bandwidths and dynamic ranges that can be required to properly handle the A/D conversion for several operation modes [Gula01]. Compared to other data conversion techniques, $\Sigma\Delta$ modulators are very suited for the implementation of multi-standard, multi-mode ADCs in highly integrated transceivers using low-cost digitally-oriented nanometer CMOS technologies. On the one hand, the key principles of $\Sigma\Delta$ modulators (oversampling and noise shaping) make them robust with respect to circuit errors. On the other, since both principles determine the dynamic range of the $\Sigma\Delta$ modulator, their variations can easily contribute to adapt the converter performance to different specifications with large hardware reuse [Burg01, Mill03, Veld03, Chri07, Crom09].

This chapter presents a triple-mode SC $\Sigma\Delta$ modulator that combines architecture- and circuit-level reconfiguration strategies in order to adapt its performance to the requirements of GSM, Bluetooth and UMTS standards. The modulator uses an expandable cascade topology with a switchable multi-bit back-end stage, reconfigures its sampling frequency and adapts the bias currents of the amplifiers in order to optimize the power consumption depending on the standard of operation. The prototype has been implemented in a 130-nm CMOS process and operates with 1.2/3.3-V supplies. It features dynamic ranges of 83.8/75.9/58.7 dB within signal bandwidths of 200 kHz/1 MHz/4 MHz with 23.9/24.5/44.5 mW power consumption, of which 9.7/10/24.8 mW are dissipated in the analog circuitry, respectively.

This chapter is organized as follows: Sect. 4.1 presents the RF SIMULINK block set that has been developed in order to extract the specifications of the $\Sigma\Delta$ modulator—which is assumed to be embedded in a zero-IF receiver—and validate the performance of the multi-mode receiver. Section 4.2 discusses the procedure followed for selecting the modulator architecture. Section 4.3 describes the SC implementation, the reconfiguration strategies and the high-level sizing of the $\Sigma\Delta$ modulator, whereas the electrical design of the circuit blocks is covered in Sect. 4.4.

A. Morgado et al., *Nanometer CMOS Sigma-Delta Modulators for Software Defined Radio*,
DOI 10.1007/978-1-4614-0037-0_4,

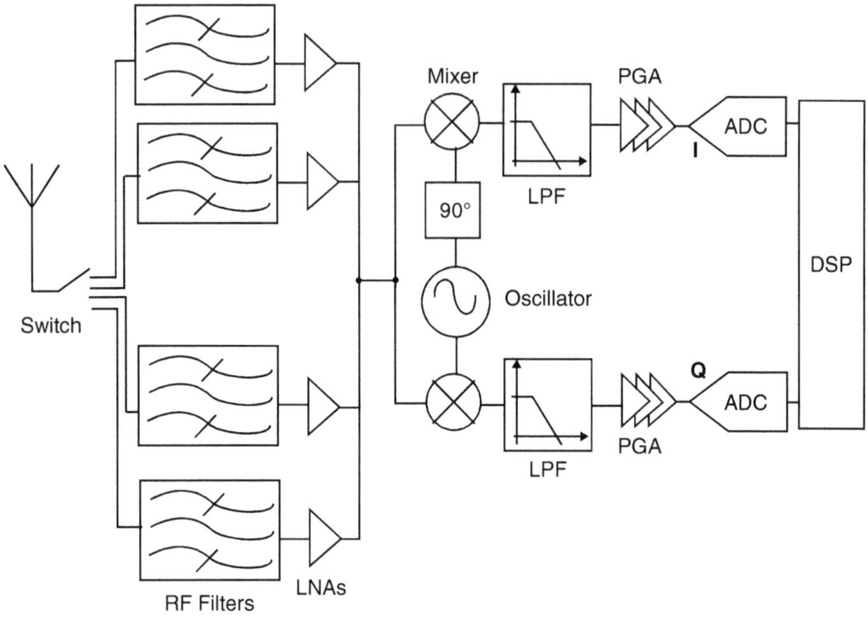

Fig. 4.1 Block diagram of a multi-standard DCR

Section 4.5 shows the chip layout, illustrating special techniques for mixed-signal systems which have been employed, as well as details on the modulator area consumption. Finally, Sect. 4.6 presents the experimental results of the prototype.

4.1 Receiver Considerations and ADC Specifications

The ΣΔ modulator presented in this chapter has been designed to meet the requirements of a multi-standard direct conversion receiver as the one shown in Fig. 4.1. This receiver architecture is commonly used in multi-standard applications because it eliminates the need for both IF- and image-reject filtering and requires only a single oscillator and mixer [Abid95]. In order to cope with the requirements of the different standards, separate (switchable) RF hardware paths (normally one per standard[1]) are used, whereas a single, digitally-programmed baseband section (from the mixer to the ADC) is implemented [Li02].

As stated in Chap. 1, a set of evaluation tests is outlined to validate the receiver performance in terms of three basic aspects: sensitivity, selectivity, and linearity [Crol97]. The first step of the systematic design methodology used in this work consists in mapping the receiver requirements onto building-block specifications (gain, DR, linearity and NF) in order to obtain the performance requirements for the ADC. This is an iterative synthesis process, generally called receiver planning

[1] Sometimes a reconfigurable LNA is employed when the standards share the same RF band [Bran05].

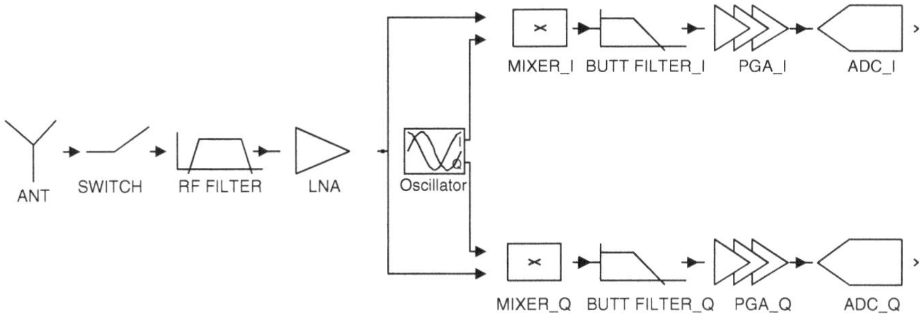

Fig. 4.2 Implementation of a multi-standard DCR using the proposed SIMULINK RF block set

Table 4.1 Input-referred requirements for the different considered standards

	GSM	Bluetooth	UMTS	WLAN
Sensitivity (S_{min})	−102 dBm	−70 dBm	−117 dBm	−65 dBm
Max. signal (S_{max})	−15 dBm	−20 dBm	−25 dBm	−30 dBm
Bandwidth	200 kHz	1 MHz	3.84 MHz	20 MHz
Interferer level	−49 dBm	−39 dBm	−46 dBm	−45 dBm
Max. out-band blocker	0 dBm	−10 dBm	−15 dBm	0 dBm
Max. in-band blocker	−23 dBm	–	−44 dBm	−30 dBm
Max. adjacent channel	−33 dBm	−27 dBm	−92.7 dBm	−65 dBm

[Crol97], that is usually accompanied by a level diagram which shows how the different signals (wanted signal and interferes) evolve along the receiver chain.

In this work, a simulation-based approach has been adopted for the receiver planning. To this purpose, the receiver front-end building blocks are modeled using MATLAB/SIMULINK [Morg09]. The behavioral models of the RF and baseband building blocks in SIMULINK take into account the operating frequency, bandwidth and amplification within the passband of the block, as well as their main non-ideal effects, including noise figure and intermodulation intercept points. In addition to these general parameters, some specific ones are also included, such as the mixer offset. In order to perform all the necessary evaluation tests, a complete DCR is implemented with this block set as shown in Fig. 4.2. Table 4.1 summarizes the specific requirements of an evaluation test of the receiver for a case study that considers four different standards, namely: GSM, Bluetooth, UMTS and WLAN [Li02, Stee07]. Besides, a complete receiver planning in which every building-block specification is a design parameter is beyond the scope of this work. Instead, fixed specifications for the blocks in the receiver chain extracted from reported radio receivers [Yoon04] were considered as shown in Table 4.2 and the ADC effective resolution was extracted from an iterative simulation-based procedure [Morg09]. The extracted ADC specifications are the starting point for the design of the $\Sigma\Delta$M in this chapter (as well as the ones in the next two chapters).

Figure 4.3 illustrates the different results that can be obtained with this tool such as a visualization of the signals (wanted and blockers), distortion and noise propagation on a direct conversion receiver, a level diagram of the different signals of interest and interferences, or a verification of the noise and intermodulation contributions

Table 4.2 Receiver building-block specifications

	SWITCH	RF FILTER	LNA	
			$Gain_{max.}$	$Gain_{min.}$
Gain (dB)	-2	-1	18	8
NF (dB)	0	0	3.5	5.0
IIP_3(dBm)	50	50	-5	0
IIP_2(dBm)	80	80	25	30

	MIXER	LPF
Gain (dB)	2	0
NF (dB)	20	25
IIP_3(dBm)	8	8
IIP_2(dBm)	55	65
Order	---	6

	PGA							
	GSM		Bluetoot		UMTS		WLAN	
	$S_{min.}$	$S_{max.}$	$S_{min.}$	$S_{max.}$	$S_{min.}$	$S_{max.}$	$S_{min.}$	$S_{max.}$
Gain (dB)	97	20	66	24	112	30	66	24
NF (dB)	10	25	10	25	10	25	10	25
IIP_3(dBm)	-40	8	-40	8	-40	8	-34	8
IIP_2(dBm)	10	65	20	60	10	65	10	65

of every building block to the NF or IIP of the overall DCR, respectively. In fact, Fig. 4.3a shows the way the required resolution for the ADC—in terms of the peak SNDR—is extracted together with the evolution of the different metrics throughout the receiver chain for WLAN. The propagation of the different test signals defined by the standard—i.e., the level diagram—is presented for UMTS in Fig. 4.3b. The evolution and contribution of NF and IIP_3 in all the building blocks that conform the receiver are shown in the remaining subfigures. Note that the simulation results of NF, IIP_2 and IIP_3 fulfill the standard requirements, as shown in Table 4.3.

Following this procedure, ADC resolutions of 13 bit for GSM, 11 bit for Bluetooth, 9 bit for UMTS and 7 bit for WLAN, along with commonly reported performances for the remaining blocks in the receiver chain, allow to fulfill the receiver requirements imposed by these four standards. The former ADC specifications for the first three standards are the starting point for the high-level exploration of ΣΔ alternatives, detailed in next section. Note that WLAN is not been included in the specifications of the presented ΣΔM because of the large operation signal bandwidth of this standard[2].

[2] Initial estimations of the power consumption showed that the power of the main building blocks in this modulator were dominated by this operation mode to such an extent that the obtained performance was really degraded. Moreover, this operation mode also dominates the power of the remaining ones because the building blocks were not flexible enough to work with such a large difference in the biasing and, consequently, in the power as well. The demanding speed requirements of this standard need a more efficient ΣΔM architecture, as those presented in Chaps. 5 and 6.

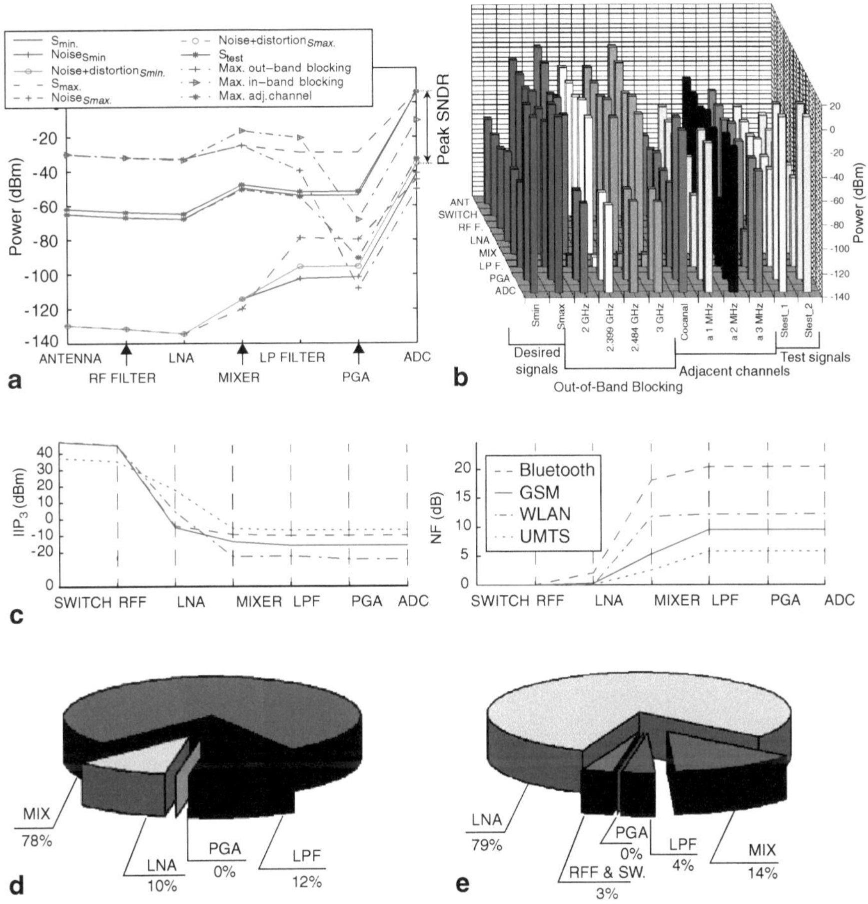

Fig. 4.3 Illustration of block set capabilities: **a** metrics propagation (WLAN), **b** level diagram (UMTS), **c** evolution of NF and IIP_3 through the receiver chain, **d** NF contribution of each block (BT), **e** IIP_3 contribution (GSM)

4.2 Selection of the Modulator Architecture

The expandable cascade described in Sect. 3.1.2 allows an architectural adaptation of the $\Sigma\Delta$M order—and, thus, of the shaping of the quantization error—according to the requirements of the corresponding operation modes. Besides, the potential use of dual quantization for the last-stage quantizer is another advantage in terms of architectural reconfiguration. These are the reasons why this architecture is selected for the targeted multi-standard applications.

Table 4.3 Summary of the simulated receiver performance

Std.	NF (dB)		IIP_3 (dBm)		IIP_2 (dBm)	
	Simulated	Specs.	Simulated	Specs.	Simulated	Specs.
GSM	9.48	10.00	−15.96	−18.00	30.39	12.50
BT	20.42	23.00	−9.93	−17.50	34.48	12.00
UMTS	5.79	9.00	−6.74	−7.00	40.71	32.00
WLAN	12.20	15.00	−24.43	−24.50	5.37	−4.00

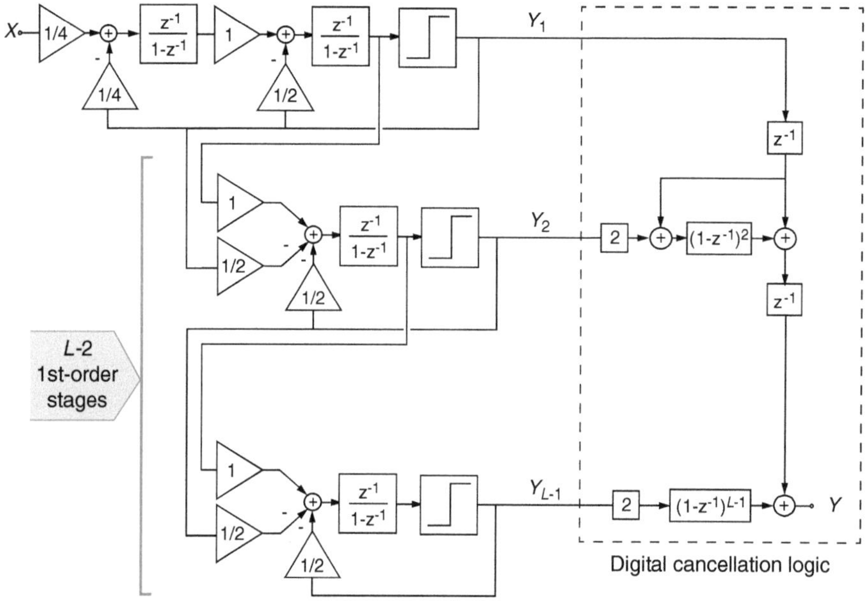

Fig. 4.4 Block diagram of the expandable cascade ΣΔM in [Mede03]

4.2.1 *Expandable ΣΔ Cascade Architecture*

The expandable cascade ΣΔM with schematic in Fig. 4.4 [Mede03] comprises a second-order stage followed by first-order stages, and can be easily extended to build a ΣΔM of a generic order L by simply adjusting the number of first-order stages. The outputs of all the stages are processed in digital domain and combined by the digital cancellation logic so that ideally only the quantization error of the last stage remains, and it is shaped by an NTF whose order equals the sum of the respective orders of all the stages in the cascade. Note that this architecture can exploit the benefits of an unconditionally stable high-order shaping thanks to its cascade structure and of a robust, linear multi-bit quantization by incorporating it only in the modulator last stage [Bran91][3]. Altogether, the main advantages of this $\Sigma\Delta$M architecture are [Mede03]:

- The systematic loss of resolution that is typically present in every cascade $\Sigma\Delta$M when compared with an ideal L-th order loop is only 6 dB.
- The modulator overload level remains constant at[4] −5.6 dBFS, regardless the order of the expandable cascade.
- The output swing required in the integrators is only the modulator full scale.
- Capacitor sharing allows to implement the integrator coefficients with only two SC branches.
- The total number of unit capacitors is only $2 \times [5 + 4(L - 1)]$, which benefits area occupation, thermal noise and amplifier dynamics.
- All first-order stages can be electrically identical[5], which considerably simplifies the electrical and physical implementation of the modulator.

Note that every cascade $\Sigma\Delta$M belonging to the family depicted in Fig. 4.4 can be univocally described by three parameters: the modulator order (L), the oversampling ratio, and the number of bits in the last stage (B). Thus, a $\{L, B, \text{OSR}\}$ triad is used to codify them.

4.2.2 *Exploration of $\Sigma\Delta$ Cascade Candidates*

The first step in the design of the reconfigurable $\Sigma\Delta$M is the exploration of the $\{L, B, \text{OSR}\}$ candidates for each standard that fulfill the corresponding standard requirements with minimum power consumption. At this step, an improved version of the analytical procedure described in [Mede03] to estimate the power consumption of cascade $\Sigma\Delta$Ms has been followed. As shown in Fig. 4.5, the procedure—based on compact expressions that contemplate both architectural and technological features—schematically consists of the following steps:

- The in-band quantization error power (P_Q) is calculated for given values of $\{L, B, \text{OSR}\}$ and of the modulator reference voltage (V_{ref}). Noise leakages due to capacitor mismatch (ΔC), finite amplifier DC gain (A_{DC}) and errors in the multi-bit DAC (INL_{DAC})—if $B > 1$—are also contemplated.

[3] The use of a multi-bit quantizer—and DAC—for only the last stage, allows a dual-quantization scheme in which the quantization noise of the overall modulator decreases while the non-linearities of the multi-bit DAC are filtered by an order equal to the number of integrators in all the precedent stages.

[4] Note that dBFS is defined as dB relative to the FS, where the FS is given by the quantizer reference voltage.

[5] If multi-bit quantization is used in the modulator last stage, its integrator coefficients are usually doubled in order to operate with an unity-gain multi-bit quantizer and make its implementation easier. This, together with the larger integrator load due to the multi-bit quantizer, normally prevents from using the same electrical design.

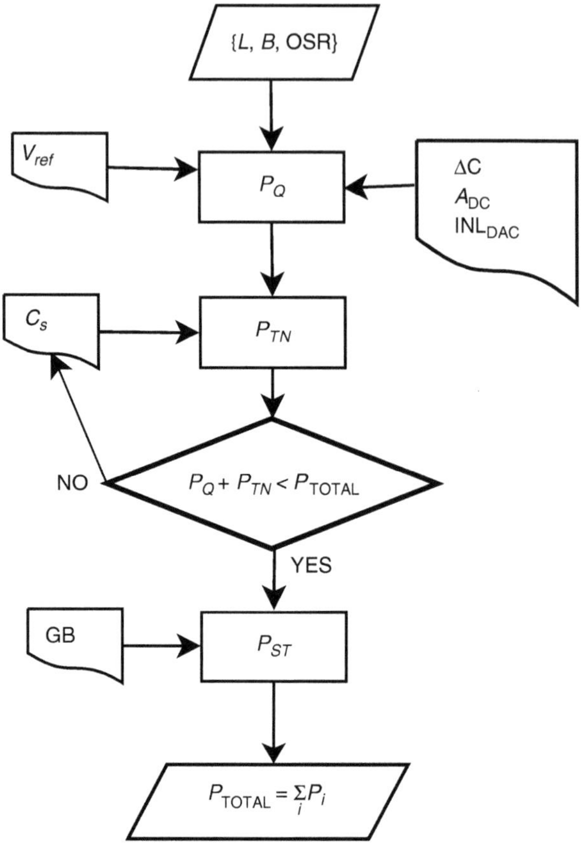

Fig. 4.5 Exploration procedure of ΣΔM candidates

- The in-band error power due to thermal noise (P_{TN}) is then considered. The value of the sampling capacitor at the modulator front-end (C_s) is selected so that $P_Q + P_{TN}$ is smaller than the maximum allowed total IBE power. P_{TN} will be mainly contributed by kT/C noise, but some room is left at this step for the contribution of the front-end amplifier noise.
- The amplifier GB is estimated so that the IBE power due to the integrator defective settling (P_{ST}) is non-limiting ($P_{ST} \ll P_Q + P_{TN}$). A linear settling model is used, considering that it takes a number $\ln(2^{\mathrm{ENOB}})$ of time constants to settle within ENOB resolution.
- The amplifier GB is then related to its power dissipation, for which the amplifier topology must be known a priori. Suitable candidates are closely related to the process technology—130-nm CMOS in our case—and its supply voltage, minimal device length, etc.
- Once the power dissipation of the front-end integrator has been estimated, that of the remaining ones—with less demanding specifications—is considered to be a

fraction of it. The overall modulator power is then basically obtained by adding all contributions, together with the dynamic power in the SC stages.

Given the targeted multi-standard application, the suitable $\{L, B, \text{OSR}\}$ triads for each standard have been explored under the following global constraints:

- The modulator reference voltage is fixed to 1.2 V in order to place the input signal level at −5.6 dBFS[6], maximizing thus the SNDR.
- Given the targeted low/medium resolutions, the explored values of L are restricted to 2, 3 and 4 for the modulator operation not to be limited by low-order quantization noise leakages.
- In order to ease the clock adaptation from one standard to another, the sampling frequency (f_s) is restricted to power-of-two divisions of a 80-MHz master clock. This limits the OSR values to explore and forces to expand the BW in UMTS from 3.84 to 4 MHz.
- The smallest unit capacitor to be considered is fixed to 0.25 pF for mismatching issues.
- In order to facilitate the circuit reconfiguration, the sampling capacitor at the modulator front-end can only take values that are multiple of the unit capacitor.

For illustration purposes, Table 4.4 shows several architectures explored for the various standards, together with the information that the former procedure provides about their high-level sizings. Shadowed cells in Table 4.4 correspond to fixed parameters for all the standards, such as the reference voltage or capacitor mismatch.

Table 4.5 summarizes the ranking of $\Sigma\Delta$Ms with the lowest estimated power after the exploration of the $\{L, B, \text{OSR}\}$ candidates for each standard. Together with the values for $\{L, B, \text{OSR}\}$, those required for f_s and C_s are also enclosed, as well as the obtained DRs and peak SNDRs. The highlighted rows in Table 4.5 correspond to the $\Sigma\Delta$Ms selected for further consideration. The rest of candidates are directly covered by the selected ones, since they just imply an increase of L or B. Thus, the selected $\Sigma\Delta$Ms at this step globally comprise: third- and fourth-order cascades; single-bit and multi-bit quantization of 2, 3 or 4 bits; sampling frequencies of 20, 40 or 80 MHz; and sampling capacitors of 0.25 or 0.5 pF. This can be implemented at circuit level by reconfiguring the third stage of the expandable cascade to either single-bit or multi-bit with programmable resolution, by dividing the master clock frequency by a factor 2 or 4, and by using switchable capacitors at the modulator front-end, respectively. Seeking for a single circuit that covers all the former possibilities can a priori be done, but such a large degree of freedom in the reconfigurability will considerably increase the circuit complexity. Thus, only one $\{L, B, \text{OSR}\}$ triad will definitively be selected for each standard. However, given that the estimated power consumptions are not very different from one case to another, the final decision was taken after extracting their complete set of building-block requirements using more accurate behavioral simulations as detailed below.

[6] Note that the gain of the PGA in Fig. 4.1 needs to be adjusted, thus, to provide this input level.

Table 4.4 Examples of high-level sizings and corresponding IBE contributions

Standard		GSM	Bluetooth	UMTS
Modulator	Topology	2-1	2-1-1	2-1-1
	OSR	100	20	10
	Clock frequency	40MHz		80MHz
	Differential reference voltage	2.4V		
Front-End Integrator	Sampling capacitor	0.25pF		
	Mimimum unit capacitor	0.25pF		
	Capacitor standard deviation	0.15%		
	Estimated switch on-resistance	230Ω	262Ω	158Ω
Amplifier	Open-loop DC gain	2000 (66dB)		
	GB (3.55-pF load)	136MHz	122MHz	202MHz
	Equivalent input noise	$9.0\text{nV}/\sqrt{\text{Hz}}$	$9.3\text{nV}/\sqrt{\text{Hz}}$	$7.3\text{nV}/\sqrt{\text{Hz}}$
A/D/A Converter	Resolution	1bit		2bit
	DAC INL	---		0.25%FS
Quantization Noise		-114.7dB	-83.4dB	-66.2dB
	Ideal quantization noise	-115.8dB	-84.0dB	-66.5dB
	Amplifier DC gain leakage	-122.1dB	-101.0dB	-92.1dB
	Capacitor mismatch leakage	-128.0dB	-93.1dB	-78.0dB
	DAC non-linearity error	---		-96.1dB
Thermal Noise		-90.0dB	-83.0dB	-80.0dB
	kT/C noise	-91.8dB	-84.8dB	-81.8dB
	Amplifier noise	-94.8dB	-87.8dB	-84.8dB
IBE Power		-90.0dB	-80.2dB	-66.0dB
Dynamic Range		88.6dB	78.8dB	64.6dB
EFFECTIVE RESOLUTION (from peak SNDR)		13.6bit	12.0bit	9.6bit

Table 4.5 Ranking of $\Sigma\Delta$Ms according to power estimations

Std.	L	B	OSR	f_S (MHz)	C_S (pF)	DR (bit)	Peak SNDR (bit)	Power (mW)
GSM	4	1	50	20	0.50	14.4	13.6	10.9
	3	2	50	20	0.50	14.4	13.6	11.7
	3	3	50	20	0.50	14.4	13.6	12.5
	4	2	50	20	0.50	14.4	13.6	13.4
	3	4	50	20	0.50	14.4	13.6	14.1
	4	3	50	20	0.50	14.4	13.6	14.2
	4	4	50	20	0.50	14.4	13.6	15.8
	3	1	100	40	0.25	14.4	13.6	17.0
	4	1	100	40	0.25	14.4	13.6	20.0
BT	4	1	20	40	0.25	12.8	12.0	18.1
	3	3	20	40	0.25	12.8	11.9	20.6
	3	4	20	40	0.25	13.0	12.2	22.5
	4	2	20	40	0.25	13.2	12.3	23.0
	4	3	20	40	0.25	13.2	12.4	23.9
	4	4	20	40	0.25	13.2	12.4	25.5
	3	1	40	40	0.25	13.4	12.6	32.1
UMTS	3	4	10	80	0.25	10.8	10.0	37.3
	4	2	10	80	0.25	10.4	9.6	38.6
	4	3	10	80	0.25	11.5	9.6	42.0

4.2.3 Final Modulator Architecture Selection

The formerly selected candidates have been extensively simulated using SIMSIDES [Ruiz05]. The architecture specifications can this way be mapped onto more refined building-block requirements such as amplifier DC gain, GB, SR, equivalent input noise, switch on-resistance, etc.

The steps followed in this process are:

- Validate that the $\Sigma\Delta$Ms selected from Table 4.5 achieve the required DR for each standard, taking into account quantization error and kT/C noise.
- Determine the maximum equivalent input noise for each amplifier that does not degrade the formerly achieved performance.
- Determine the required amplifiers dynamics (GB and SR), taking into account settling errors during both the integration and sampling phases [Rio06].
- Refine the DC gain and SR requirements at the front-end integrator in order to limit the generated distortion near the modulator overload level.

At this step different amplifiers are considered in each integrator in order to gain insight on their individual needs. Once the final modulator architecture is selected for each standard, we will try to cover the global amplifier specifications using reconfigurable amplifiers (in terms of bias currents and/or transistors sizings). Switches,

Table 4.6 Amplifier requirements after fine tuning of the different {*L*, *B*, OSR} candidates

Std.	{*L*, *B*, OSR}	Integrator	Amplifier input noise (nV/Hz$^{1/2}$)	Amplifier DC gain	Eq. capacitive load (pF)	GB (MHz)	SR (V/μs)
GSM	{3, 1, 100}	#1	7.0	1500	3.5	11.9	55.1
		#2	135.0	250	5.7	11.7	47.0
		#3	237.5	250	5.7	5.4	21.7
	{3, 2, 50}	#1	6.0	700	3.7	18.1	68.1
		#2	32.5	400	5.7	17.3	47.0
		#3	225.0	250	16.0	2.3	25.1
	{4, 1, 50}	#1	6.0	800	3.7	16.7	67.9
		#2	12.5	250	5.7	24.2	43.4
		#3	125.0	250	5.7	7.1	61.5
		#4	212.5	250	5.7	6.5	18.1
BT	{3, 3, 20}	#1	8.0	2000	3.5	61.8	88.2
		#2	26.0	550	5.7	70.8	108.5
		#3	49.0	350	16.0	40.1	193.4
	{4, 1, 20}	#1	7.0	2000	3.5	56.8	117.6
		#2	12.0	350	5.7	70.8	108.5
		#3	19.0	400	5.7	33.4	271.2
		#4	40.0	300	5.7	12.9	135.6
UMTS	{3, 4, 10}	#1	18.0	2500	3.5	107.6	150.9
		#2	12.0	500	5.7	178.8	289.1
		#3	35.0	400	16.2	130.7	492.6
	{4, 2, 10}	#1	11.0	1800	3.5	118.3	184.9
		#2	18.0	400	5.7	172.9	271.1
		#3	45.0	450	5.7	227.4	361.4
		#4	35.0	450	16.2	65.2	193.2

which will not be reconfigured from one standard to another, are sized at this step considering their slow-down effect on the integrators dynamics [Yu99] and their induced dynamic distortion [Wu98]. They exhibit an on-resistance around 260 Ω, which does not compromise performance in the different standards and avoids the use of clock-boosting techniques.

The requirements of the selected ΣΔMs after the fine tuning process are summarized in Table 4.6, in terms of the amplifier equivalent input noise, DC gain and dynamics for each integrator. Figure 4.6 depicts the SNDR curves obtained by behavioral simulation for the modulator sizings in Table 4.6. The ΣΔMs exhibit peak SNDRs larger than 81 dB for GSM, 71 dB for Bluetooth and 58 dB for UMTS. Based on the results in Table 4.6, especially on those related to the amplifier dynamics, the finally selected {*L*, *B*, OSR} triads are {3, 1, 100}, {4, 1, 20} and {4, 1, 10} for GSM, Bluetooth and UMTS, respectively.

4.3 SC Implementation of the Reconfigurable ΣΔ Modulator

Figure 4.7 shows the block diagram of the multi-mode ΣΔM, which adapts its topology, conversion BW and operating frequency in order to fulfill the requirements of the three standards, 13/11/9 bit effective resolution within 0.2/1.0/4.0 MHz of BW. For the standard with the largest SNDR (GSM) a third-order shaping and high oversampling are employed, whereas the modulator uses a fourth-order shaping and

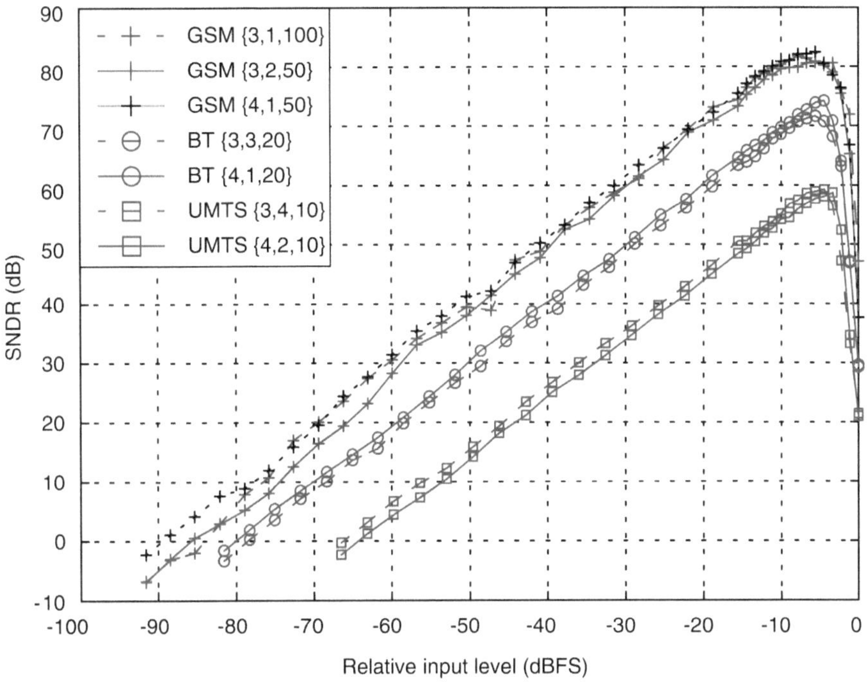

Fig. 4.6 SNDR curves obtained by behavioral simulation for the ΣΔMs with parameters in Table 4.6

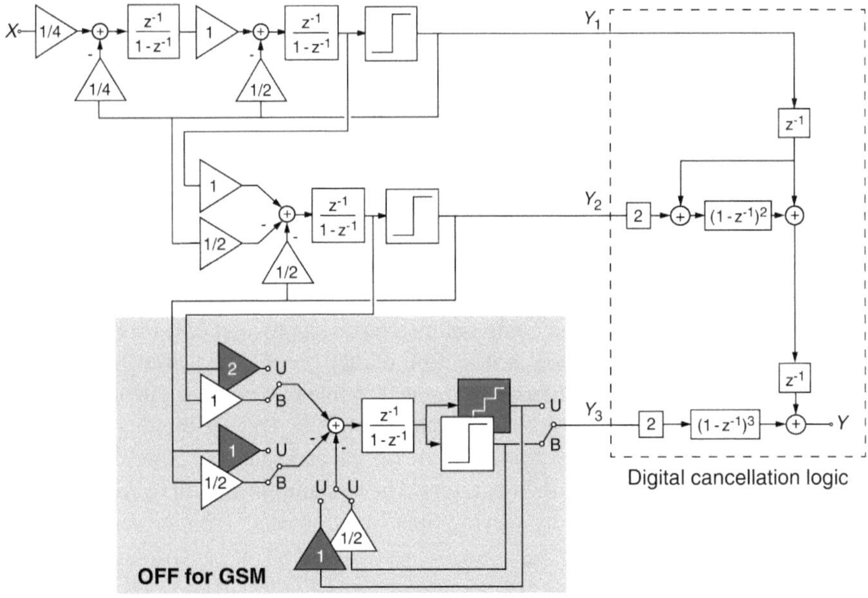

Fig. 4.7 Conceptual block diagram of the multi-mode ΣΔM

Table 4.7 Timing of the ΣΔM

	ϕ_1	ϕ_2	ϕ_1	ϕ_2
INTEGRATORS	sample	integrate	sample	integrate
COMPARATORS	regenerate refresh output	No change (NC)	regenerate refresh output	NC
ADC	regeneration refresh output	NC	regeneration refresh output	NC
DAC	refresh output	NC	refresh output	NC

2 bits in the last-stage for the standard with the largest BW (UMTS) due to the limited oversampling. Note that the clock frequency is doubled for this mode. A fourth-order shaping is also used for Bluetooth but, contrary to UMTS, the last-stage quantizer is reconfigured as single-bit.

The fully-differential SC implementation of the multi-mode ΣΔM is shown in Fig. 4.8. The formerly described reconfiguration strategies are handled at the circuit level with three control signals: two of them (*GSM* and *SB* in Fig. 4.8) are used to power on/off the last-stage of the cascade and to adapt its quantizer resolution, whereas the third one (not shown in Fig. 4.8) controls the division of the master clock frequency. A sampling capacitor of 0.25 pF is used for all standards, thus eliminating the need for switchable capacitor arrays at the modulator front-end. This value makes the ENOB to be limited by kT/C noise for GSM, whereas it is dominated by quantization noise for UMTS.

The timing scheme of the ΣΔ modulator is illustrated in Table 4.7. The modulator operation is controlled by two non-overlapped clock phases, ϕ_1 and ϕ_2. The integrator input signals are sampled during phase ϕ_1 and then integrated together with the corresponding feedback signals during phase ϕ_2. The comparators and the ADC are activated at the end of ϕ_2 (using $\overline{\phi}_2$ as strobe) to avoid any possible interference of the integrator transient response at the beginning of sampling. The operation over time of each block for the conversion of an input sample is summarized in Table 4.7. Note that the timing guarantees a single delay per clock cycle.

Considering the SC implementation in Fig. 4.8, the modulator specifications of ENOB and BW for each standard have been mapped onto electrical requirements of the amplifiers, switches and comparators, considering worst cases for speed and for thermal noise. In this case, a new fine tuning process has been followed in order to fulfill the various standard requirements while covering the global amplifier specifications with only two reconfigurable amplifiers. One of the reconfigurable amplifiers is to be used in the first stage of the cascade and the other one will be employed in the third and fourth integrators. The two amplifiers will be reconfigured from one standard to another by varying their bias currents.[7] This strategy helps to

[7] Note that the sizing of the amplifiers is not modified with the operation mode. This would have entailed a more complex transistor-level design.

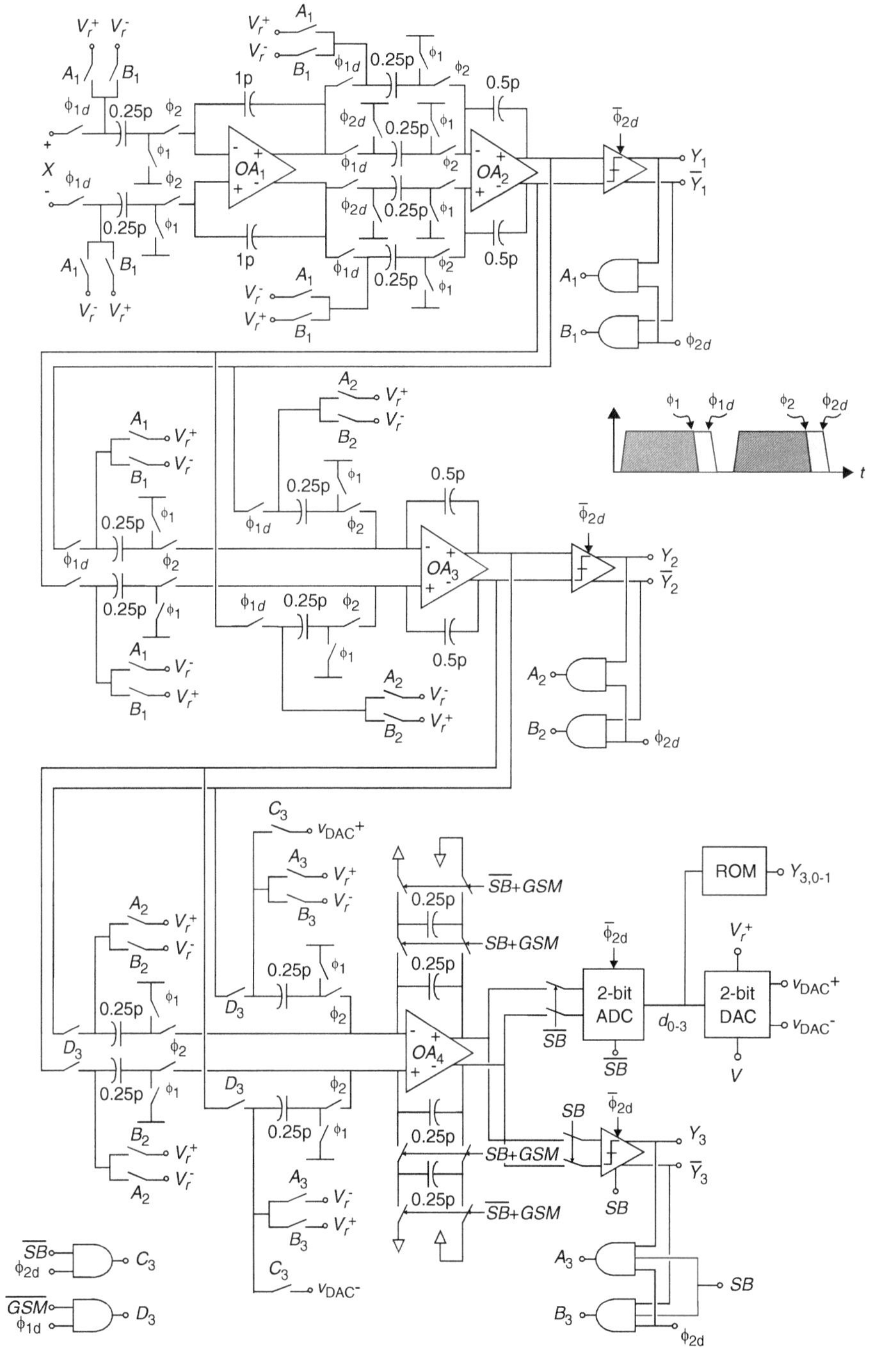

Fig. 4.8 SC schematic of the reconfigurable multi-mode ΣΔM

Table 4.8 Requirements of amplifiers and comparators in the reconfigurable ΣΔ modulator

		Amplifier					Comparator	
Std.	**Integrator**	**Input Noise ($nV/Hz^{1/2}$)**	**DC gain (dB)**	**Eq. load (pF)**	**GB (MHz)**	**SR (V/μs)**	**Hysteresis (mV)**	**Offset (mV)**
GSM	#1	55.4	57.9	3.5	21.1	90.4	30	40
	#2			5.7	13.1	88.9		
	#3	85.0	58.1	5.7	8.6	36.1		
BT	#1	28.3	55.7	3.5	93.4	271.2	30	40
	#2			5.7	57.9	266.8		
	#3	30.0	54.0	5.7	49.6	108.5		
	#4							
UMTS	#1	15.8	50.0	3.5	193.9	628.6	30	40
	#2			5.7	120.2	621.6		
	#3	25.6	53.3	5.7	316.4	324.9		
	#4			16.2	111.7	173.7		

considerably reduce the complexity of the multi-mode cascade ΣΔM and to shorten the design cycle of its cells.

Table 4.8 summarizes the amplifier and comparator requirements when operating in the various standards. These requirements define the starting point for the electrical synthesis of the modulator blocks, which is described in the next section.

4.4 Electrical Design of the Modulator Blocks

4.4.1 Amplifiers

All amplifiers in the multi-standard ΣΔM use the folded cascode topology that is depicted in Fig. 4.9, since the required DC gains are not high, whereas the speed specifications are quite demanding when operating in UMTS mode. The differential input pair uses pMOS transistors, which affects power consumption but enables to

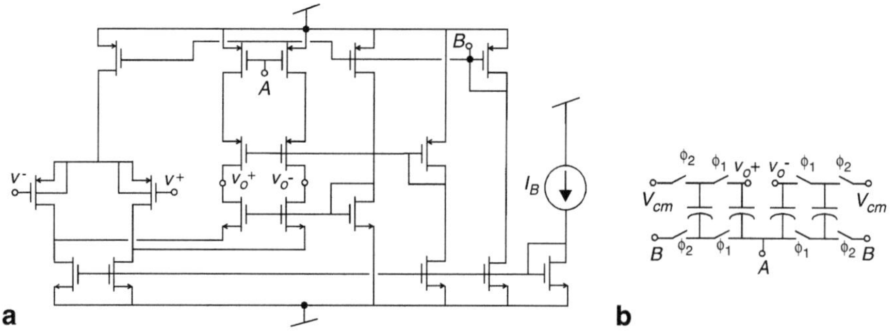

Fig. 4.9 Folded cascode amplifier: **a** core, **b** SC common-mode feedback net

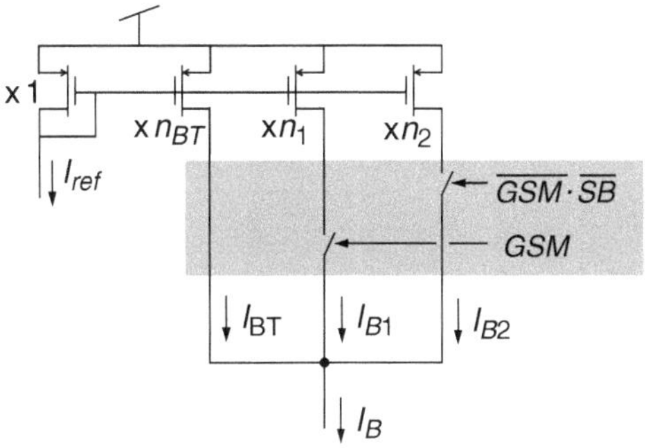

	(GSM,SB)	I_B
BT mode	(0,1)	I_{BT}
GSM mode	(1,X)	$I_{BT} + I_{B1}$
UMTS mode	(0,0)	$I_{BT} + I_{B2}$

Fig. 4.10 Adaptation of the bias current of an amplifier

cancel the body effect as an strategy to reduce the impact of substrate noise coupling. Minimum-length transistors have been avoided in the amplifier input pair and in the current mirrors in order to reduce 1/f noise and mismatching effects.

The four amplifiers (OA_1 to OA_4 in Fig. 4.8) share the same transistor sizes, with the only difference being a larger differential input pair in the second and third amplifiers (OA_2 and OA_3) in order to obtain a faster response. The bias current of each amplifier is adapted from one standard to another according to the scheme illustrated in Fig. 4.10. This figure shows that the currents I_{B1} and I_{B2}—with current mirroring ratios n_1 and n_2—are added to the current tail I_{BT}—defined as $n_{BT} \times I_{ref}$—according to the table enclosed in Fig. 4.10 in order to generate the output current I_B. Depending on the amplifier and on the operation mode, the bias currents vary from 16 to 80 μA.

Table 4.9 summarizes the electrical parameters of each amplifier on the different operation modes obtained after full transistor sizing. Results correspond to the worst-case value of each individual parameter obtained from a corner analysis, considering fast and slow device models, ±10% variation in the 3.3-V nominal supply and temperatures in the range [−40 °C, +85 °C]. Note that the electrical results in

Table 4.9 Worst-case results for the amplifiers obtained by electrical simulation

	GSM		Bluetooth			UMTS			
	OA_1	OA_2, OA_3	OA_1	OA_4	OA_2, OA_3	OA_1	OA_4	OA_2	OA_3
DC gain	61.6dB	64.2dB	62.7dB		64.6dB	58.8dB		57.3dB	59.9dB
GB	244MHz	143MHz	183MHz	105MHz	130MHz	326MHz	102MHz	251MHz	232MHz
Phase margin	72°	76°	72°	79°	76°	72°	84°	76°	
SR	253V/μs	102V/μs	144V/μs	80V/μs	85V/μs	474V/μs	141V/μs	352V/μs	289V/μs
Eq. load	0.64pF	1.33pF	0.63pF	1.14pF	1.32pF	0.64pF	2.15pF	1.34pF	
Output swing	±1.83V	±1.92V	±2.10V		±2.01V	±1.40V		±1.02V	±1.21V
Input cap.	0.29pF	0.44pF	0.29pF		0.43pF	0.30pF		0.45pF	
Output cap.	31fF	38fF	29fF		39fF	31fF		39fF	24fF
Eq. input noise	11.8nV/Hz$^{1/2}$	10.4nV/Hz$^{1/2}$	13.6nV/Hz$^{1/2}$		11.0nV/Hz$^{1/2}$	10.5nV/Hz$^{1/2}$		7.8nV/Hz$^{1/2}$	8.2nV/Hz$^{1/2}$
Bias current	28μA	24μA	16μA		20μA	52μA		80μA	66μA
Power	2.32mW	1.99mW	1.32mW		1.65mW	4.31mW		6.63mW	5.47mW

Fig. 4.11 Illustration of the non-linearity of the amplifiers DC gain

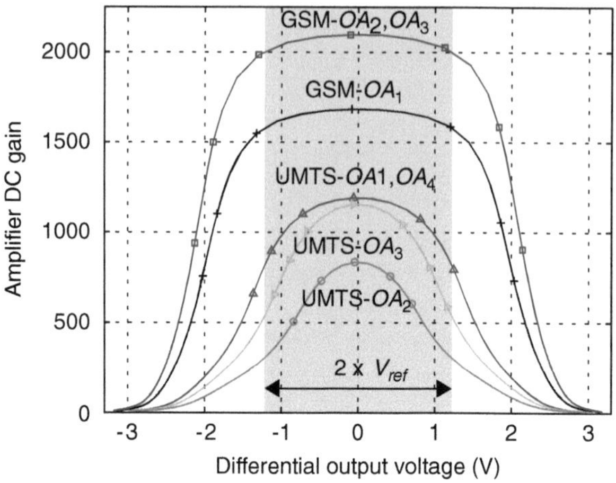

Table 4.9 cover those obtained in the high-level sizing of the multi-standard ΣΔM (Table 4.8), the main difference being the value of the equivalent integrator loads, which are finally considerably smaller than initially estimated at the high level.

Special attention has been paid to the non-linearity of the amplifier DC gains, caused by the variation of the amplifier output conductance when its output voltage swings. As illustrated in Fig. 4.11 for several amplifiers at the nominal case, it translates into a reduction of the open-loop DC gain as the output voltage moves away from the quiescent point, so that small-gain regions are often visited during the normal operation of the modulator (shadowed areas in Fig. 4.11). This reflects in Table 4.9 as a considerable reduction of the amplifiers' output swings for UMTS (the standard with the largest bias currents), which must be taken into accounted in behavioral simulations. Non-linearities introduced by the front-end amplifier for GSM must be also carefully considered. In order to accurately account for them in behavioral simulations, we have resorted to a table look-up procedure using amplifier DC curves obtained by electrical simulation, whose data are included in behavioral simulations through a fast-convergence iterative procedure [Ruiz05, Rio06].

4.4.2 Capacitors

All integrators in the reconfigurable SC ΣΔM use 0.25-pF sampling capacitors. This value has been fixed considering the trade-off between the modulator operation for GSM—in which thermal noise[8] and mismatch can seriously limit the required resolution (13 bit)—and that for UMTS—in which the required BW (4 MHz) in

[8] Note that the most important contribution to the overall modulator thermal noise comes from the first integrator since thermal noise of the remaining integrators suffers a filtering whose order depends on the location of the corresponding integrator in the modulator.

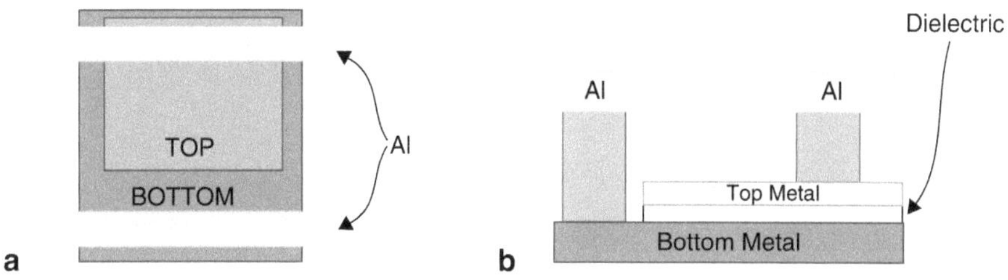

Fig. 4.12 MiM capacitor: **a** top view, **b** side view

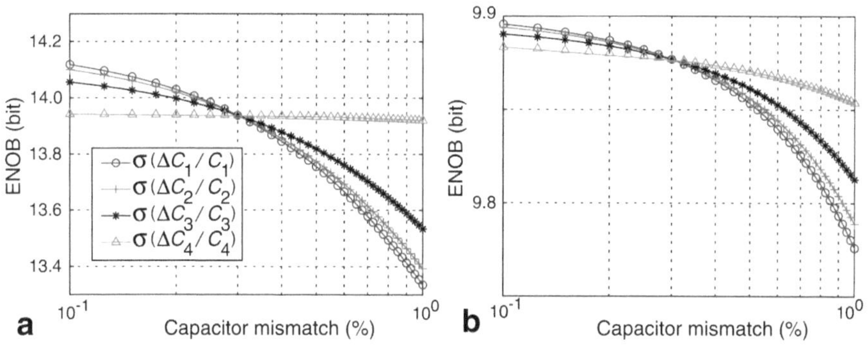

Fig. 4.13 ENOB versus capacitor mismatches: **a** Bluetooth, **b** UMTS

combination with large capacitive loads at the integrators can seriously impact power consumption.

Capacitors use metal-insulator-metal (MiM) structures available in the intended technology, which allows thin inter-metal oxide between two metals. The location of these additional metal layers are above standard metals—except Aluminium (Al)—for lower parasitic capacitance and less substrate loss. Aluminium is used to connect the top and bottom plates of the capacitors as illustrated in Fig. 4.12. With these structures, the 0.25-pF unit capacitor occupies approximately 21.5 μm × 17.5 μm, its size being 29.5 μm × 24.5 μm when contacts to metal 8 are symmetrically placed in the four corners of the structure to enable easy contacts to lower metals and, thus, from the capacitor metal plates to the corresponding connections. These capacitors exhibit a good matching (<0.3%)-enough for meeting specifications-and very small bottom-plate parasitics (<5%). Note that Fig. 4.13 shows the influence of capacitor mismatching for BT and UMTS, illustrating that mismatch leakage noises for standard values of technological mismatches are below the required resolutions. Note that a standard deviation of a capacitance ratio—defined as $\sigma(\Delta C_i/C_i)$ in Fig. 4.13 with *i* referring to the corresponding integrator—of 0.3% is assumed for all capacitors except the one that is varied in Fig. 4.13.

The fully-differential implementation of the reconfigurable modulator requires only 2 × 17 unit capacitors, which helps to reduce the area dedicated to the implementation of integrator coefficients. Figure 4.14 shows the symbolic layout of

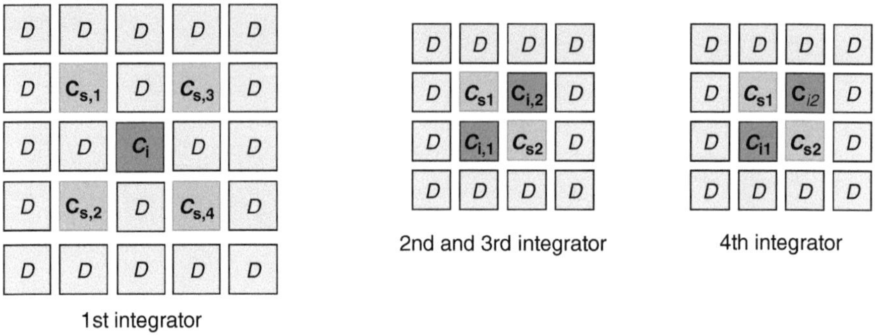

Fig. 4.14 Representation of the layout of unit capacitors (*D* stands for dummy)

capacitors implementing the integrators weights in the ΣΔ modulator. Note that only capacitors in the first integrator actually use a common-centroid topology, whereas the matching of the remaining weights is just based on closely-placed unit capacitors.

4.4.3 Switches

All the switches in the ΣΔ modulator are of CMOS type and their designs have been tackled with two main considerations in mind. First, the on-resistance of the switches can heavily affect the integrators' dynamics and slow their transient responses down [Rio06]. Second, its value can be highly dependent on voltage, so that the non-linearity of the input sampling switches can introduce dynamic distortion at the ΣΔM front-end—the more evident, the larger the input signal frequency [Yu99]. Among the solutions to these problems, resorting to larger aspect ratios increases capacitive parasitics, whereas including clock boosting increases complexity and diminishes the robustness of the final IC [Wu98].

Behavioral simulation results show that switch on-resistances of about 260 Ω can be tolerated in combination with the amplifier dynamics with no significant response degradation. This value can be obtained in the used technological process employing standard-threshold CMOS transmission gates, with no need for clock boosters. The sizes of the pMOS and nMOS transistors have been selected equalizing their transconductances, in order to keep the total on-resistance as linear as possible. Figure 4.15a shows its nominal DC curve obtained by electrical simulation.

In order to accurately evaluate the induced dynamic distortion, the non-linear sampling process at the modulator front-end has been electrically simulated using the differential circuit shown in Fig. 4.15b. Distortion is mainly determined by the switches that are directly connected to the input ($S_{1p,\,n}$), given that the remaining ones ($S_{2p,\,n}$) are connected to the central voltage, which is approximately constant. Electrical simulations have been performed for an input sinewave with amplitude near the modulator overload level and with frequency at 1/3 of the considered standard bands. Especial attention has been paid to GSM—whose results are depicted in

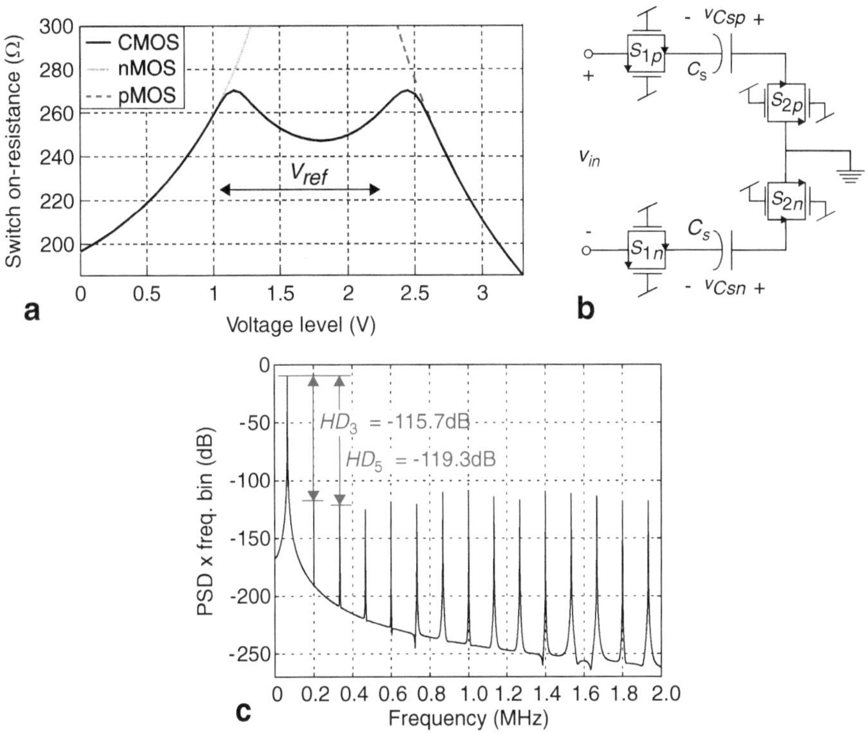

Fig. 4.15 CMOS switch: **a** on-resistance, **b** circuit for evaluating the distortion introduced by the input sampling switches, **c** distortion for a $0.63\,V_{pd}$ @66.7 kHz input tone (GSM mode)

Fig. 4.15c for a $0.63\,V_{pd}$ input sinewave, in which the signal BW is narrow but the required linearity is relatively high, and to UMTS, in which the required resolution is low but the signal band is wide. The resulting harmonic distortion in all simulated cases is considerably lower than that required for the $\Sigma\Delta$ modulator operation on the corresponding standard.

4.4.4 Comparators

The specifications for the comparators in the single-bit modulator stages are quite relaxed in terms of resolution, but demanding in terms of speed according to Table 4.8. Offsets and hysteresis around 30–40 mV can be tolerated, but the maximum comparison time must be around 1/4 of the clock period—3 ns for the 80-MHz clock used for UMTS. For this reason, the latched topology shown in Fig. 4.16 has been adopted [Wang00]. It includes a pre-amplifying stage to attenuate the impact of common-mode interferences on the comparator and of kick-back noise on the integrator. As

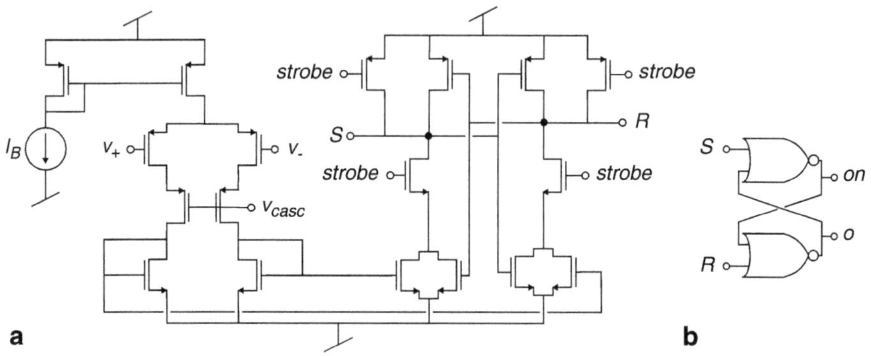

Fig. 4.16 Comparator: **a** pre-amplifier and regenerative latch, **b** set-reset latch

Table 4.10 Worst-case results for the comparator obtained by electrical simulation

Hysteresis	0.6 mV	Offset	35.3 mV
Low-to-high resolution time	2.10 ns	High-to-low resolution time	2.11 ns
Power consumption	0.50 mW		

shown in Fig. 4.8, the CMOS latch uses $\bar{\phi}_{2d}$ as strobe in order to start the comparator regenerative decision before the beginning of the integration phase (ϕ_1).

Monte Carlo and corner analyses have been performed to electrically characterize the comparator after full transistor sizing. Table 4.10 summarizes its worst-case performance for a 0.2-pF load, with data shown being the statistical results considering the mean values plus 3 sigmas.

4.4.5 *Multi-bit Quantizer*

A 2-bit quantizer is employed on UMTS mode in the last modulator stage to digitize the fourth-integrator output and convert it back to analog domain (signals $Y_{3,0-1}$ and $v_{\mathrm{DAC}^{+,-}}$ in Fig. 4.8, respectively). The 2-bit A/D/A converter is implemented with a flash quantizer and a resistor-ladder DAC, as shown in Fig. 4.17.

The quantizer has a differential flash architecture and compares the integrator output ($V_i^+ - V_i^-$ in Fig. 4.17) with voltage tabs generated in the resistor ladder. Comparators in the quantizer are similar to those described above, but with an extra differential pair at the pre-amplifying stage in order to operate with fully-differential inputs. The thermometer code at the comparators outputs is then translated into a 1-of-4 code (d_{0-3}) that controls the resistor-ladder DAC. The latter uses 6 segments of 400-Ω unit resistors laid with unsalicided p+ poly that are connected between the differential reference voltages ($V_r^+ = 2.25$ V and $V_r^- = 1.05$ V), thus providing a differential full scale of 1.2 V with 500-μA current consumption.

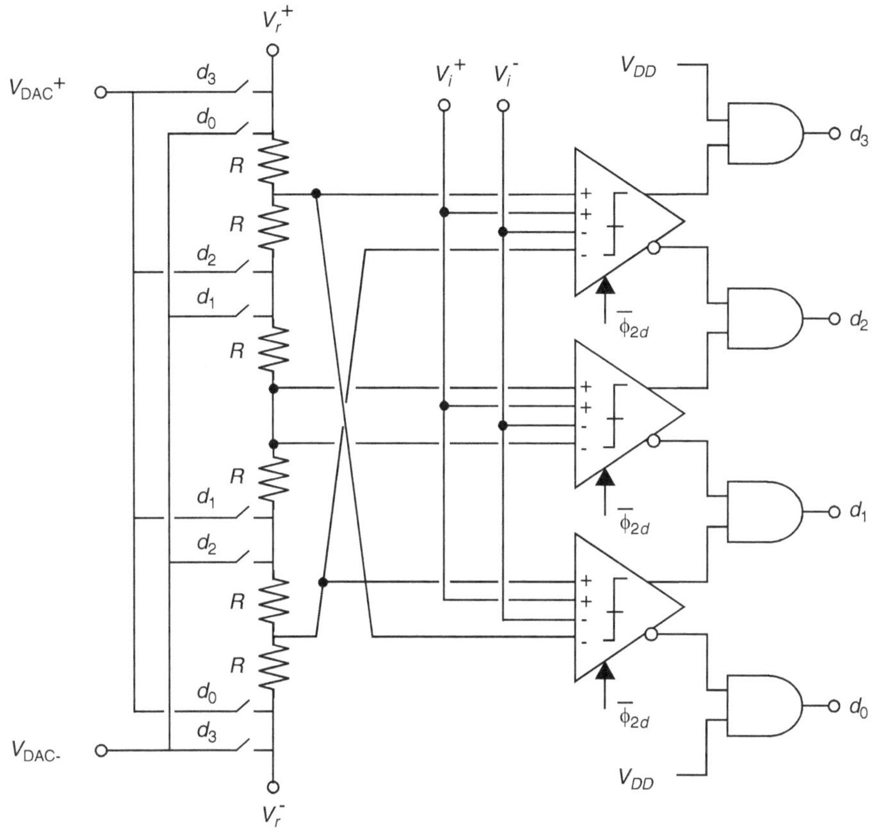

Fig. 4.17 2-bit quantizer in the third modulator stage (active for UMTS mode)

4.4.6 Auxiliary Blocks

The auxiliary blocks required for the modulator operation are included on-chip, namely the clock phases, common-mode and reference voltage generators, as well as the master current generator.

4.4.6.1 Clock Phase Generation, Buffering and Distribution

Figure 4.18 shows the schematic of the clock phase generator. Two non-overlapped clock phases—ϕ_1, ϕ_2—that control the sampling and integration operations in the $\Sigma\Delta$ modulator are obtained from an external 80-MHz clock signal. Delayed versions of these phases—ϕ_{1d}, ϕ_{2d}—are generated in order to attenuate signal-dependent charge injection [Lee85]. The complementary versions of all these phases are also generated—$\overline{\phi}_1$, $\overline{\phi}_2$, $\overline{\phi}_{1d}$, $\overline{\phi}_{2d}$—for the control of the CMOS switches. The clock

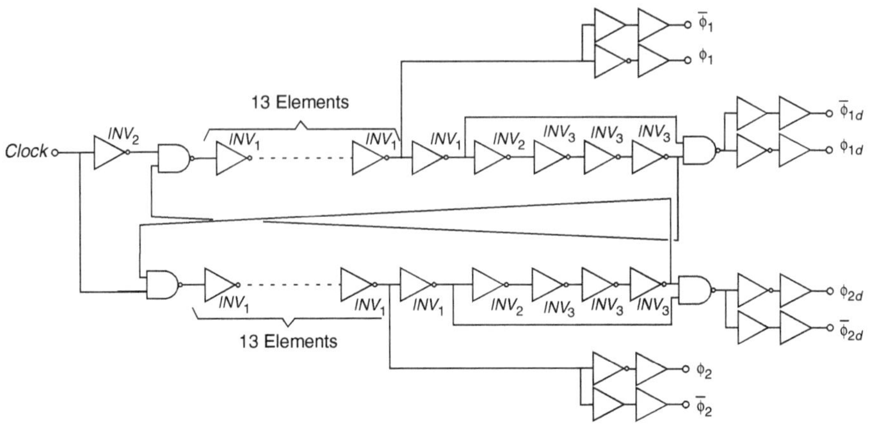

Fig. 4.18 Clock phase generator

Table 4.11 Loads corresponding to each clock phase

	$\overline{\phi}_{1d}$	ϕ_{1d}	$\overline{\phi}_{2d}$	ϕ_{2d}	$\overline{\phi}_1$	ϕ_1	$\overline{\phi}_2$	ϕ_2
C_L(fF)	342.8	247.1	188.0	926.0	670.0	222.0	670.8	222.0

phases generator includes a divide-by-2 module that is active for GSM and Bluetooth in order to make the ΣΔ modulator operate internally at 40 MHz.

All generated phases must drive a load given by, on the one hand, the input capacitance of the controlled switches and, on the other, by that of the digital gates needed to enable DAC operations (see A_i, B_i, C_3 and D_3 generation in Fig. 4.8). This can be translated into an equivalent capacitive load to be driven by all generated phases. Table 4.11 shows the loads corresponding to all generated phases. However, the clock phases need to be routed through the whole chip. Indeed, a long U-shaped bus—whose size is shown in Fig. 4.19a—is used for the clock phases distribution[9]. Figure 4.19b shows a 3D representation of a standard portion of the clock phases routing. It can be observed that each phase is routed together with the complementary (ϕ_i and $\overline{\phi}_i$, respectively) in metal 4, while both phases are surrounded by ground strips of metal 4 in parallel (*GND* in Fig. 4.19). The whole bus is covered by the same ground above and below the routed phases with plates at metal 6 and 2, respectively. Note that Fig. 4.19c depicts a symbolic layout view of the distribution of the phases, showing the sizes of metal strips and their distances. Figure 4.19d shows a lumped LCR parasitics model for the actual transmission line used for routing the phases. To create this model, two scenarios have been considered in a electromagnetic field solver. The first one—depicted in Fig. 4.19e—considers a simple metal strip line

[9] Due to the layout separation into different regions in order to reduce switching noise coupling, the clock phases generator—located in the digital region—is placed far away from the switches where the clock phases must be routed.

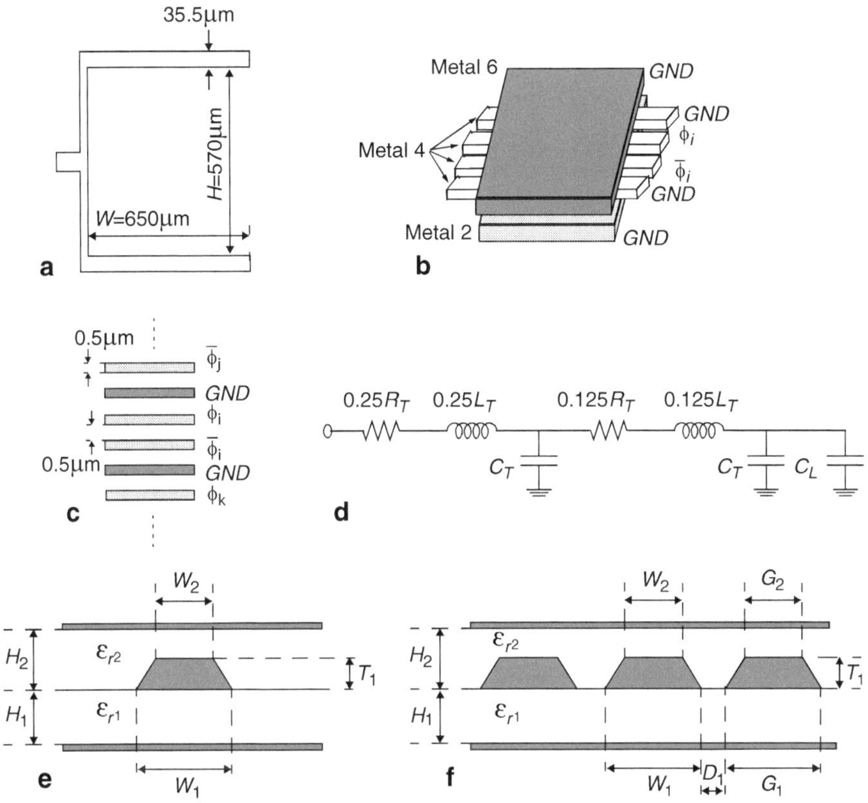

Fig. 4.19 Phases bus: **a** size, **b** phases isolation, **c** layout distribution of phases, **d** parasitics electromagnetic extracted model. Electromagnetic model scenarios: **e** embedded strip line, **f** coplanar embedded strip line

Table 4.12 Parasitics for the lumped LCR model in Fig. 4.19d

L_T (nH)	C_T (pF)	R_T (Ω)
0.362	0.132	170.2

embedded into two infinite ground planes while the second one—see Fig. 4.19f—accounts for two additional strip lines surrounding the main line (coplanar strip line). The technological information for the variables shown in both figures—together with the distances between all strip lines and their sizes—are given to the field solver in order to obtained two lumped models as the one shown in Fig. 4.19d for each scenario. Finally, the worst-case value for each independent parameter in the model is further considered. Note that the total bus length is taken as $H + W/2 = 0.935$ mm—see Fig. 4.19a. Table 4.12 shows the value of the elements finally employed for Fig. 4.19d model.

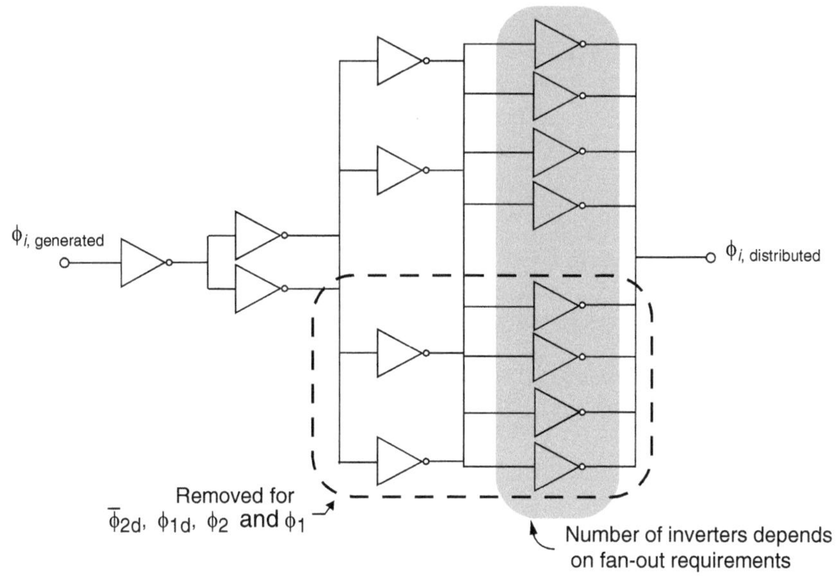

Fig. 4.20 Clock phase driver tree

Table 4.13 Number of inverters in the last column of Fig. 4.20

ϕ_{2d}	$\overline{\phi}_{2d}$	ϕ_{1d}	$\overline{\phi}_{1d}$	ϕ_2	$\overline{\phi}_2$	ϕ_1	$\overline{\phi}_1$
8	3	4	5	4	7	4	7

In order to avoid a substantial corruption of the clock phases—and thus of the modulator performance—after routing and delivery to the corresponding switches and digital gates, buffers to drive the corresponding loads are needed. Figure 4.20 shows the buffer tree employed for each clock phase where there is a set of levels with increasing number of inverters in parallel whose values depend on the required fan-out. In order to extract the necessary number of inverters for each clock phase, first the addition of C_L—with values given in Table 4.11—and C_T is normalized to 8—maximum number of inverters at the last level. This way, Table 4.13 is obtained. In addition, the whole modulator has been extensively simulated including the obtained buffer tree in the previous step and the lumped LCR structure in Fig. 4.19d for each phase and no appreciable degradation in performance has been observed compared to a structure that does not account for routing parasitics. Note that even when less than 5 inverters are required[10], the last column is still used in order to keep the same delay for each phase.

The resulting non-overlapping time, and the falling and rising phase delays are approximately 0.24 ns, and 0.20 and 0.11 ns, respectively when accounting for the

[10] In this case, the same functionality can be obtained removing the last level of inverters in the tree and taking the phase from the previous level. However, the critical path would change.

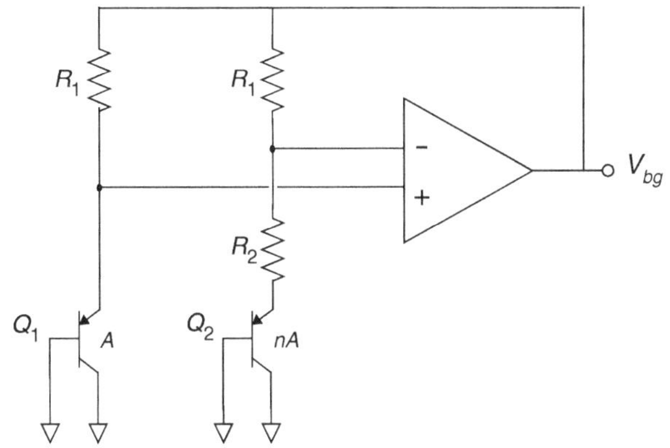

Fig. 4.21 Scheme of the bandgap

buffer tree and the LCR lumped model together with the clock phase generator. The effective operation time for the sampling phase is 94.4 and 95.6% of half the clock period for 80 and 40 MHz, respectively, the respective values for the integration phase being 88.0 and 95.2%.

4.4.6.2 Bandgap

A bandgap is used for generating a 1.2-V temperature-independent voltage (V_{bg}) that is later employed in the reference voltage generator and the master current generator. Figure 4.21 shows the schematic used for this purpose. Note that the circuit in the figure makes use of lateral bipolar transistors—available in standard CMOS technologies. For zero temperature coefficients, $(1+R_1/R_2)\cdot\ln(n)$ must be roughly 17.2 [Raza00]. This relation between pnp transistor units and resistors is obtained in practice by choosing $R_1 = 5.5\,\text{k}\Omega$, $R_3 = 1.125\,\text{k}\Omega$ and $n = 24$.

4.4.6.3 Reference Voltage Generator

The reference voltages required for the modulator operation—namely, $V_{ref} = V_r^+ - V_r^- = 1.2\,\text{V}$—can be obtained from the 1.2-V bandgap voltage V_{bg} as

$$V_r^+ - V_r^- = V_{bg} \tag{4.1}$$

Figure 4.22 shows that references V_r^+, V_r^- are obtained from a fully-differential amplifier in inverting configuration with gain 1. Note that a buffer is needed for taking the bandgap voltage V_{bg}. The fully-differential amplifier V_{ref}-*OP* consists of an asymmetric OTA with static CMFB net.

The main design considerations for the generation of these references are a fast settling [Pies02] and a low output impedance in the V_r^+, V_r^- lines in order to avoid dynamic distortion at the integrators [Ribn91]. A slow reference buffer approach has been adopted together with an external large capacitor, so that the output impedance

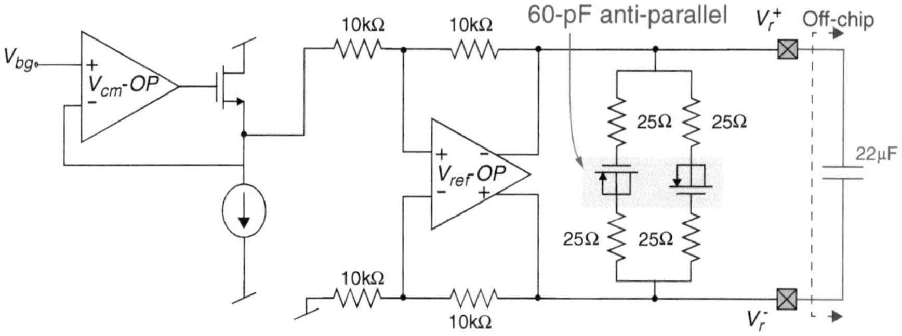

Fig. 4.22 Reference voltage generator

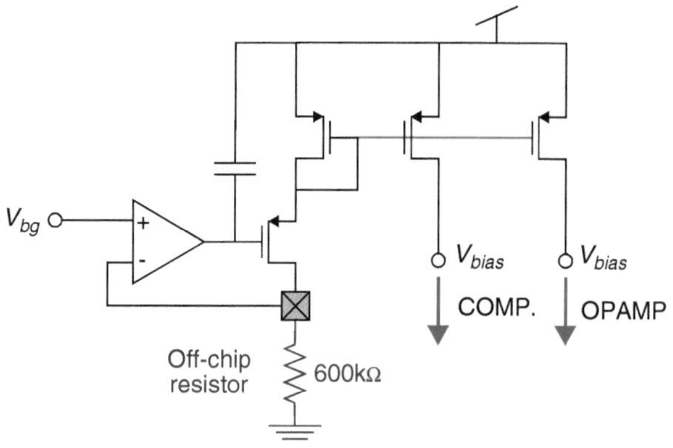

Fig. 4.23 Master current generator

is below 4 Ω (required for GSM) along the signal band (up to 4 MHz for UMTS). The external capacitor is connected between the reference voltages, valued according to the (pad + wire + lead + pin) parasitics, so that the spurious components around half the sampling frequency are removed from the differential reference voltage. Besides, a damping network is used to remove references ringing.

4.4.6.4 Master Current Generator

The bias currents needed in the ΣΔ modulator are all internally generated, as illustrated in Fig. 4.23, from a single master current of 2 μA, provided by an external 600-kΩ resistor. This current is mirrored and properly scaled to bias all the amplifiers in the integrators, the pre-amplifying stages of the comparators, and the common-mode and reference voltage generators. Note that the adaptation of the amplifiers' bias currents from one standard to another is based on switches as illustrated in the scheme in Fig. 4.10. All the current tails of that schematic are generated using the scheme shown in Fig. 4.23.

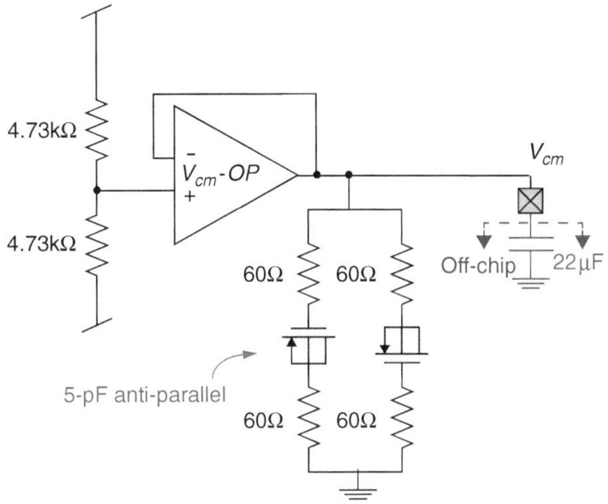

Fig. 4.24 Common-mode voltage generator

4.4.6.5 Common-Mode Voltage Generator

The common mode $V_{cm} = 1.65$ V is half of the 3.3-V supply voltage. Therefore, this voltage is

$$V_{cm} = \frac{1}{2}(V_{DD} - V_{ss}) \tag{4.2}$$

The circuit in Fig. 4.24 has been used to implement the required ratio of 1/2 in a robust manner. This analog central voltage V_{cm} is obtained from a simple resistor ladder, whereas a single-ended folded-cascode OTA is used to buffer the voltage to the $\Sigma\Delta$M. Note that a large external capacitor is necessary together with an on-chip damping network to maintain stable V_{cm} despite all switching activity due to the dynamic loading depending on clock phases.

4.4.7 Complete Modulator Simulation Results

The performance of the multi-standard $\Sigma\Delta$M has been evaluated through several transient electrical simulations at transistor level in CADENCE-SPECTRE [Cade08]. The worst-case thermal noise corresponding to the modulator operation on each standard has been incorporated into transient simulations by means of an input equivalent noise source, with noise sequences generated in MATLAB.

Figure 4.25 shows the modulator output spectra for each operation mode for a 0.63 V_{pd} input tone, amplitude which is close to the modulator overload level (−5.6 dBFS). Note that no harmonics are visible from the spectra obtained by electrical simulation, what validates the careful analysis performed for the impact of non-linearities on amplifiers and switches.

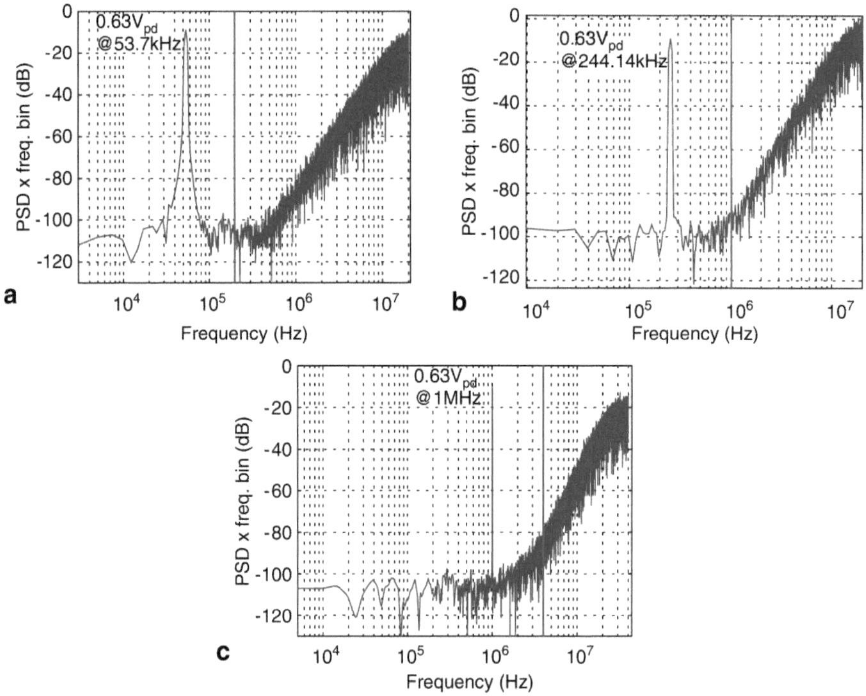

Fig. 4.25 Modulator output spectrum for a maximum amplitude input tone for: **a** GSM, **b** Bluetooth, **c** UMTS

Table 4.14 Electrical simulation results for the reconfigurable ΣΔM (0.63 V_{pd} input tone)

	GSM		Bluetooth		UMTS	
	Behavioral Simulation (worst)[a]	Electrical Simulation (typ.)[b]	Behavioral Simulation (worst)[a]	Electrical Simulation (slow)[d]	Behavioral Simulation (worst)[c]	Electrical Simulation (slow)[d]
SNDR	79.3dB	80.0dB	67.2dB	70dB	56.3dB	56.0dB
ENOB	12.9bit	13.0bit	10.9bit	11.3bit	9.1bit	9.0bit
BW	200kHz		1MHz		4MHz	

a Worst-case value for each amplifier parameter (Table 4.9) and maximum thermal noise.
b Typical MOS models and worst-case equivalent thermal noise.
c Worst-case value for each amplifier parameter (Table 4.9) and maximum capacitor values.
d Slow MOS models, maximum temperature and worst-case equivalent thermal noise.

Figure 4.26 shows the modulator SNDR curve obtained for each standard by incorporating into behavioral simulations the worst-case electrical parameters of the modulator blocks after their complete designs at the transistor level.

Table 4.14 summarizes the electrical simulation results of the modulator performance, which are in good agreement with those obtained by behavioral simulation.

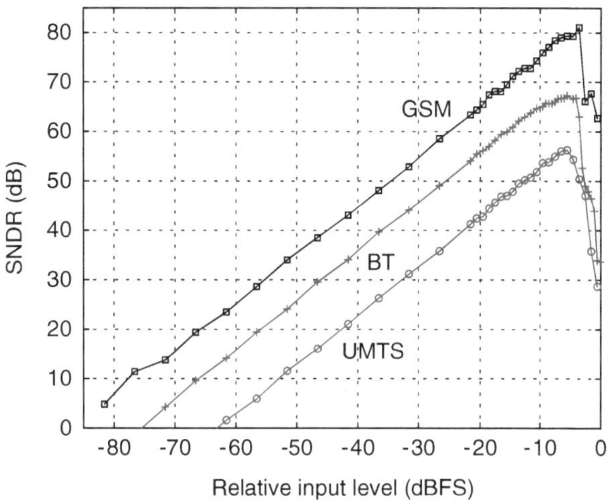

Fig. 4.26 Worst-case SNDR curves obtained by behavioral simulation after full transistor sizing of the $\Sigma\Delta M$ blocks

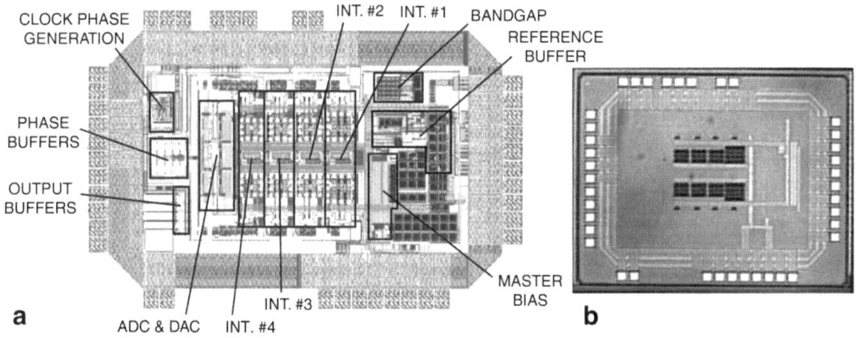

Fig. 4.27 Prototype implementation: **a** layout, **b** microphotograph

4.5 Layout and Area Distribution

The reconfigurable $\Sigma\Delta$ modulator has been implemented in a 130-nm 1-poly 8-metal digital CMOS process. The main parts of the prototype layout are depicted in Fig. 4.27a, whereas Fig. 4.27b shows its microphotograph[11]. The floorplan of the chip includes separate analog, mixed and digital supplies, as well as guard-rings surrounding each section of the circuit. The modulator occupies 2.68 mm^2 including bonding pads (0.76-mm^2 active area). In addition, several details on the layout implementation of the chip are given below.

Figure 4.28 shows the area consumption computation divided into all main parts in the modulator. Special attention has been paid to the area division of the integrators, digital cells and analog auxiliary blocks—such as the bandgap, the references buffers

[11] Only MiM capacitors and wide routing lines with stacked metal layers are visible after passivation.

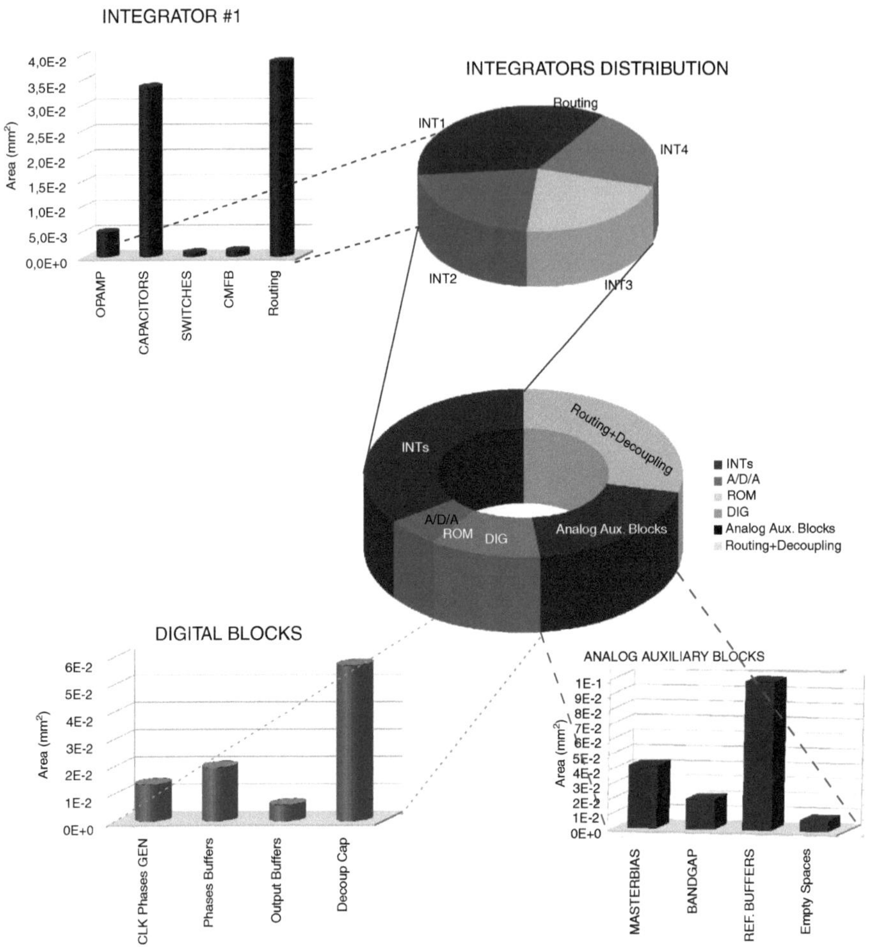

Fig. 4.28 Area distribution of this prototype

and the master current generator. Note that part of the empty spaces of the modulator building blocks are used to place decoupling capacitors or for routing.

The practical implementation of the layout follows the approach proposed in [Rio06]. However, a set of building blocks have been laid out in a particular manner in order to reduce coupling among all the different regions. This is summarized as follows:

- The layout is divided into different regions, corresponding to the analog, mixed-signal and digital parts of the circuit. As stated before, separate analog, mixed and digital supplies are employed for this purpose. In addition, large guard rings (with dedicated pads and pins) surround each section of the circuit. This well-known strategy—commonly employed in high-performance mixed-signal chips—helps to reduce switching noise coming from the digital parts of the chip. Besides, decoupling capacitors are extensively used throughout the chip for each supply and its corresponding ground. Note that reference and common-mode buffers make use of decoupling as well.

- The chip is fully differential, what, together with an extensive use of layout symmetry and common-centroid techniques, helps to reduce common-mode interferences.
- CMOS switches are implemented with single-finger transistors what helps to avoid any crossing among digital and analog signals. This is done, in practice, by routing the clock phase of the NMOS transistor in metal 3 in parallel to the gate of the PMOS transistor of the transmission gate in polysilicon layer, where the complementary phase goes. Note that analog signals are connected to the diffusions of the transistors so there is no vertical cross coupling with fast clock phases routed through the transistors gates. This is convenient to avoid switching noise coupling in the analog signals.
- All digital clock phases, DAC control signals (A_i, B_i, C_3 and D_3 in Fig. 4.8, for $i = 1$, 2 and 3) and the complementaries of all the former signals are routed together in a clock phase bus (see Sect. 4.4.6). This bus is already routed twice for enabling the top and bottom CMOS switches. As shown in Fig. 4.19c, two metal strips of a dedicated ground are placed at both sides of each of these signals and the corresponding complementary for isolation purposes. In fact, 0.5-μm spacing is kept among all paths (from signal to signal or from ground to path) to reduce cross coupling.
- Not only digital signals are routed based on a bus configuration. Analog supplies together with the bias currents of the amplifiers and comparator preamplifiers conform two buses that surround the amplifiers in the chip layout. Besides, the reference and common-mode voltages go through a fully-differential bus inside the area devoted to capacitors and switches.

Note that some auxiliary blocks such as the bandgap, and the common-mode and reference voltage generators might not be used if the modulator is embedded in a complete communication system or, at least, their functionality could be shared by all the blocks of the overall transceiver. Moreover a dedicated voltage DC source or, alternatively, the same one that provides the supply voltages[12] could have been employed for the generation of these references, as usually done for many ADC prototypes. In case these blocks are not computed, the area occupation lowers to only 0.65 mm^2.

4.6 Experimental Results

The $\Sigma\Delta$M samples has been packaged in a 44-pin CQFP and mounted on one 4-layer PCB. A first order RC anti-aliasing filter whose corner frequency can be adjusted

[12] In this case, a voltage adaptation of the 3.3-V supply to the 1.2-V reference will be needed.

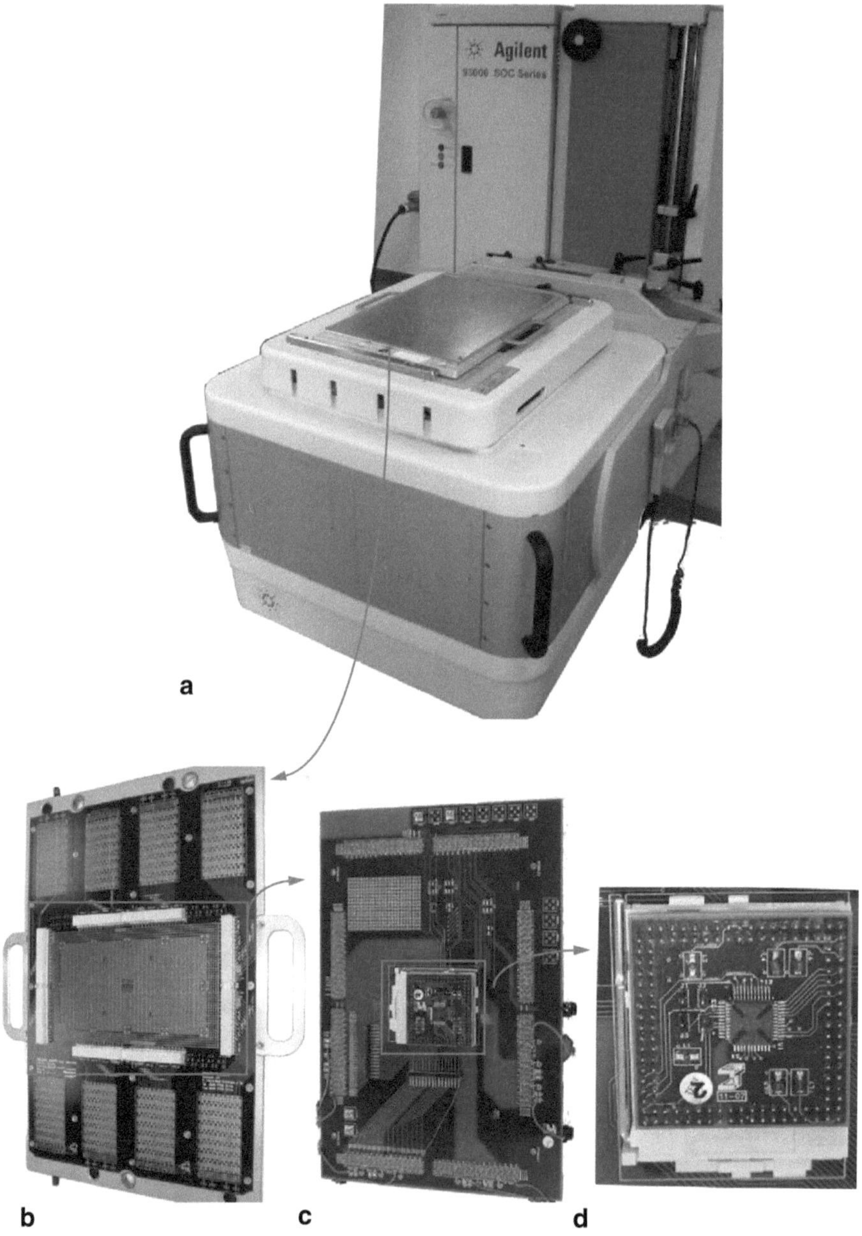

Fig. 4.29 First test set-up: **a** 93000 SoC test unit, **b** generic mother board, **c** specific mother board, **d** main PCB (rotated views)

depending on the operation mode has been included on the PCB. The thermal noise generated by this passive filter is well below specifications.

Two test set-ups have been used to measure the performance of the chip. The original set-up, displayed in Fig. 4.29, uses three PCBs that are needed to connect

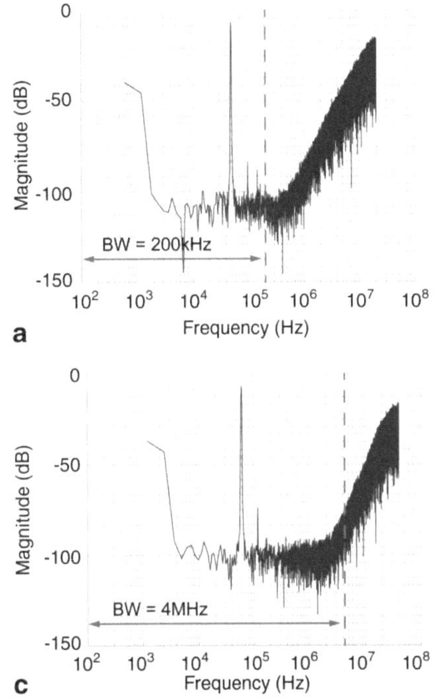

Fig. 4.30 Measured spectra in the first test set-up for: **a** GSM, **b** Bluetooth, **c** UMTS

the chip to an Agilent 93000 SoC test unit that generates the input, the supplies, the control and clock signals, and acquires the output bit streams of the modulator stages. After the bit-stream acquisition, data are transferred to a workstation, in which the digital cancellation logic is applied by software and the overall modulator output is post-processed. Figure 4.30 shows the measured 65536-point Hanning-windowed FFTs of the modulator output, for the different standard configurations, considering a 58.6-kHz input sinewave with an amplitude of −9.5 dBFS. A certain amount of non-linearity is appreciable in the measured spectra, which is originated in the input signal conditioning at the PCB level. An SR770 FFT spectrum analyzer, that is capable to process a single-tone signal with spurious below 90 dBc [Stan92], was used to measure the input signal generated by the Agilent 93000 SoC test unit. A similar amount of non-linearity to that in Fig. 4.30 was observed.

The measured SNDR curves with this test set-up are depicted in Fig. 4.31 for each operation mode. The prototype achieves a dynamic range of 80.7 dB for GSM, 75.0 dB for Bluetooth, and 57.3 dB for UMTS, whereas the corresponding peak SNDRs are 74.0, 68.4, and 52.8 dB. Note from Fig. 4.31 that the SNDR drops prematurely due to the harmonic content of the input signal given to the prototype, with the measured reductions of the peak SNDRs being 1.0 bit for GSM, 0.7 bit for Bluetooth and UMTS when compared to the transistor-level simulation results in Table 4.14.

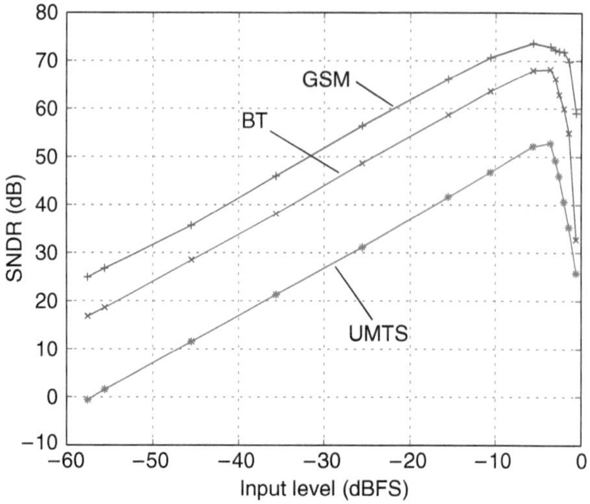

Fig. 4.31 Best SNDR plots extracted from the first test set-up

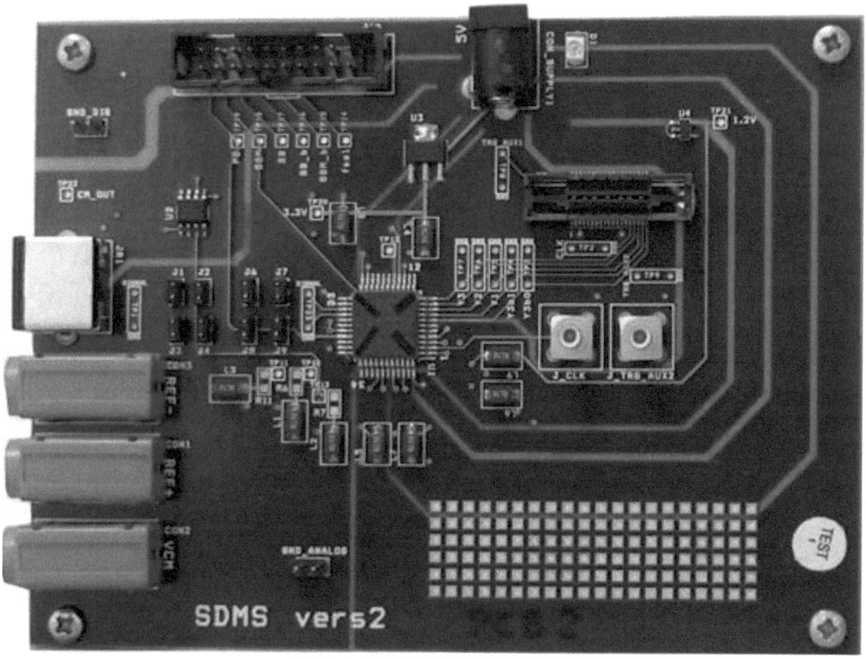

Fig. 4.32 Photograph of the PCB for the second (and definitive) test set-up

Figure 4.32 shows the PCB used for the second experimental set-up that is used to remove the induced distortion. Here, an Agilent A3631A unit generates the voltage supply, while the input is generated by the ultra-low distortion Tektronix SG5010 audio oscillator. An SRS CG635 is employed for the clock signal and an Agilent logic analyzer 16823B is used to acquire the output bit streams of the modulator

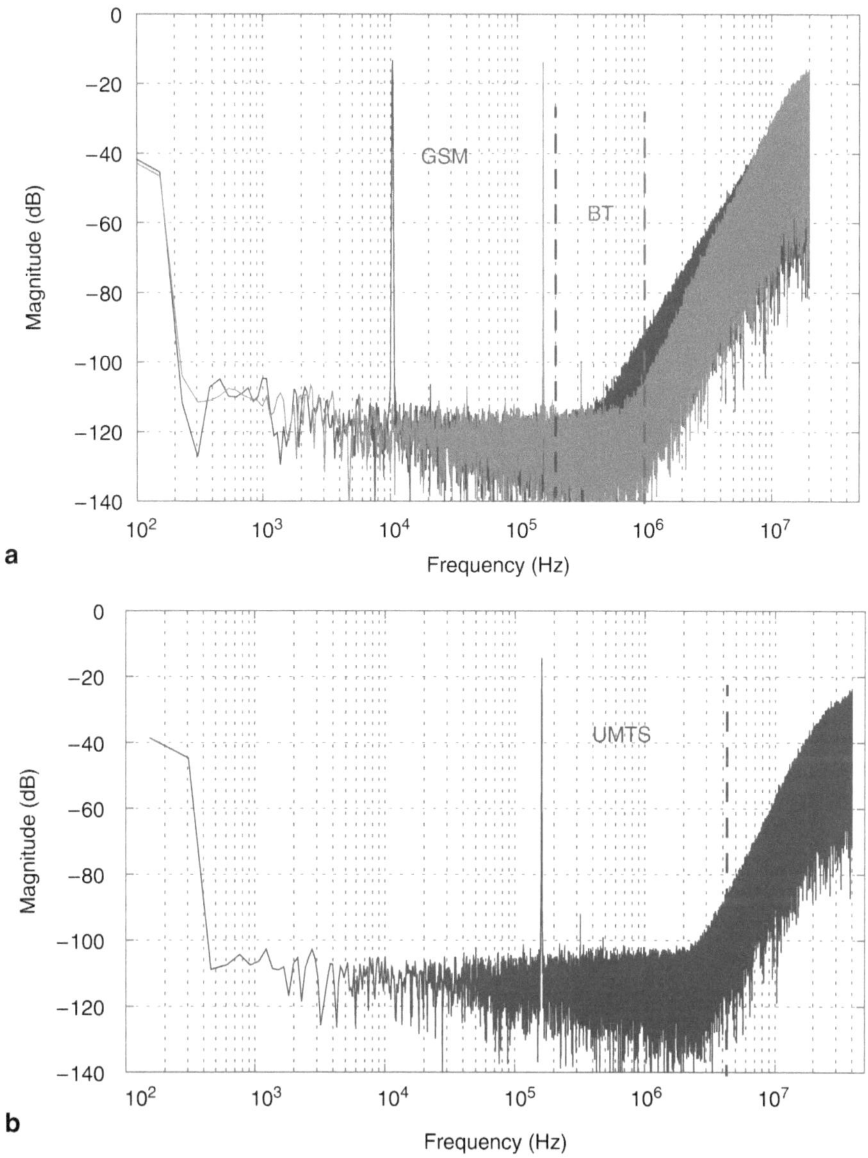

Fig. 4.33 Measured spectra for: **a** GSM and Bluetooth, **b** UMTS

stages and to generate the control signals. A pattern generator, that is part of the logic analyzer, is used for this last purpose. After bit-stream acquisition, data are transferred to a workstation to be post-processed by software.

Figure 4.33 shows the measured 64 k-point Hanning-windowed FFT of the modulator output for each standard, and for a single-tone input with −10-dBFS amplitude.

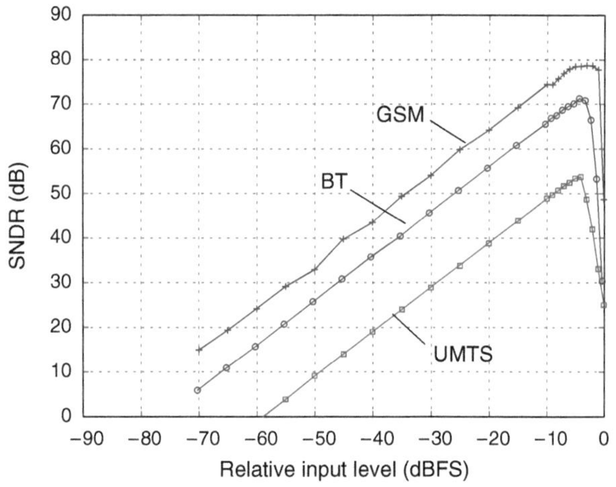

Fig. 4.34 Measured SNDR *curves*

The reconfiguration of the quantization noise shaping is clearly visible, together with the domination of white noise for GSM and of quantization noise for UMTS within the corresponding bandwidths. The spectra show less harmonic content compared to those in Fig. 4.30. This translates into larger peak SNDRs. The measured SNDR curves are depicted in Fig. 4.34 for each operation mode. The prototype achieves a dynamic range of 83.8 dB for GSM, 75.9 dB for Bluetooth and 58.7 dB for UMTS, whereas the corresponding peak SNDRs are 78.7, 71.3 and 53.7 dB. There is an SNDR improvement of 3.7, 3.1 and 0.9 dB for GSM/BT/UMTS when compared to the first experimental set-up. Therefore, this is the test set-up selected for the actual experimental characterization and, eventually, for the comparison with the state of the art.

Table 4.15 summarizes the most significant features of the measured performance for the multi-mode $\Sigma\Delta$M prototype, together with the power dissipation of the different sections. The modulator power consumption is 23.9/24.5/44.5 mW for GSM/Bluetooth/UMTS, of which 9.7/10.0/24.8 mW are dissipated in the analog circuitry.

Experimental results are in good agreement with the transistor-level simulations results provided by Table 4.14. Table 4.15 shows that there is an SNDR reduction of only 1.3 and 2.3 dB for GSM and UMTS, respectively. By contrast, there is an SNDR improvement of 1.3 dB for Bluetooth due to the use of multi-bit quantization in the last stage of the cascade. A 1.5-dB SNDR improvement has been measured when a 2-bit last-stage quantizer is used for BT rather than its nominal resolution of only one bit. Although an approximate in-band quantization error reduction of 6 dB is expected when increasing the resolution of the last-stage quantizer in one bit, this does not translate into a direct SNDR improvement because of the thermal noise contribution to the IBE power.

Table 4.15 Summary of measured performance

	GSM	BT	UMTS
Modulator architecture	2-1	2-1-1(2b)	
OSR	100	20	10
BW (MHz)	0.2	1	4
Sampling frequency (MHz)	40		80
DR (dB)	83.8	75.9	58.7
Peak SNDR (dB)	78.7	71.3	53.7
Process	0.13-μm 1P8M CMOS		
Core area	0.65/0.76[a] mm^2		
Power consumption[b] (mW)	23.9	24.5	44.5
Analog core @ 3.3V	9.7	10.0	24.8
Reference buffer @ 3.3V	19.0	18.9	20.0
Mixed section @ 3.3V	9.6	10.0	13.9
Digital section @ 3.3V	4.1		5.5
Digital section @ 1.2V	0.6		1.2

a This value considers the area occupied by the bandgap and the reference buffer.
b Digital output buffers and the reference buffer are not included.

4.7 Summary

This chapter introduces, firstly, a tool to extract the specifications of an ADC embedded in a DRC for multi-standards applications. Besides, this tool allows the user to include the performance of each RF and analog baseband building block and, later, verify the receiver metrics.

Based on the information provided by this tool, a multi-mode $\Sigma\Delta$ modulator has been presented that efficiently adapts its performance to the required specifications of GSM, Bluetooth and UMTS standards. The chosen architecture is an expandable $\Sigma\Delta$M made up of a first second-order stage followed by first-order stages where dual quantization can be used for the last one.

The modulator is capable to vary the order of the loop filter, the sampling frequency, the number of bits in the internal quantizer and the bias currents of its amplifiers. The selection of the modulator topology for each standard of operation and the design of its main blocks relies on a top-down CAD methodology that combines simulation and statistical optimization at different levels of the modulator hierarchy. Furthermore, the topology of the different blocks is described and their designs are presented at transistor level.

The prototype has been fabricated in a 130-nm CMOS technology and features dynamic ranges and peak SNDRs of 83.8/75.9/58.7 dB and 78.7/71.3/53.7 dB within 200 kHz/1 MHz/4 MHz, respectively. As will be stated in Chap. 7, this prototype achieves a competitive performance, while covering a wide conversion region of the resolution-bandwidth plane.

Note, however, that 4G receivers require a wider conversion region—especially in terms of bandwidth—and a larger number of operation standards to be converted in order to fuel more connectivity capabilities, necessary for this kind of multi-mode wireless transceiver [Bran05]. To this purpose, next chapter will show an integrated solution that covers the additional conversion of GPS, DVB-H and WiMAX operation standards.

Chapter 5
A 1.2-V 90-nm CMOS Reconfigurable 2-2 Cascade SC ΣΔM for GSM/BT/GPS/UMTS/DVB-H/WiMAX

THE RECEPTION OF WIRELESS COMMUNICATION STANDARDS, such as WiMAX or WLAN, together with those intended for cellular communication in the same multi-mode chipset is usually considered the main achievement in the evolution towards 4G transceivers [Bran05]. If additional wireless standards, such as low-rate low-distance standards (ZigBee or Bluetooth), positioning systems (GPS) or digital television (DVB-H), are included, the functionality of these types of devices will be boosted.

This chapter shows the design, implementation and testing of a reconfigurable ΣΔ modulator intended for 4G wireless telecom systems. The chip enables the A/D conversion of GSM, BT, GPS, UMTS, DVB-H and WiMAX wireless standards for a direct conversion receiver, while both the system architecture and the power consumption of the modulator building blocks are adapted to the corresponding operation modes. The modulator uses a cascade topology made up of unity-STF stages and can adjust its loop-filter order to the operation mode by switching off the back-end stage when needed. Furthermore, the modulator sampling frequency and the bias currents of the opamps can be adjusted to reduce the overall power consumption as well. The prototype is implemented in a 90-nm CMOS technology using a low supply voltage of 1.2 V. Experimental measurements feature a dynamic range of 78/70/71.5/66/62/52 dB within 100 kHz/500 kHz/1 MHz/2 MHz/4 MHz/10 MHz, respectively.

The chapter is divided as follows. Section 5.1 describes the employed architecture together with its system-level parameters, such as the in-loop coefficients. Section 5.2 presents the SC implementation of the modulator and discusses specific topological issues. The transistor-level design of the embedded modulator building blocks is then discussed in Sect. 5.3. Section 5.4 details the layout of the chip and describes the main points in the PCB design phase. Finally, Sect. 5.5 provides the experimental results.

A. Morgado et al., *Nanometer CMOS Sigma-Delta Modulators for Software Defined Radio*,
DOI 10.1007/978-1-4614-0037-0_5,

Table 5.1 Modulator specifications

Standard	GSM	BT	GPS	UMTS	DVB-H	WiMAX
SNDR (dB)	75	70	65	60	55	50
BW (MHz)	0.1	0.5	1.0	2.0	4.0	10.0

5.1 Modulator Architecture

Compared to the prototype presented in the previous chapter, three additional operation modes are included; namely: GPS, DVB-H and WiMAX. Table 5.1 shows the specifications of all the operation modes considered for the ΣΔM presented in this chapter. The values of these specifications are extracted as an average of those obtained from the SIMULINK toolbox described in Sect. 4.1 and those from open literature. The functionality and conversion capabilities of this chip are expanded compared to the one in Chap. 4, especially in terms of bandwidth. To do so, the prototype incorporates a number of architectural improvements and reconfiguration techniques. As well as presenting the chosen ΣΔM architecture, this section will also discuss the selection of its architectural parameters such as the values of the in-loop coefficients.

5.1.1 Architecture Selection

The following points are important when selecting the architecture:

- Efficient operation for all the modes in the modulator (see specifications in Table 5.1). Thus, low-bandwidth modes require high resolution, while high-speed operation is needed for the modes with the largest bandwidths.
- Low-power requirements for the main building blocks in the modulator, such as the opamps.
- This modulator will be implemented in a 90-nm CMOS technology with a low supply voltage of 1.2 V, so low-voltage operation must also be taken into account in the architectural selection.
- Easy architectural reconfiguration.

Figure 5.1 shows the block diagram of a ΣΔ modulator that presents a good trade-off for all the aforementioned architectural requirements. This architecture is a cascade with unity-STF stages—due to the use of feed-forward paths—that conform the overall modulator. The quantization error of the first stage is extracted and later transferred to the second stage through only one inter-stage node [$I_1(z)$]. These types of topologies—and their advantages—are discussed in Sect. 3.1.3. The main benefits of these architectures, with the previously mentioned architectural requirements in mind, can be listed as follows [Silv01, Silv04a, Hamo06, Yao06]:

- Reduced sensitivity to the non-idealities of the building blocks, especially opamp non-linearities and large-signal parameters of amplifiers such as SR. This point covers the first of the architectural requirements listed above (i.e., efficient operation for all modes) because the high robustness against the building block

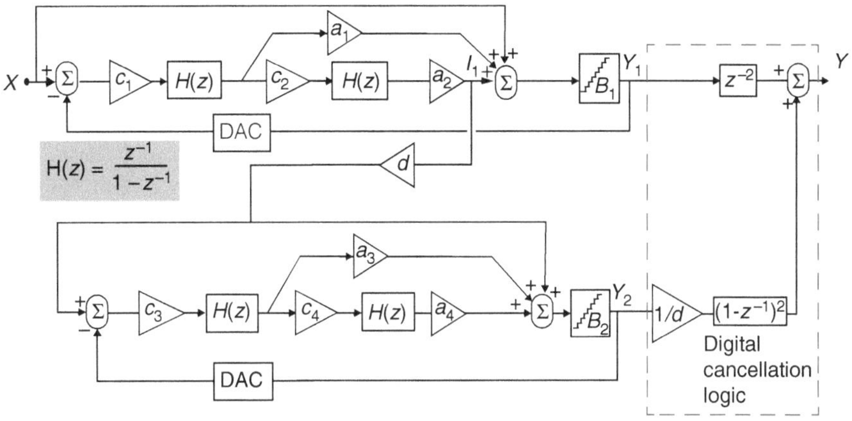

Fig. 5.1 Block diagram of the reconfigurable ΣΔM architecture

non-linearities helps to increase resolution (specially for low-bandwidth modes whose DRs are usually dominated by thermal noise and peak SNDRs by distortion). Low sensitivity to SR also leads to an efficient operation for high-speed modes. In addition, the second architectural requirement (i.e., low power for the modulator building blocks) also benefits from the low opamp requirements, mainly in terms of DC gain, gain non-linearities and SR.

- Low internal node swings in the modulator or, analogously, reduced output swing requirements for the opamps[1]. This simplifies the design of the opamps and, potentially, leads to a lower power operation[2] (see the second architectural requirement). It also fuels low-voltage operation of the opamps in the modulator because of the increase in the output voltage headroom (third architectural requirement).
- Overload levels are higher in comparison to those of conventional ΣΔMs, as a consequence of the previous feature. Larger resolutions can therefore be anticipated (see the first architectural requirement).
- Only one DAC is needed for the in-loop feedback operation. If the resolution of the quantizers is adapted to the operation mode[3], the circuit complexity will be reduced. This feature fulfils the fourth architectural requirement (easier architectural reconfiguration).
- Simplicity for conformation of cascade topologies. The fourth architectural requirement is also covered by this because the back-end stage can be either connected or disconnected more easily. This leads to a simpler in-loop order adaptation.

The topology in Fig. 5.1 therefore satisfies all the aforementioned architectural requirements in terms of the efficient operation of a reconfigurable modulator with low

[1] Note that these requirements also depend on the resolution of the embedded quantizers and the values of the in-loop coefficients.

[2] Consider the case in which, for instance, a two-stage opamp is needed rather than a single-stage one because of the opamp output swing requirements.

[3] This reconfiguration strategy is not used in this ΣΔM but in that presented in Chap. 6.

power and low supply in a multi-mode scenario. This architecture also has a number of drawbacks [Ghar06, Tang07], namely:

- Analog adders are needed for the implementation of the feed-forward paths and thus for the USTF in this modulator architecture. If a passive adder is used, there will be a scaling factor implicit on the addition that needs to be compensated by a voltage scaling in the quantizers references. An active addition[4], on the contrary, demands the use of an amplifier. This entails higher power consumption and can potentially lead to an inefficient A/D conversion.
- Timing issues in the architecture. This is especially problematic when two paths are used for the inter-stage connection in cascades made up of USTF stages. Fortunately, the use of only one inter-stage path eliminates these issues (see Sect. 3.3).

5.1.2 System-Level Parameters

There are many parameters to be explored for the architecture in Fig. 5.1 in a multi-mode application. In the first place, the OSR, the in-loop order (L), the resolution of the embedded quantizers (B_i), the reference voltage (V_{ref}) and the in-loop coefficients of the architecture (c_i and a_j)[5] need to be considered. In theory, all these parameters could be varied independently[6] and according to the requirements of each mode. However, such a scenario would lead to an extremely complex circuitry.

A number of reconfiguration strategies are included, taking advantage of the multimode capabilities of the topology in Fig. 5.1. First, the OSR—which is the most commonly employed reconfiguration strategy [Rosa09]—is adapted to the mode. Second, L can be 4 but also 2 by switching off the last stage. The latter case corresponds to a 2-2 cascade operation (as shown directly in Fig. 5.1) with the following NTF:

$$\mathrm{NTF}(z) = \frac{1}{d} \cdot (1 - z^{-1})^4 \tag{5.1}$$

while a second-order single-loop ΣΔM, in which the first stage in Fig. 5.1 is the only one working, results in

$$\mathrm{NTF}(z) = (1 - z^{-1})^2 \tag{5.2}$$

Both cases lead to STF $(z) = 1$. Note from Fig. 5.1 that the digital cancellation logic is only employed in the 2-2 cascade configuration. The final values of these two design

[4] The ΣΔM presented in Chap. 6 uses this approach. More details on these kinds of analog adders can be found there. By contrast, the ΣΔM in Fig. 5.1 employs passive adders for power saving.

[5] Note that some of these parameters—such as OSR, V_{ref} or B_i—can be considered closer to the electrical implementation than to the architecture itself. However, their values will influence the performance of the overall modulator and could be reconfigured as well.

[6] Of course, the integrator coefficients (c_i) and the feed-forward coefficients (a_j) are related as described in (3.8) for every stage of the block diagram in Fig. 5.1. This will be addressed later.

parameters (L and OSR) will be given in Sect. 5.1.4 for each mode. They will be selected based on behavioral simulations.

Many system characteristics are influenced by the values of B_1 and B_2; e.g., the power of the in-band quantization error, the output swing requirements of the opamps, the modulator's robustness against its building blocks non-linearities and opamp large-signal parameters such as SR, the possible need for DEM techniques, and the power and area consumption in the modulator. Fixed values of 3 levels for the quantizers of both stages in this $\Sigma\Delta$M are used to simplify its circuit-level implementation. Quantization errors, opamp output swings and SR, power and area consumption are therefore all reduced compared to a single-bit solution [Paul87]. Furthermore, an almost inherently linear DAC can be obtained with only one extra switch in the SC feedback branch of the D/A conversion, so no DEM is required [Lewi92, Balm00, Reut02]. The resulting electrical implementation of the $\Sigma\Delta$M is thus simplified.

The reference voltage is taken as the supply voltage (1.2 V) to maximize the modulator SNDR. The output swing requirements of the opamps must therefore be lower than V_{ref}, contrary to several reported modulators with a reference voltage lower than the supply voltage. The selection of a reference voltage equal to the supply makes the low-voltage operation of conventional cascade $\Sigma\Delta$Ms difficult, at least for medium-to-large input levels. For example, the reference voltage of the chip presented in Chap. 4 is 27.5% lower than the supply voltage. This problem is solved here with the combined use of a unity-STF architecture, multi-bit quantizers and a proper selection of the in-loop coefficients as shown below.

5.1.3 Modulator In-Loop Coefficients

Not only the low-voltage operation of the modulator—provided by the opamp output swing requirements—depends on the in-loop coefficients but also many other metrics of the modulator performance, namely:

- Stability and output swing requirements of the opamps. The modulator's overloading, the robustness against its building block non-linearities and the opamp SR depend on the in-loop coefficients in a similar way.
- The equivalent loads of the opamps. Opamp dynamics are also determined by the in-loop coefficients in the modulator. The design of this modulator has taken this into account and low values of the equivalent capacitances were obtained, as shown in Sect. 5.3.1.
- Thermal or kT/C noise. Its corresponding power can be affected by the coefficients under certain circumstances. Let us assume, for instance, that a minimum capacitance of 50 fF is set for matching and there is an integration capacitance of 100 fF. An integrator coefficient of 2 leads to a sampling capacitance of 50 fF, while a coefficient of 0.5 results in a sampling capacitor of 200 fF. The latter solution presents less thermal noise but larger capacitances are needed. Thus, there is a trade-off between coefficients and capacitor values for matching, kT/C noise and opamp dynamics.
- Area, because the number of unit capacitors in the modulator is directly related to the values of its coefficients.

Table 5.2 Relation of the in-loop coefficients of the block diagram in Fig. 5.1. [Silv04a]

First-stage coefficients	Second-stage coefficients
$a_1 = \frac{2}{c_1}$	$a_3 = \frac{2}{c_3}$
$a_2 = \frac{1}{c_1 \cdot c_2}$	$a_4 = \frac{1}{c_3 \cdot c_4}$

The same NTFs in (5.1) and (5.2) and a unity STF are respectively obtained for the architecture in Fig. 5.1 for a 2-2 cascade and the first-stage in a single-loop configuration, when the in-loop coefficients in Fig. 5.1 maintain the relations given by Table 5.2. These relations define the feed-forward factors resulting from a scaling on the integrator coefficients, as demonstrated in Sect. 3.1.3.

5.1.3.1 Exploration of the In-Loop Coefficients

An exploration of all the modulator in-loop coefficients that takes into account all the aforementioned restrictions might lead to a very large design space. A complete exploration should also look at other parameters that are closer to the electrical level, such as the integrator sampling capacitors and opamp GB and SR. Given the complexity of this analysis, simplifications are needed for a practical exploration of this design space at the architectural level. To do so, the following set of restrictions is considered:

- Only in-loop coefficients that are power of 2 are considered under the relation given by Table 5.2 (see Sect. 3.1.3). This allows an easy implementation of the modulator in-loop coefficients, while limiting the design space to be explored. Furthermore, the inter-stage factor d is also restricted in the same way. The digital coefficient $1/d$ (located at the output of the back-end stage) can therefore be made with just shifting operations in a register of the digital cancellation logic in Fig. 5.1.
- Sizing parameters of the modulator building blocks, such as the values of the sampling capacitors in the modulator, opamp GB and SR, are not taken into account at this point although a set of strategies will be used later to reduce the IBE power related with these metrics, such as large OSR to reduce kT/C noise or low capacitor values to diminish the opamp equivalent loads. In fact, only the opamp output swing requirements and the number of unit capacitors (N) will be considered at this level.

SIMSIDES [Ruiz05] is used to perform the behavioral simulations necessary for the exploration of the design space under the above-mentioned restrictions. First, a single-loop configuration of the ΣΔM in Fig. 5.1 is considered; i.e., the back-end stage is assumed off and the number of unit capacitors (inter-stage path included) is not considered either. As an illustration of the results of this procedure, Table 5.3 is included. In this table, the values of c_1 and c_2 are taken from 0.25 to 4, while the values of a_1 and a_2 are extracted from Table 5.2. The necessary unit capacitors N (in a single-ended form) together with the required minimum output swings for the opamps (OS_i), normalized to the reference voltage, are given in Table 5.3. Shadowed

Table 5.3 In-loop coefficients and opamps output swings in the first stage of Fig. 5.1

c_1	c_2	a_1	a_2	OS_1(%)	OS_2(%)	N
0.25	0.25	8	16	23.6	3.6	35
0.25	0.5	8	8	23.1	7.0	25
0.25	1	8	4	22.4	12.6	20
0.25	2	8	2	21.9	22.4	19
0.25	4	8	1	20.9	37.3	20
0.5	0.25	4	8	44.2	7.0	21
0.5	0.5	4	4	43.8	13.4	15
0.5	1	4	2	42.8	25.2	12
0.5	2	4	1	40.6	44.7	12
0.5	4	4	0.5	39.4	73.1	15
1	0.25	2	4	82.1	13.9	14
1	0.5	2	2	80.4	26.7	10
1	1	2	1	77.9	50.6	8
1	2	2	0.5	76.4	90.6	10
1	4	2	0.25	72.4	150.3	14
2	0.25	1	2	140.2	28.1	12
2	0.5	1	1	137.8	54.5	9
2	1	1	0.5	138.1	101.6	9
2	2	1	0.25	132.7	182.8	12
2	4	1	0.13	127.3	304.7	18
4	0.25	0.5	1	228.0	57.5	14
4	0.5	0.5	0.5	227.5	110.8	13
4	1	0.5	0.25	214.9	207.3	14
4	2	0.5	0.13	206.9	369.0	19
4	4	0.5	0.06	189.0	599.2	29

areas in the table correspond to those coefficients where the output swing is lower than the reference voltage. The lower the integrator coefficients, the less output swing requirements. As a rule of thumb, the farther c_i and a_j are from 1, the larger the number of employed unit capacitors. There are some exceptions to this rule, mainly due to the existence of an implicit scaling of a feed-forward coefficient on the inter-stage connection of the cascade in Fig. 5.1. A clear example of this is given for the selected in-loop coefficients in this design (more details can be found below). Values greater than 1 for c_1 entail output swings larger than V_{ref} and, thus, greater than the 1.2-V supply in this design, so these values will not be further considered. If $c_1 = 1$, the requirement on the front-end amplifier output swing is very large, while $c_1 = 0.5$ does not impose such a high value for the first-amplifier output swing requirements but twice that of $c_1 = 0.25$. This is why $c_1 = 0.25$ is selected. Note from Table 5.3 that the set of coefficients $\{c_1, c_2, a_1, a_2\} = \{0.25, 1, 8, 4\}$—highlighted in the table—is a good trade-off, in which the opamps output swings are limited to low values and N is not high.

All the in-loop coefficients of the scheme in Fig. 5.1 have now been considered. Table 5.4 illustrates the behavioral simulation results when coefficients c_3, c_4, a_3 and a_4 are also included. However, these coefficients are limited to $c_1 = 0.25$, and

Table 5.4 In-loop coefficients and opamps output swings in Fig. 5.1

c_1	c_2	a_1	a_2	c_3	c_4	a_3	a_4	OS_1 (%)	OS_2 (%)	OS_3 (%)	OS_4 (%)	N
0.25	0.25	8	16	0.25	0.25	8	16	23.4	3.7	19.9	6.5	66
0.25	0.25	8	16	0.25	0.5	8	8	23.4	3.7	19.6	14.1	63
0.25	0.25	8	16	0.25	1	8	4	23.4	3.7	19.9	27.8	65
0.25	0.25	8	16	0.25	2	8	2	23.4	3.7	19.4	50.1	72
0.25	0.25	8	16	0.25	4	8	1	23.4	3.7	18.9	79.2	86
0.25	0.25	8	16	0.5	0.25	4	8	23.4	3.7	41.7	17.2	66
0.25	0.25	8	16	0.5	0.5	4	4	23.4	3.7	45.2	38.1	63
0.25	0.25	8	16	0.5	1	4	2	23.4	3.7	45.87	82.1	67
0.25	0.25	8	16	0.5	2	4	1	23.4	3.7	44.7	152.8	78
0.25	0.25	8	16	0.5	4	4	0.5	23.4	3.7	47.5	245.3	70
0.25	0.5	8	8	0.25	0.25	8	16	23.1	7.3	20.1	5.1	47
0.25	0.5	8	8	0.25	0.5	8	8	23.1	7.3	21.1	9.9	44
0.25	0.5	8	8	0.25	1	8	4	23.1	7.3	20.4	18.6	45
0.25	0.5	8	8	0.25	2	8	2	23.1	7.3	20.1	33.2	50
0.25	0.5	8	8	0.25	4	8	1	23.1	7.3	17.2	53.4	60
0.25	0.5	8	8	0.5	0.25	4	8	23.1	7.3	42.7	11.3	46
0.25	0.5	8	8	0.5	0.5	4	4	23.1	7.3	44.2	23.9	43
0.25	0.5	8	8	0.5	1	4	2	23.1	7.3	43.7	49.6	45
0.25	0.5	8	8	0.5	2	4	1	23.1	7.3	41.2	89.4	52
0.25	0.5	8	8	0.5	4	4	0.5	23.1	7.3	40.6	140.2	48
0.25	1	8	4	0.25	0.25	8	16	22.6	13.9	19.2	4.3	41
0.25	1	8	4	0.25	0.5	8	8	22.6	13.9	19.2	7.8	37
0.25	1	8	4	0.25	1	8	4	22.6	13.9	18.1	14.4	35
0.25	1	8	4	0.25	2	8	2	22.6	13.9	17.2	24.6	39
0.25	1	8	4	0.25	4	8	1	22.6	13.9	16.1	37.3	47
0.25	1	8	4	0.5	0.25	4	8	22.6	13.9	39.6	8.6	36
0.25	1	8	4	0.5	0.5	4	4	22.6	13.9	39.2	17.4	33
0.25	1	8	4	0.5	1	4	2	22.6	13.9	40.1	33.6	34
0.25	1	8	4	0.5	2	4	1	22.6	13.9	37.8	59.0	39
0.25	1	8	4	0.5	4	4	0.5	22.6	13.9	36.5	91.5	37
0.25	2	8	2	0.25	0.25	8	16	22.9	26.8	20.9	3.8	50
0.25	2	8	2	0.25	0.5	8	8	22.9	26.8	20.1	7.0	44
0.25	2	8	2	0.25	1	8	4	22.9	26.8	19.5	12.6	41
0.25	2	8	2	0.25	2	8	2	22.9	26.8	18.1	20.9	41
0.25	2	8	2	0.25	4	8	1	22.9	26.8	17.4	31.7	48
0.25	2	8	2	0.5	0.25	4	8	22.9	26.8	41.5	7.2	36
0.25	2	8	2	0.5	0.5	4	4	22.9	26.8	39.8	14.2	32
0.25	2	8	2	0.5	1	4	2	22.9	26.8	39.0	26.4	30
0.25	2	8	2	0.5	2	4	1	22.9	26.8	37.7	43.7	34
0.25	2	8	2	0.5	4	4	0.5	22.9	26.8	34.3	68.6	33
0.25	4	8	1	0.25	0.25	8	16	22.9	51.6	21.4	3.6	71
0.25	4	8	1	0.25	0.5	8	8	22.9	51.6	21.2	6.5	61
0.25	4	8	1	0.25	1	8	4	22.9	51.6	20.9	11.6	56
0.25	4	8	1	0.25	2	8	2	22.9	51.6	19.6	18.6	55
0.25	4	8	1	0.25	4	8	1	22.9	51.6	18.8	28.8	56
0.25	4	8	1	0.5	0.25	4	8	22.9	51.6	43.2	6.9	45
0.25	4	8	1	0.5	0.5	4	4	22.9	51.6	41.9	12.7	39
0.25	4	8	1	0.5	1	4	2	22.9	51.6	42.4	22.6	36
0.25	4	8	1	0.5	2	4	1	22.9	51.6	39.4	39.0	36
0.25	4	8	1	0.5	4	4	0.5	22.9	51.6	38.1	57.2	39

$c_3 = 0.25$ or 0.5 in Table 5.4 to show simulation results with low requirements for the opamps output swing. Furthermore, an inter-stage scaling factor $d = 1$ is selected because it entails no reduction in the power of the in-band quantization error[7]—see $1/d$ term in (5.1), while the output swings in the back-end stage in the modulator can be controlled with its coefficients[8]. In addition, higher values of d impose a reduction in the last-stage coefficients. The rows in Table 5.4 are shadowed for the set of coefficients in which all opamp output swings are lower than the reference voltage. A last-stage set of coefficients identical to that initially considered for the first stage—$\{c_1, c_2, a_1, a_2\} = \{c_3, c_4, a_3, a_4\} = \{0.25, 1, 8, 4\}$—represents a good trade-off in terms of the requirements of the opamp output swing and the total number of unit capacitors. Note from Table 5.4 that only 14 unit capacitors are needed for the implementation of the last stage in-loop coefficients, while the first stage requires the use of 22 unit capacitors. This reduction in the number of in-loop coefficients in the last stage benefits from the fact that the inter-stage node $I_1(z)$ is connected from the output of the second integrator to the input of the third integrator and to the second-stage analog adder with effective coefficients $d \cdot a_2 \cdot c_3 = 1$ and $d \cdot a_2 = 4$, respectively. The in-loop coefficients of this analog adder are now $\{d \cdot a_2, a_3, a_4\} = \{4, 8, 4\} = 4 \cdot \{1, 2, 1\}$; i.e., this set of coefficients can be simplified to only $\{1, 2, 1\}$ with an unit capacitor that is 4 times larger than the one originally considered. This results in area saving (because of the lower N) or fewer capacitance mismatches (due to the use of larger unit capacitors). In addition these coefficients fit well with the recommendations previously given for the selection of topological coefficients. To sum up, the in-loop coefficients $\{c_1, c_2, c_3, c_4\} = \{0.25, 1, 0.25, 1\}$ are those finally selected for this $\Sigma\Delta$M, with output swings lower than 0.35 V (almost 30% of the reference voltage) for $V_{ref} = 1.2$ V and 2×35 unit capacitors.

Figure 5.2 also illustrates the required output swings for the opamps (OA_i where i stands for the opamp number in the modulator), considering two different sets of coefficients in Table 5.4. The first set is the one selected for the modulator in Fig. 5.1 ($\{c_1, c_2, c_3, c_4\} = \{0.25, 1, 0.25, 1\}$), while the second set provides output swings for the fourth opamp greater than V_{ref}. Figure 5.2a shows that the opamp output voltages are always below 0.35 V—and most of the occurrences are inside the range of ± 0.2 V—for the selected coefficients. Figure 5.2b shows the opamp output swing requirements for the in-loop coefficients $\{c_1, c_2, c_3, c_4\} = \{0.25, 0.25, 0.5, 2\}$. A maximum required voltage of roughly ± 1.9 V is needed for the last case.

The topology in Fig. 5.1 presents constant opamp output swings for almost the whole range of input levels, contrary to conventional cascades, in which the output swing is proportional to the input amplitude [Nam05]. In fact, Fig. 5.3 depicts the requirements of the opamp output swing when varying the input amplitude. Note that the output swing requirements remain constant for input amplitudes lower than -3 dBFS.

---

[7] Section 4.2.1 illustrates that the systemic power loss of the in-band quantization error in the $\Sigma\Delta$M presented in Chap. 4 was 6 dB.

[8] High values of d lead to either very low values of c_3 and c_4 or large output swings for the last-stage opamps.

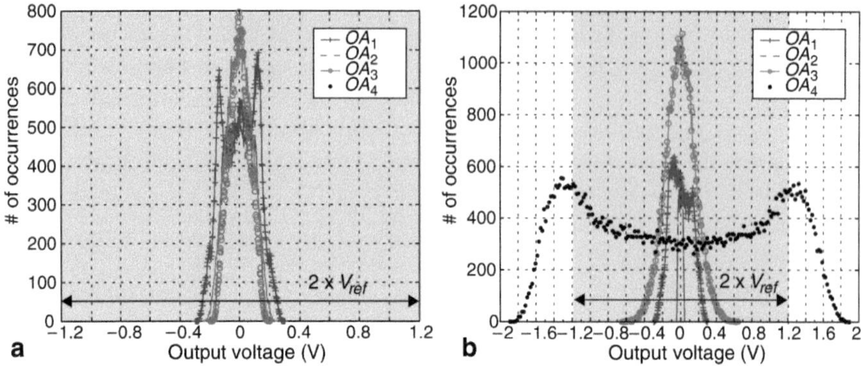

Fig. 5.2 Opamp output swing requirements for: **a** $\{c_1, c_2, c_3, c_4\} = \{0.25, 1, 0.25, 1\}$, **b** $\{c_1, c_2, c_3, c_4\} = \{0.25, 0.25, 0.5, 2\}$

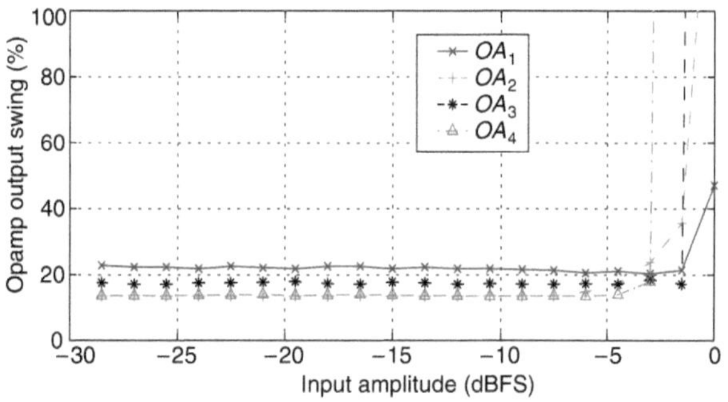

Fig. 5.3 Opamps output swing requirements versus input amplitude for the selected in-loop coefficients ($\{c_1, c_2, c_3, c_4\} = \{0.25, 1, 0.25, 1\}$)

5.1.4 Initial Modulator Sizing

The selection of the ΣΔM architecture and the coefficients employed have been detailed above. However, there are still a number of high-level parameters to be chosen that will have a strong impact on the final performance of the overall modulator. Among them, the oversampling ratio and the modulator order (L)—either a second-order or a 2-2 cascade—are the main ones to be considered for each operation mode.

Large OSRs are used to reduce thermal noise in modes that require high-to-medium resolution. However, the larger the OSR, the greater the sampling frequency. This approach is therefore only valid for low-to-medium bandwidths. The maximum sampling frequency has been limited to 240 MHz as a design choice. In practice, larger values of the sampling frequencies are difficult to implement in SC ΣΔMs due to the large opamp dynamic requirements, resulting in excessive power consumption even when it is feasible technologically speaking. Alternatively, L can be increased to obtain enough resolution since broadband standards are limited to low OSRs. Narrowband modes—that use larger OSRs—can work with lower orders (a

Table 5.5 Initial modulator high-level sizing parameters

Standard	GSM	BT	GPS	UMTS	DVB-H	WiMAX
L		2			4	
OSR	200	80	60	30	15	12
f_s (MHz)	40	80		120		240

second-order single-loop $\Sigma\Delta$M) since thermal noise is limiting the resolution when compared to the contribution of the quantization error to the overall IBE power. Behavioral simulations, based on SIMSIDES [Ruiz05], were performed with these considerations in mind. In addition, a first estimation of the main circuit non-idealities was included in the behavioral simulations and a safe margin was taken in order to fulfill the specifications given in Table 5.1.

The results of behavioral simulations, while also taking into account the aforementioned criteria, lead to the values of L, OSR and f_s presented in Table 5.5. The values for OSR decreased with bandwidth and required ENOB. A single-loop topology was used for bandwidths in the range from 100 kHz to 1 MHz and a cascade from there up to 10 MHz of bandwidth. Although these parameters are found to be a good trade-off for the respective operation modes in the modulator, the reconfiguration strategies implemented in this $\Sigma\Delta$M enables a flexible selection of their values in each of the operation modes.

This high-level sizing of the modulator is the starting point for the SC implementation of the prototype. This will be described in the next section, which will also deal with specific issues of the implementation of the architecture in Fig. 5.1 at this level, such as a necessary scaling factor for the quantizer voltage references and the timing of this architecture. Finally, a high-level synthesis is shown. It will be the starting point for the later transistor-level design of the modulator building blocks.

5.2 SC Implementation and Related Issues

Figure 5.4 shows the conceptual SC schematic of the $\Sigma\Delta$M in Fig. 5.1. The modulator consists of two stages, each of them including two integrators, one analog adder and a 1.5 bit—or 3-level—quantizer to implement the main path, whereas the DAC[9] performs the in-loop feedback operation. Note that the digital cancellation logic (see Fig. 5.1) is applied by software and is therefore not implemented on-chip.

Four different amplifiers (OA_i in the figure) are used to implement the four integrators in the cascade. Note that the required analog additions are based on passive SC networks. The desired integrator analog coefficients $\{c_1, c_2, c_3, c_4\} = \{0.25, 1, 0.25, 1\}$—and the subsequent feed-forward coefficients $\{a_1, a_2, a_3, a_4\} =$

[9] The DAC operation is already performed at the sampling capacitors of the front-end integrator controlled by digital signals A_i, B_i and C_i for each stage i. Note that the digital gates that generate these control signals are grouped together in the DAC block in Fig. 5.4; though, formally this block should be denoted as DAC control-signal generator.

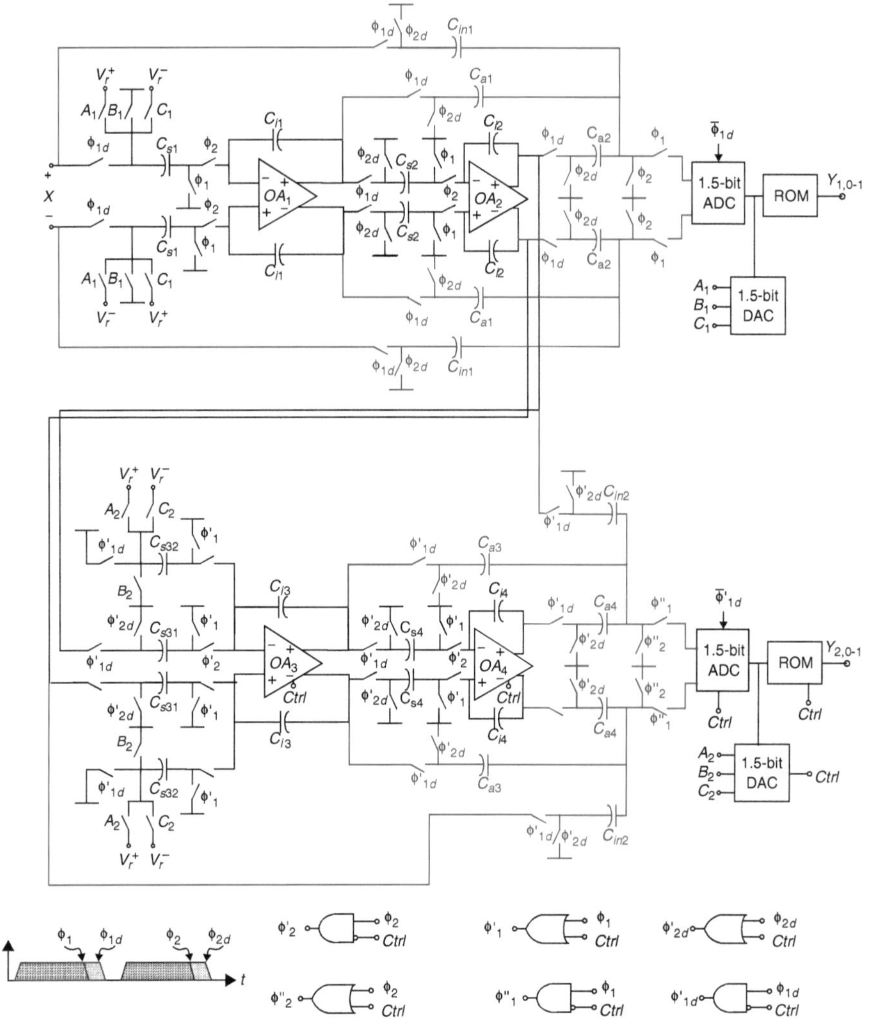

Fig. 5.4 SC schematic of the reconfigurable multi-standard $\Sigma\Delta$M

{8, 4, 8, 4}—are implemented as capacitor ratios with selected values listed in Table 5.6. Capacitor values are scaled depending on the integrator location. The main purpose here is to reduce the opamp dynamic requirements by diminishing the equivalent loads of the opamps, while keeping mismatches and thermal noise contributions under control.

The main architectural reconfiguration strategy implemented in this $\Sigma\Delta$M is the adaptation of the order (L) that can be configured to be either 2 or 4 with the control signal *Ctrl* (see Fig. 5.4). Thus, *Ctrl* switches all the sampling capacitors off in the last stage by connecting all of them to the common-mode voltage when $L = 2$ and the

Table 5.6 Capacitor values in Fig. 5.4

Capacitor	C_{s1}		C_{i1}	C_{s2}	C_{i2}	C_{in1}	C_{a1}	C_{a2}
Capacitance (fF)	150		600	100	100	50	200	100
Capacitor	C_{s31}	C_{s32}	C_{i3}	C_{s4}	C_{i4}	C_{in2}	C_{a3}	C_{a4}
Capacitance (fF)	400	100	400	100	100	200	400	200

back-end stage is not used. The integration and sampling operations in this $\Sigma\Delta$M are controlled by two non-overlapped clock phases (ϕ_1 and ϕ_2, respectively), and their delayed and complementary versions. However, modified versions of these clock phases are needed in order to avoid floating nodes at the capacitors. This way, the appearance of dielectric relaxation [Fata90] is also avoided. To this purpose, phase signals ϕ_i', ϕ_{id}' and ϕ_i'' are generated as shown at the bottom of Fig. 5.4. Digital OR or AND operations are applied to the corresponding phase: *Ctrl* signal and its complementary version ($\overline{Ctrl}$), respectively. If $Ctrl = $ '1', the last stage switches will be activated for ϕ_1', ϕ_{2d}' and ϕ_2'', whereas these switches will be deactivated when they are controlled by ϕ_2', ϕ_{1d}' and ϕ_1''. In addition, the activation of the *Ctrl* signal involves the turning off of the modulator building blocks of the last stage—such as opamps and comparators—by means of a power-down control in order to save power consumption.

5.2.1 *Practical Implementation of Analog Adders*

A SC passive architecture for the analog adders is the best solution in terms of speed requirements when compared to an active solution since the latter entails the use of amplifiers [Tang07]. This leads to a lower-power solution, specially for high-speed operation modes [Gagg04]. However, the passive approach involves a number of design constraints, namely:

- Input signal is compressed while being added. This effect is due to the implicit capacitor divider present in the analog adder. This issue and its related solutions will be discussed later.
- Contrary to amplifier-based active solutions, there is no virtual ground. The influence of parasitics is therefore greater.
- Note that, regardless of the selected approach—active or passive—, capacitor matching plays an important role that will be evaluated in Sect. 5.3.2.

5.2.1.1 Quantizer Scaling Factor

The use of passive SC analog additions involves a reduction of the full scale at the quantizer input and, consequently, a compression of the references and the quantizers

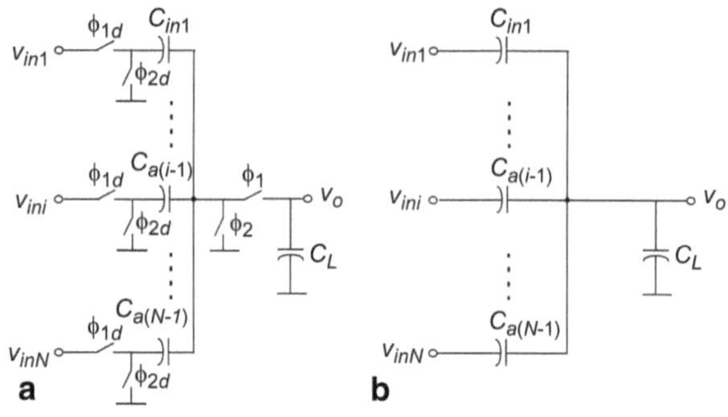

Fig. 5.5 Analog adder: **a** SC implementation, **b** configuration during sampling phase

voltage levels to be used [Silv04b, Fuji06, Tang07]. Fig. 5.5 shows the schematic of a generic analog adder (Fig. 5.5a) and its equivalent circuit during the sampling phase (Fig. 5.5b), this phase being the one used to perform the actual analog addition. If charge redistribution is considered for the capacitor divider shown in Fig. 5.5b, it can be shown that the output voltage of the analog adder[10] is given by:

$$v_o = \frac{C_{in1} \cdot v_{in1}}{C_{in1} + \sum\limits_{i=2}^{N} C_{a(i-1)} + C_L} + \frac{\sum\limits_{i=2}^{N} C_{a(i-1)} \cdot v_{ini}}{C_{in1} + \sum\limits_{i=2}^{N} C_{a(i-1)} + C_L} \tag{5.3}$$

where C_L is the capacitive load due to the input parasitic capacitances of the comparators in the quantizer.

Let us assume that the only input signal is v_{in1}. Then, there is a scaling factor α over that input signal, defined as follows

$$\alpha = \frac{C_{in1}}{C_{in1} + \sum\limits_{i=2}^{N} C_{a(i-1)} + C_L} \tag{5.4}$$

Note that, ideally, the only signal path for the modulator input signal in the first-stage analog adder is the one that corresponds to the v_{in1} node due to the unity-STF implementation of the modulator. In fact, particularizing for the first-stage analog adder in Fig. 5.4, the modulator input signal is scaled by a factor

$$\alpha = \frac{C_{in1}}{C_{in1} + C_{a1} + C_{a2} + C_L} \tag{5.5}$$

where the capacitor load is given by the parasitics at the input of the in-loop quantizer. If—in a first approximation—this capacitance is considered negligible, this factor can be taken as a ratio between feed-forward coefficients, given by:

$$\alpha = \frac{1}{1 + a_1 + a_2} \tag{5.6}$$

[10] This voltage is also the input of the embedded quantizer directly connected to the analog adder. Therefore, this voltage is compared with the quantizer voltage tabs.

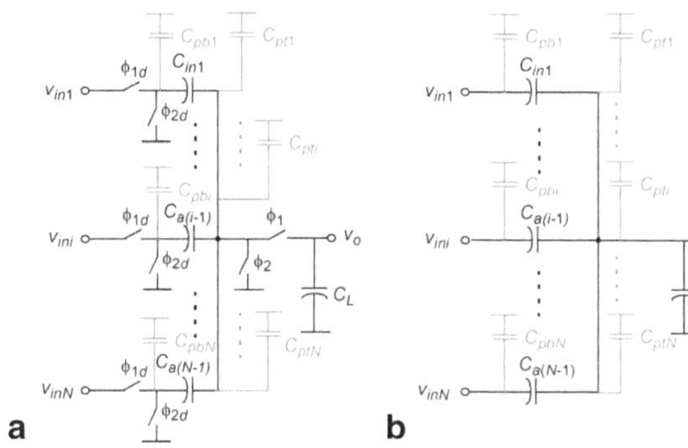

Fig. 5.6 Analog adder including parasitics: **a** SC implementation, **b** sampling phase

However, given the low values of the feed-forward capacitors in Table 5.6, this approximation is not recommended for this specific case. Parasitic capacitances at the input of the embedded quantizer therefore need to be included. Note that the 1.5-bit (or 3-level) quantizer is implemented with two comparators. Therefore, the value of the analog adder load is extracted by adding the input parasitic capacitance of the two preamplifying blocks at the comparator input (explained in detail in Sect. 5.3.4). Actual preamplifier parasitic capacitances are estimated as 45 fF based on AC simulations; thus, the scaling factor is

$$\alpha = \frac{C_{in1}}{13 \cdot C_{in1} + C_L} = 0.0676 \tag{5.7}$$

The presence of parasitics could modify the required value of α. In order to analyze this, Fig. 5.6 shows the analog adder schematic with the inclusion of bottom- and top-plate parasitic capacitances[11], respectively denoted as C_{pbi} and C_{pti} in the figure. In the actual addition, the influence of C_{pbi} can be neglected since—due to Kirchhoff's second law—these capacitors are shorted to the corresponding input voltage v_{ini} when performing the addition operation. However, Fig. 5.6b shows that all capacitors C_{pti} are in parallel configuration with C_L. C_L must therefore be updated now to $C_L + \sum C_{pti}$ in (5.4). Taking this into account, and including the parasitic capacitances of the switches with values shown in Sect. 5.3.3, α changes to 0.0655.

A preamplifying stage just before the comparator input with a gain equal to $1/\alpha$ (23.7 dB) can be included to compensate for signal attenuation. However, this preamplifier gain must be stable with any kind of variation such as those deriving from temperature, supplies ringing and technological variations. Furthermore, this solution would demand a power-hungry amplifier with very large output swing and high-speed operation. An SC active implementation of the analog adder could have similar requirements in terms of power but no compression over the output voltage of the analog adder, so it could be considered as a feasible alternative. However, there is still a power-saving strategy, in which the attenuation factor α is not compensated but the reference voltages of the quantizer resistor ladder are scaled by α. This scaling

[11] Note that these parasitic capacitances account not only for the parasitics of the sampling capacitors but also for the parasitic capacitances of the associated switches.

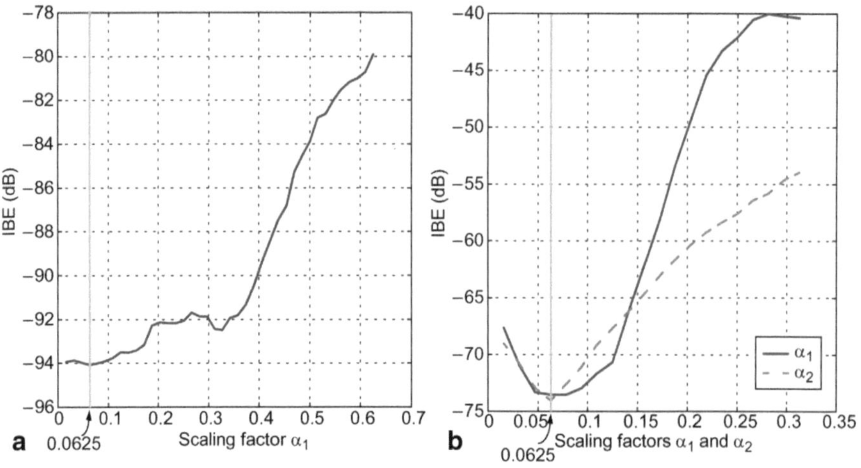

Fig. 5.7 Quantizer references compression versus IBE power: **a** GSM ($L=2$ and $OSR=200$), **b** WiMAX ($L=4$ and $OSR=12$)

factor is performed via an on-chip resistor ratio. Nevertheless, a factor $\alpha=0.0655$ is difficult to implement accurately following this procedure. For instance, this value can be implemented as a 131-to-2000 resistor ratio. In practice, this ratio would require a huge amount of elements and area. A scaling factor of 0.0625 is selected because of its simpler implementation and closeness to the former value; indeed, only a 2-to-32 ratio is needed.

Let us consider the 2-2 cascade ΣΔM with SC schematic in Fig. 5.4. Now, there is more than one passive analog adder. Thus, the analytical approach described above for extracting the scaling factor of the first-stage analog adder (α_1)—denoted above as α because only one analog adder was considered—is not directly applicable to that of the second stage (α_2). The reason is that the first-stage analog adder presents a capacitor divider directly over the modulator input signal, while the second-stage analog adder, on the contrary, processes the quantization error of the first-stage quantizer in the cascade. The evolution of the IBE power with the scaling factors α_1 and α_2 is simulated with SIMSIDES [Ruiz05]. The simulation results are depicted in Fig. 5.7a, b for GSM and WiMAX operation modes, respectively. This figure shows that, fortunately, the same value of 0.0625 for both scaling factors gives the minimum IBE power in GSM and WiMAX. In addition, this result simplifies circuit implementation because only one resistor ladder is needed. However, this small scaling factor imposes strict requirements on the comparators input offset [Silv04b]. If there were no compression over the quantizer full scale, the maximum allowable comparator offset with no degradation in the power of the in-band quantization error would be the differential reference ($2 \cdot V_{ref}$) divided by the number of elements of the quantizer (K) and 2; i.e., half LSB of the A/D conversion. However, when compression is needed, the maximum comparator offset requirement is directly multiplied by this

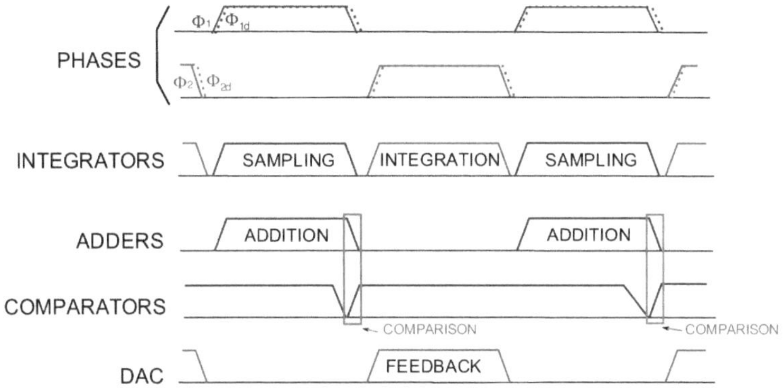

Fig. 5.8 Timing operation of the modulator building blocks in Fig. 5.4

scaling factor $\alpha = \alpha_1 = \alpha_2 = 0.0625$ [Nam05, Tang07]; in our case:

$$\text{Offset} < \frac{2 \cdot \mathrm{V_{ref}}}{2 \cdot K} \cdot \alpha = \frac{\mathrm{V_{ref}}}{K} \cdot \alpha = 25\,\mathrm{mV} \tag{5.8}$$

this maximum offset level of 25 mV constituting an additional constraint for the comparator design. This issue can be relieved by design—as done here and explained in Sect. 5.3.4—and/or using auto-zeroing [Enz96a, Figu09].

5.2.2 *Timing*

In order to build the cascade topology in Fig. 5.1, the input of the first-stage embedded quantizer and that of the DAC could have been subtracted to extract the corresponding quantization error and give it to the back-end stage, as usually done in conventional $\Sigma\Delta$Ms (see, for example, the scheme in Fig. 4.7) [Mats87, Long88, Chou89, Rebe90]. However, this solution is not feasible in the SC implementation of Fig. 5.1 (see Fig. 5.4) due to timing problems related with the feed-forward paths addition of the two stages in a cascade modulator. This would create a synchronization issue as described in Sect. 3.3.2, and, consequently, both the NTF and STF would change. A unity STF in the last stage can therefore only be used when the quantization error from the previous stage is given through only one path and this signal is thus available for the last-stage analog addition operation on time, as already done in the scheme of Fig. 5.1 and in its SC implementation of Fig. 5.4.

Figure 5.8 summarizes the distribution of the timing operations in the modulator. During sampling phase (ϕ_1):

- Integrators sample their input signals.
- Analog adders perform the actual addition via a capacitor divider.
- At the very end of this phase, when the value of the addition is stable at the input of the comparator preamplifier, the comparator regenerative latch is triggered to perform the comparison. This is done with the rising edge of phase $\bar{\phi}_{1d}$.

Table 5.7 In-band thermal noise power

Standard	GSM	BT	GPS	UMTS	DVB-H	WiMAX
$P_{TN,\mathrm{kT/C}}$ (dB)	−86.5	−83.4	−82.4	−82.7	−77.0	−72.5
$P_{TN,\mathrm{Opamp+kT/C}}$ (dB)	−84.2	−77.1	−73.9	−75.5	−69.8	−65.2

During integration phase (ϕ_2), the following operations are performed:

- The embedded quantizer retains the same value taken from comparison in the previous phase ϕ_1.
- DAC performs the feedback operation.
- Integrators integrate the difference between the previously sampled signal and that fed back by the DAC.

5.2.3 *High-Level Synthesis*

The values of the sampling capacitors are scaled[12] according to their location on the modulator chain since the thermal noise contribution of each integrator—through the kT/C noise of its sampling capacitor—is high-pass filtered with an order equal to the number of the preceding integrators in the modulator. This is done to reduce the opamps' equivalent loads at each phase. However, a minimum capacitive value of 50 fF was selected to avoid appreciable SNDR degradations due to capacitor mismatches. The first-integrator sampling capacitor is set to 150 fF, which together with an OSR of 200 is enough to keep the power of the thermal noise contribution below the IBE power required for GSM mode. Table 5.7 shows the in-band thermal noise power, including the budget devoted to the kT/C noise given by the sampling operation of all capacitances in the modulator and the equivalent input noise of the opamps. Initially, very large values of $80\mathrm{nV}/\sqrt{\mathrm{Hz}}$ are considered in all the modulator opamps for GSM, BT and GPS, with $40\mathrm{nV}/\sqrt{\mathrm{Hz}}$ for UMTS, DVB-H and WiMAX. These opamp input equivalent noise values are used to extract the values in Table 5.7 by behavioral simulations in SIMSIDES [Ruiz05]. However, only kT/C noise will be considered for the rest of the behavioral simulations in this section[13].

The dynamic requirements of the opamps are also simulated in SIMSIDES. These requirements are included in the SIMULINK model of the modulator through a set of values for the amplifier GB, SR, the equivalent loads, and switches on-resistances[14]. Note from Table 5.6 that very small capacitors are used in order to reduce the equivalent capacitor values (see Sect. 5.3.1).

Unity-STF ΣΔM topologies lead to the use of low-to-medium opamp gains with no appreciable SNDR degradation [Hamo06, Yao06]. This is specially beneficial

[12] Scaling refers here to the reduction of the capacitor value for minimum opamp loads up to the point at which thermal noise and mismatching start playing an important role.

[13] The thermal noise power extracted from this procedure, is later included in the modulator electrical model in SPECTRE via an input source. The data provided to this source is taken from a MATLAB routine that generates a temporal white-noise sequence for the corresponding noise PSD.

[14] The switches on-resistances slow down the dynamic operation of the opamps [Rio06].

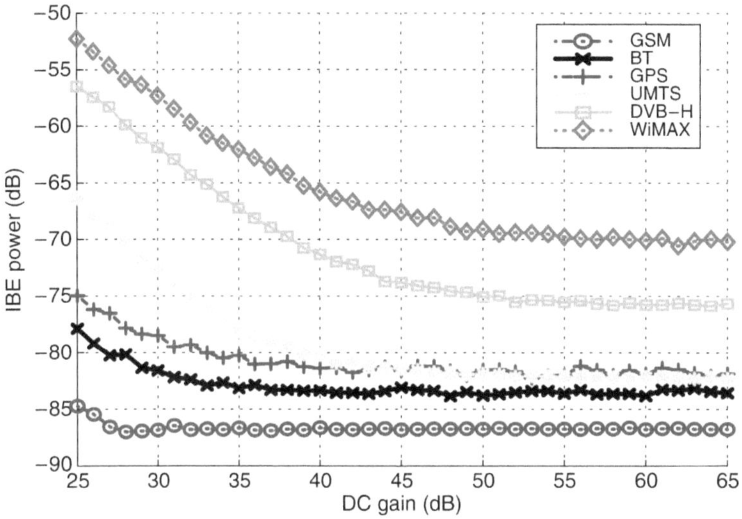

Fig. 5.9 Evolution of the in-band error power with the DC gain of the modulator opamps

with technological scaling, where the transistor intrinsic gain and, subsequently, that of the amplifier get lower and lower [Sans06]. Figure 5.9 illustrates this by depicting the IBE when the DC gain of all opamps in the modulator is varied. Note from Fig. 5.9 that only 2 or 3 dB of degradation in the IBE—compared to its minimum value—can be observed for DC gains of approximately 35 and 45 dB in the single-loop and cascade configurations, respectively.

Amplifiers gain non-linearities and SR are not considered at this step of the modulator synthesis since their effects are highly attenuated by the employed topology shown in Fig. 5.1 [Silv01, Silv04a, Silv04b, Goth02]. Note similar behavioral simulations, like the ones shown in the case study in Sect. 3.2 (see, for example, Fig. 3.22). These effects will be evaluated when the electrical building blocks are already designed.

5.3 Electrical Design of the Building Blocks

This section gives details on the design of the amplifiers and their capabilities to adapt their performance, the design of switches, capacitors and comparators. It then goes on to describe the implementation of the 1.5-bit quantizer and the electrical design of other auxiliary blocks, such as the clock phase generator.

5.3.1 Amplifiers

Figure 5.10 depicts the schematic of the amplifier. It consists of a folded-cascode topology in order to boost its speed operation—necessary for wideband modes such as WiMAX, while medium-to-low DC gains are needed. All the opamps in the

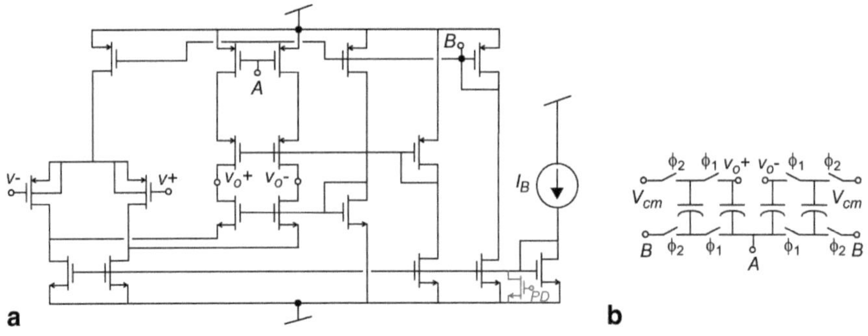

Fig. 5.10 Folded cascode amplifier: **a** core, **b** SC common-mode feedback net

modulator (OA_1-OA_4 in Fig. 5.4) share the same transistor sizes, the only difference being that the bias current of each amplifier can be adjusted independently of the others. Power-down transistors, shown in Fig. 5.10a in grey, are used in OA_3 and OA_4 in order to turn them off when the back-end stage is not needed. Note that a PMOS input differential pair is employed to cancel substrate body effect out and thus make the amplifier less sensitive to switching noise coupling coming from the modulator digital part [Arag99, Char01]. In addition, an SC common-mode feedback net is used to stabilize the amplifier output common-mode voltage as shown in Fig. 5.10b.

5.3.1.1 Adaptation of Amplifier Performance

The performance of each amplifier in the ΣΔM is adapted from one operation mode to another by varying its bias current (I_B in Fig. 5.10) independently of that of the others. Figure 5.11 illustrates the flexibility of the opamps, showing the adaptability of GB (Fig. 5.11a), output swing (OS) (Fig. 5.11b) and finite DC gain (Fig. 5.11c) as a function of the consumed power. These simulations include the values of the best, nominal and worst cases that result from a corner analysis, considering fast and slow device models, $\pm 5\%$ variation in the 1.2-V supply and temperatures in the range $[-40°\text{C}, +85°\text{C}]$. The power consumption of every opamp for each operation mode is directly related to its bias current, which also determines the corresponding SR and non-linear behavior. Non-linearities will be studied further below on, where opamp non-linear gain is examined as a function of the bias current.

Table 5.8 presents the resulting equivalent capacitive loads of the integrators during both sampling and integration phases. These capacitive loads—denoted as $C_{eq,s}$ and $C_{eq,i}$, respectively—are defined in [Rio06] as

$$C_{eq,s} = C_p + (C_L + C_{sn1} + \cdots + C_{snj})\left(1 + \frac{C_p}{C_i}\right)$$
$$C_{eq,i} = C_p + C_{s1} + \cdots + C_{si} + C_L\left(1 + \frac{C_p + C_{s1} + \cdots + C_{si}}{C_i}\right) \quad (5.9)$$

where C_p and C_L, C_i, C_{snj} and C_{si} stand for the input and output parasitic capacitances of the opamp, the integration capacitance, the sum of all the capacitances of the

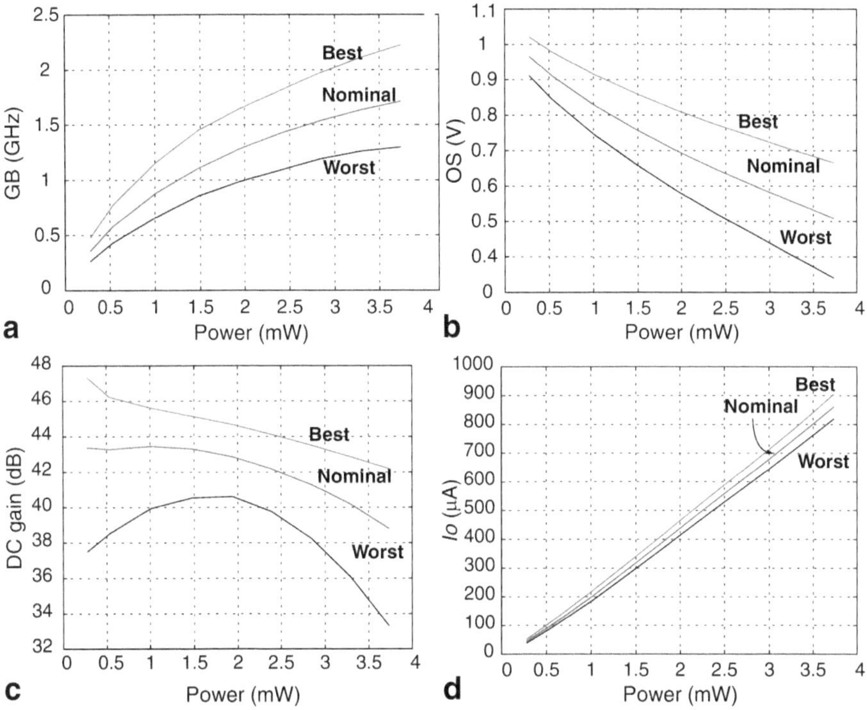

Fig. 5.11 Illustrating opamp performance adaptability on: **a** GB, **b** output swing, **c** DC gain, **d** output current

Table 5.8 Equivalent capacitor loads for each integrator

Eq. capacitance	*L*	Int. #1	Int. #2	Int. #3	Int. #4
$C_{eq,s}$	4	0.825	2.250	0.892	0.750
$C_{eq,i}$		0.360	0.390	0.755	0.390
$C_{eq,s}$	2	0.825	0.750		
$C_{eq,i}$		0.360	0.390		

sampling circuits connected to the corresponding integrator (i.e., SC branches of the following integrators and analog adders) and the sampling capacitors of the integrator, respectively.

The capacitor values given in Table 5.6, assuming parasitics capacitors of 100 fF at the opamp input and of 40 fF at the output, are used to obtain the data given in Table 5.8. Note that the modulator order also determines the equivalent sampling load of the second integrator since its output is connected to one of the sampling capacitors of the third integrator and another of the last-stage analog adder only in a MASH configuration, $C_{eq,s}$ being 3 times greater in the latter case compared to the single-loop configuration.

The equivalent loads for the second integrator in the 2-2 cascade $\Sigma\Delta$M are used to illustrate opamp dynamics performance as a function of bias current, as shown in

Table 5.9 Results for the amplifiers performance obtained by electrical simulation

I_B (μA)	5	10	20	30	40	50	60
DC gain (dB)	45.8	45.6	45.2	44.7	43.9	43.1	42.1
gm **(mA/V)**	0.77	1.27	1.95	2.43	2.82	3.13	3.38
GB$_i$ (MHz)	314	518	796	992	1151	1277	1379
GB$_s$ (MHz)	54	90	138	172	199	221	239
Phase margin	62.7°	65.6°	68.7°	70.4°	72.2°	73.1°	74.0°
Io **(μA)**	43	94	198	304	410	518	625
SR$_i$ (V/μs)	48	106	225	349	472	598	724
SR$_s$ (V/μs)	5	12	25	39	53	67	81
$C_{eq,i}$ **(pF)**				0.39			
$C_{eq,s}$ **(pF)**				2.25			
Output swing (V)	±0.96	±0.91	±0.82	±0.76	±0.69	±0.64	±0.59
Input cap. (fF)	74	77	79	81	82	83	84
Output cap. (fF)	26	20	14	12	10	9	8
Eq. input noise	9.6nV/Hz$^{1/2}$	8.0nV/Hz$^{1/2}$	6.9nV/Hz$^{1/2}$	6.5nV/Hz$^{1/2}$		6.3nV/Hz$^{1/2}$	
Power (mW)	0.27	0.53	1.00	1.50	1.90	2.35	2.80

Table 5.9. To this purpose, the on-resistance of the switches for the integrator whose equivalent loads are computed (R_{on}) and that of the next sampling SC branches—with a total sampling capacitance of C_{ns}—connected to the integrator output node (R_{non}) are taken into account in the GB and SR calculation, giving rise to the following figures—which take into account the opamp GB and SR, the on-resistances of the switches and the operation clock phases (both sampling and integration)—defined in [Rio06] as follows

$$\mathrm{SR}_i = \left(1 + \frac{C_p + C_s}{C_i}\right)\frac{I_o}{C_{eq,i}} \qquad \mathrm{SR}_s = \left(1 + \frac{C_p}{C_i}\right)\frac{I_o}{C_{eq,s}} \tag{5.10}$$

$$\mathrm{GB}_{s,on} = \frac{\mathrm{GB}_s}{1 + \mathrm{GB}_s \cdot 2R_{non}C_{ns}} \qquad \mathrm{GB}_{i,on} = \frac{\mathrm{GB}_i}{1 + \mathrm{GB}_i \cdot 2R_{on}C_s} \tag{5.11}$$

with GB_i and GB_s defined as

$$\mathrm{GB}_i = \frac{g_m}{C_{eq,i}} \qquad \mathrm{GB}_s = \frac{g_m}{C_{eq,s}} \tag{5.12}$$

Table 5.9 also includes other simulation results such as DC gain, transconductance (gm), current at the opamp output branch (I_o), output swing, input and output parasitic capacitances, equivalent input thermal noise and, finally, power consumption. Table 5.10 shows a polynomial approximation for the main parameter results of the amplifier as a function of the bias current I_B. Once the opamp has been already designed, these equations can be included in behavioral models in order to perform very fast parametric simulations with only one variable (I_B) per amplifier and used for verification purposes in the design phase. This procedure will be updated with opamp non-linearities further on.

Finally, it should be noted that the I_B of each amplifier—and comparator preamplifier—is adapted through an adjustable off-chip network of potentiometers. This will be addressed in Sect. 5.3.7.

Table 5.10 Polynomial approximation of the main amplifier features for the nominal case

C_{out} (fF)	$1.19 \cdot 10^{27} \cdot I_B^6 - 4.63 \cdot 10^{23} \cdot I_B^5 + 7.01 \cdot 10^{19} \cdot I_B^4 - 5.22 \cdot 10^{15} \cdot I_B^3 + 2.00 \cdot 10^{11} \cdot I_B^2 - 3.88 \cdot 10^6 \cdot I_B + 43.1$
C_{in} (fF)	$-9.47 \cdot 10^{17} \cdot I_B^4 + 2.58 \cdot 10^{14} \cdot I_B^3 - 2.34 \cdot 10^{10} \cdot I_B^2 + 9.33 \cdot 10^5 \cdot I_B + 68.71$
A_{DC} (dB)	$-7.49 \cdot 10^8 \cdot I_B^2 - 2.92 \cdot 10^4 \cdot I_B + 46.37$
I_0 (μA)	$6.10 \cdot 10^9 \cdot I_B^2 + 1.02 \cdot 10^7 \cdot I_B - 6.26$
gm (mA/V)	$4.11 \cdot 10^{12} \cdot I_B^3 - 1.08 \cdot 10^9 \cdot I_B^2 + 1.02 \cdot 10^5 \cdot I_B + 0.24$
OS^+ (V)	$-4.03 \cdot 10^{11} \cdot I_B^3 + 9.65 \cdot 10^7 \cdot I_B^2 - 1.16 \cdot 10^4 \cdot I_B + 1.02$
OS^- (V)	$4.03 \cdot 10^{11} \cdot I_B^3 - 9.65 \cdot 10^7 \cdot I_B^2 + 1.16 \cdot 10^4 \cdot I_B - 1.02$

5.3.1.2 Opamp Non-linear Gain

Figure 5.12 shows the evolution of the amplifier output with the input voltage for different bias currents. A zoom over the central region covers an output voltage range slightly larger than the maximum given by the behavioral simulations in Fig. 5.2a (±0.35 V). It can be observed that an increase of I_B entails a degradation of the amplifier linearity. However, the reduced amplifier output swings (0.35 V in the worst case) lead to high linearity in the whole range of output voltages for bias currents below 50 μA. This is better observed in Fig. 5.13 where the opamp gain is depicted versus its output voltage.

In general, a set of several input voltages should be considered in order to depict SNDR curves for all operation modes, while also exploring the different values of

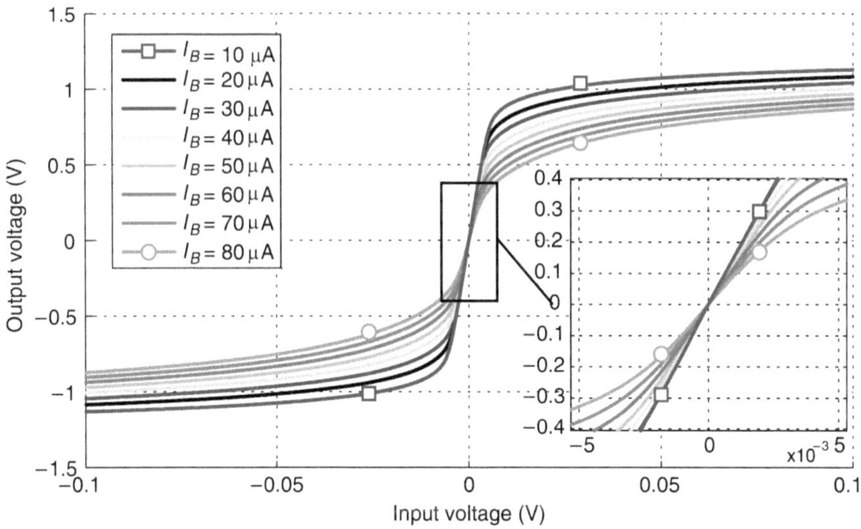

Fig. 5.12 DC transfer characteristic

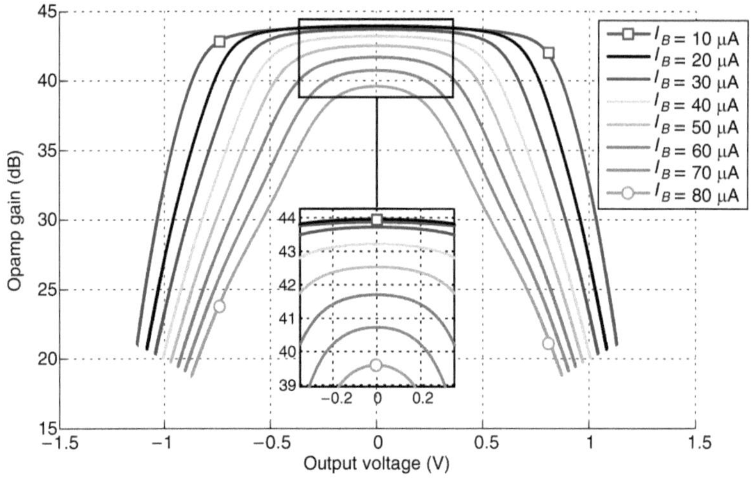

Fig. 5.13 Opamp gain versus its output voltage

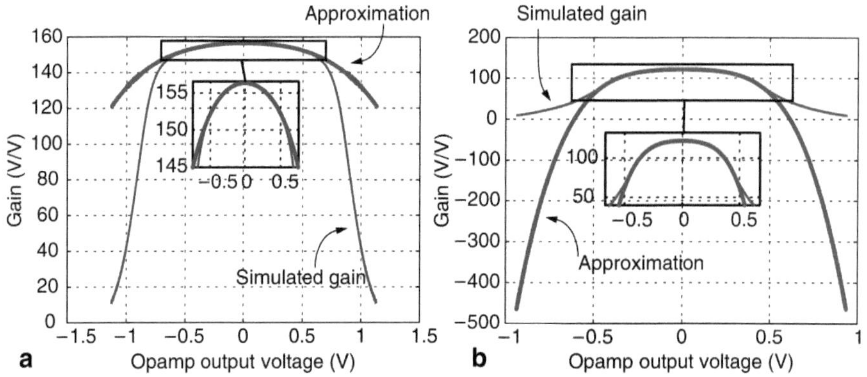

Fig. 5.14 Fourth-order polynomial approximation to opamp gain for: **a** $I_B = 10$ μA, **b** $I_B = 60$ μA

I_B for the four amplifiers in the modulator one by one. Such a huge design space cannot be explored based on transient simulations in order to have a reasonable time to market for a product based on these systems. Exploration must be based on behavioral simulations because of the simulation speed improvement of the last case [Wolf97]. The impact of gain non-linearities can therefore be included in behavioral simulations for evaluating their effect in very fast simulations. In our case, SIMSIDES [Ruiz05] is used, which allows the designer to include opamp gain non-linearities with a fourth-order polynomial equation as defined in (3.23).

A fitting function in natural units is used for the gain approximation based on (3.23) for the whole amplifier output swing obtained by behavioral simulations (± 0.35 V). Figure 5.14 illustrates this for $I_B = 10$ μA and $I_B = 60$ μA.

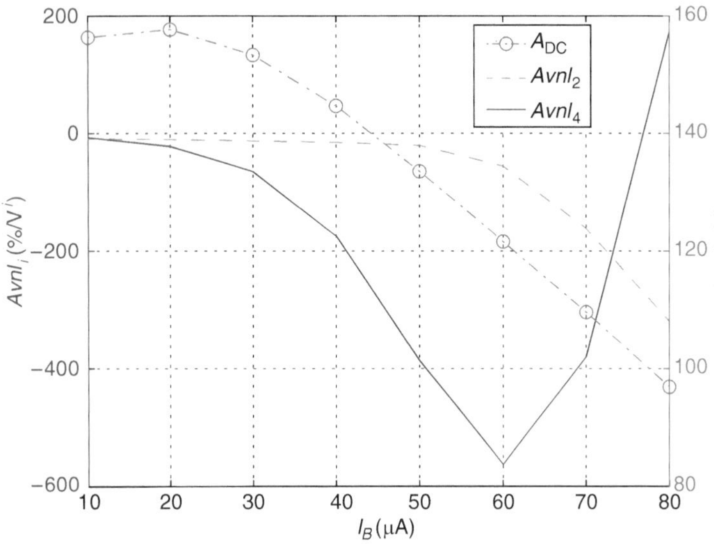

Fig. 5.15 Gain coefficients variation with bias currents

Table 5.11 Polynomial approximation for coefficients in (3.23)

Av	$2.02 \cdot 10^{-4} \cdot I_B^3 - 3.77 \cdot 10^{-2} \cdot I_B^2 + 1.07 \cdot I_B + 149.38$
$Avnl_2$ (%)	$6.530305 \cdot 10^{-8} \cdot I_B^6 - 1.602393 \cdot 10^{-5} \cdot I_B^5 + 1.48553 \cdot 10^{-3} \cdot I_B^4$ $-6.72407 \cdot 10^{-2} \cdot I_B^3 + 1.56177 \cdot I_B^2 - 17.6448 \cdot I_B + 65.3041$
$Avnl_4$ (%)	$-4.6137 \cdot 10^{-7} \cdot I_B^6 + 1.1775 \cdot 10^{-4} \cdot I_B^5 - 1.1329 \cdot 10^{-2} \cdot I_B^4$ $+ 5.2459 \cdot 10^{-1} \cdot I_B^3 - 12.5180 \cdot I_B^2 + 143.2500 \cdot I_B - 61.0220$

Figure 5.15 shows the evolution of A_{DC} and $Avnl_i$—respectively defined as opamp DC gain and non-linear gain factors in (3.23)—with I_B. Note that odd-order non-linear coefficients are zero in the approximation; in other words, the approximation consists of even-order polynomials. Table 5.11 gives the extracted symbolic expressions to obtain A_{DC} and $Avnl_i$ as a function of I_B. Once this methodological procedure has also been completed for the amplifier non-linear gain, the main amplifier features obtained by electrical simulation at the transistor level can be approximated with the equations in Tables 5.10 and 5.11. They can then be included in behavioral simulations, with I_B being the only parameter to be swept for each amplifier.

Behavioral and transistor-level simulations suggest that good results are obtained with bias currents of 42.5 μA, and 34 and 19 μA for the first-stage amplifiers, and the third and fourth amplifiers, respectively in WiMAX mode. The bias currents of the amplifiers in the remaining modes are always below 40 μA. Furthermore, Fig. 5.13 shows that low currents need to be selected for high-linear opamp response but low speed (see Table 5.9). High currents, however, are used for higher opamp speed requirements but worse linearity. This trade-off is convenient for our purposes since, on the one hand, low currents are used for narrowband standards where high linearity and low speed are needed. Besides, power consumption is reduced in this case. On the other hand, wideband standards, in which speed is the main feature to be exploited and the power budget is greater, non linearities are not so influential given the lower resolution requirements.

5.3.2 *Capacitors*

Metal-oxide-metal (MoM) structures were used to implement the capacitors in the modulator for several reasons. First, this type is more area efficient than other solutions such as MiM capacitors. Second, this capacitor structure is compatible with standard digital CMOS processes—unlike, for instance, MiM capacitors that need specific layers and, thus, additional technological steps during manufacture, available only in the mixed-mode/RF process of the employed technology. MoM capacitors rely on coupling capacitance between metal fingers running in parallel. Metal layers are stacked from metal 1 to upper metal layer to increase capacitance. The 90-nm technology used for this prototype allows the use of a minimum number of metal layers of 3 and a maximum number of 6. In this design, all the 6 available metals are used for maximum capacitance density. The width and spacing of the metal fingers are fixed to the minimum dimensions; i.e., 0.14 μm, in order to achieve maximum capacitance density. The number of fingers per metal layer is limited to an even number in order to obtain a symmetric capacitor.

This kind of capacitor exhibits good matching (less than 0.1% for the selected unit capacitors) and small bottom-plate parasitics (less than 5%). Only 2×35 unit capacitors are needed—as suggested by Table 5.4—to implement the $\Sigma\Delta$ modulator. Figure 5.16 illustrates how the unit capacitors are arranged. The arrangement is based on common-centroid arrays and closely placed units with the extensive use of dummies to ensure similar surroundings for each unit capacitor[15].

To evaluate the influence of capacitance mismatches, behavioral simulations are conducted with SIMSIDES [Ruiz05]. Figure 5.17 illustrates how the SNDR curves change for a 100-run Monte Carlo simulation performed for the architecture in Fig. 5.4 when a 0.1% standard deviation is considered for all the capacitors that conform the actual SC implementation. The impact on the peak SNDR is also zoomed in the figure. Figure 5.18 shows the variation of the peak SNDR against capacitor mismatches for GSM and WiMAX modes. As expected, the impact of capacitor mismatches is greater for WiMAX—in which quantization error dominates and noise leakages influence the performance of the cascade—than for GSM—in which thermal noise dominates the IBE of the single-loop modulator.

[15] In order to keep a perfect common-centroid structure for the first analog adder, the unit element of C_{a1} that is farthest from C_{in1} in the same row should have been placed at the location of C_{in1}, and C_{in1} at the gravity center of the structure. However the approach in Fig. 5.16 is recommended because C_{a1} is made up of 8 elements and C_{in1} of only 1, and placing C_{in1} at the actual structural center will entail an increase of the structure and, thus, in the area as well. Therefore, C_{in1} is placed as close as possible to the gravity center of the structure, while C_{a1} only has 1 out of 8 elements that is not correctly placed for a common-centroid distribution.

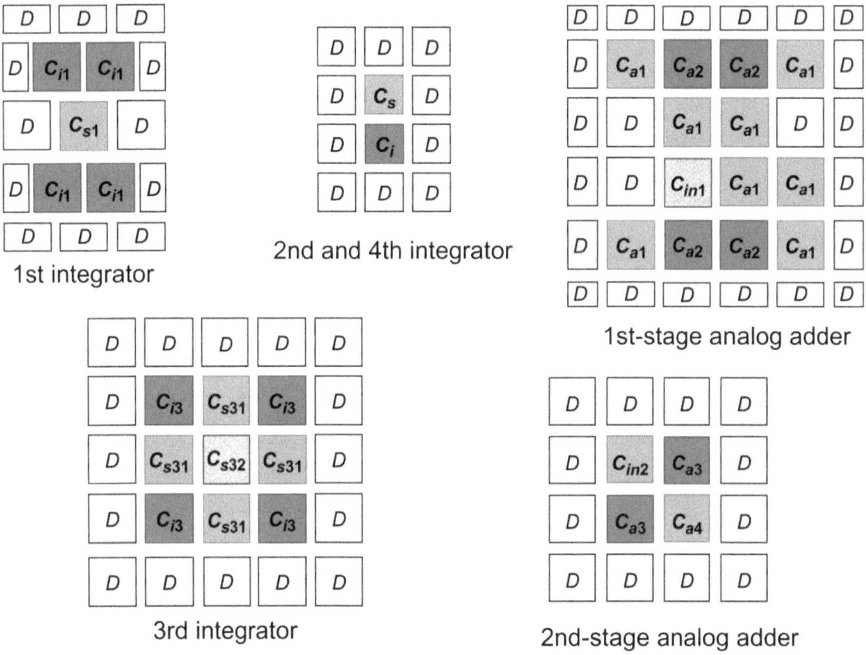

Fig. 5.16 Representation of the layout of unit capacitors (*D* stands for dummy)

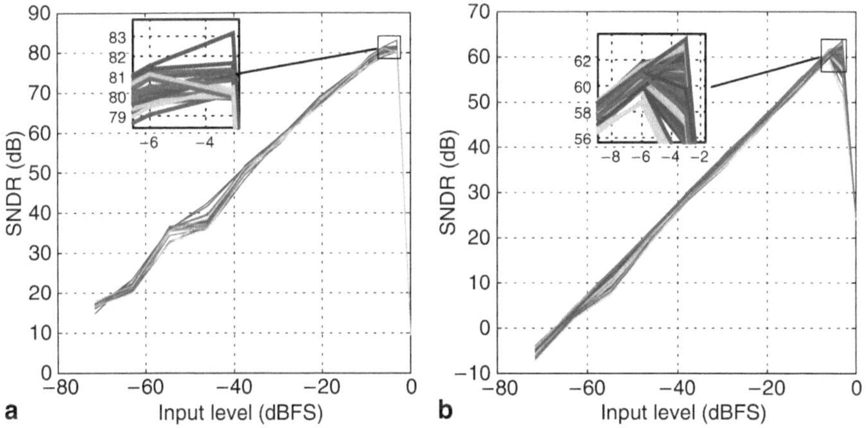

Fig. 5.17 SNDR variation with a 0.1% standard deviation in all capacitors for: **a** GSM, **b** WiMAX

5.3.3 Switches

All switches used in the modulator are standard-threshold CMOS transmission gates, with no need for boot-strapping. They have been designed taking into account the trade-off between the non-linear on-resistance and its associated drain/source parasitic capacitances (C_{dd}/C_{ss}). Figure 5.19 shows the on-resistance variation with the DC voltage level. A ratio of $W_{\mathrm{PMOS}}/W_{\mathrm{NMOS}} = 3.125$ leads to a resistance plot

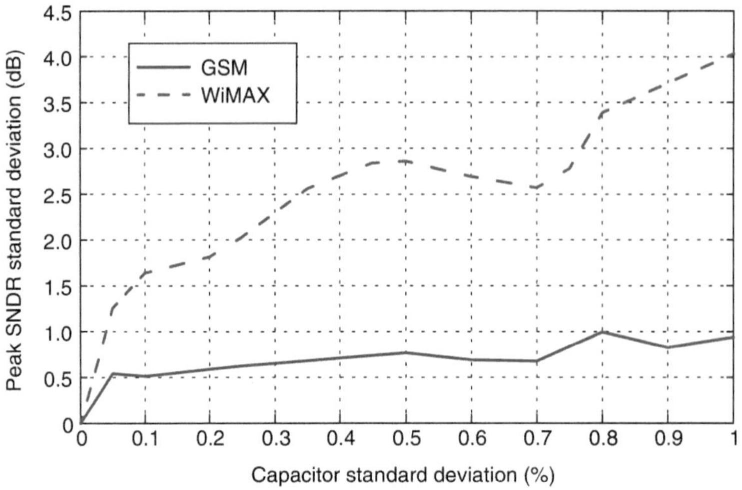

Fig. 5.18 Effects of capacitor mismatches on SNDR

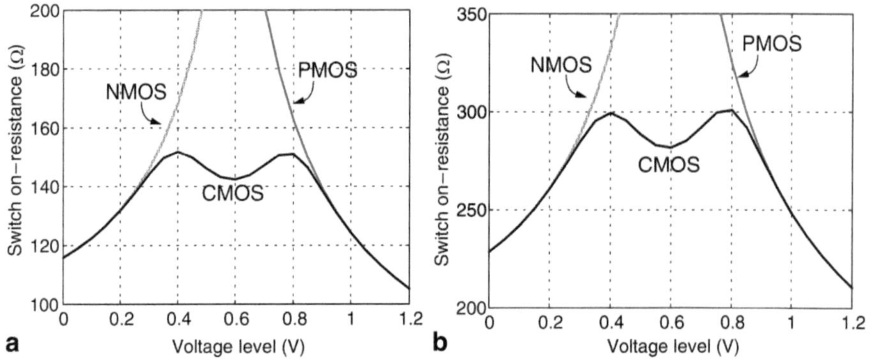

Fig. 5.19 On-resistance versus voltage level: **a** $W_{PMOS} = 12\,\mu m$, **b** $W_{PMOS} = 6\,\mu m$

centered on the common-mode voltage (0.6 V). A switch on-resistance with a mean value of 147 Ω can be tolerated in the first, third and fourth integrators, while values in the range of 285 Ω can be used for the second integrator and analog adders to give low parasitic capacitances. All switch transistor sizing and the corresponding electrical simulation results are summarized in Table 5.12.

Switch linearity has been evaluated following the procedure described in Sect. 4.4.3. To this purpose, the schematic in Fig. 4.15b is simulated at transistor-level, considering an input sinewave with an amplitude equal to the reference level and a frequency at 1/3 of the considered standards BW. Figure 5.20 depicts the results for GSM, where high linearity is required within a narrow BW, and WiMAX, where resolution is low but the signal band is wide. Switches with $W_{PMOS} = 12\ \mu m$ are considered for these simulations and the obtained THD is equal to −125.9 dB for GSM and −82.5 dB for WiMAX. Thus, the resulting harmonic distortion is well below that required by the modulator operation.

Table 5.12 Switch sizing and electrical simulation results

	Int. #1	Int. #2	Int. #3	Int. #4	Adder #1	Adder #2
W_{PMOS} (μm)	12	6	12		6	
W_{NMOS} (μm)	3.84	1.92	3.84		1.92	
L_{CMOS} (nm)	80					
R_{on} (Ω)	147	285	147		285	
C_{dd} (fF) C_{ss} (fF)	30.63	15.46	30.63		15.46	

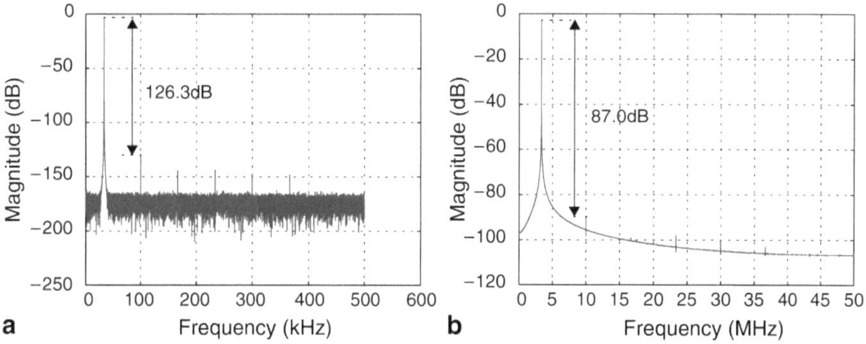

Fig. 5.20 Distortion due to non-linear sampling for an input tone of 0-dBFS input power: **a** GSM, **b** WiMAX

5.3.4 *Comparators*

The comparator comprises three blocks; namely, a preamplifying first stage, a regenerative latch and a set-reset latch to store the voltage given by the former latch. The schematics of the preamplifiers and the regenerative latch are different from those used in the comparators in Chap. 4. The current regenerative latch operates faster (approximately 10 times), as can be observed by comparing the electrical simulation results in Sect. 4.4.4 and those presented in this section.

Figure 5.21 depicts the scheme of the preamplifying stage. Its design aims to fulfill several goals; namely, to obtain a high DC gain—since this helps to reduce the comparator input referred offset—, a low kick-back noise on the modulator and high speed, while retaining a low parasitic input capacitance. This capacitance plays an important role in the quantizer scaling factor α as shown in (5.7). In fact, the larger its value, the lower α and, consequently, the more demanding the comparator offset requirements too. This last trade-off shows that augmenting the preamplifier DC gain should not be done exclusively by increasing the size of the input differential pair transistor (M_1 in the figure)—in order to increase gm—for a given bias current (I_{B}), because this would imply an increment of the preamplifier input parasitic capacitance as well. It is therefore very important to obtain a high output resistance. To this

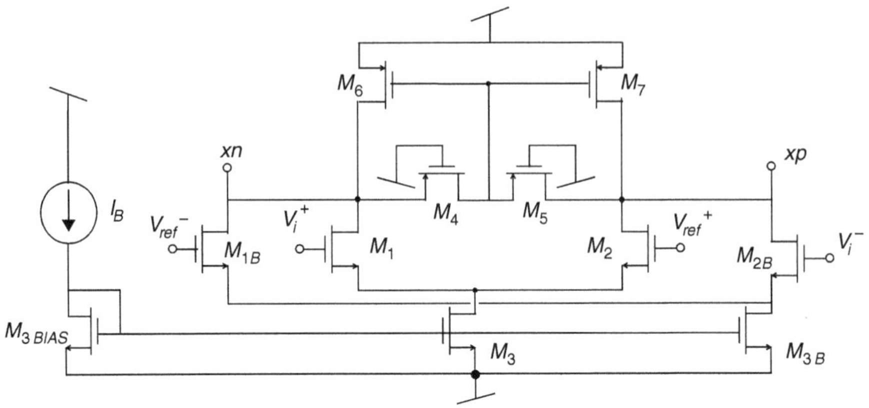

Fig. 5.21 Comparator preamplifier schematic

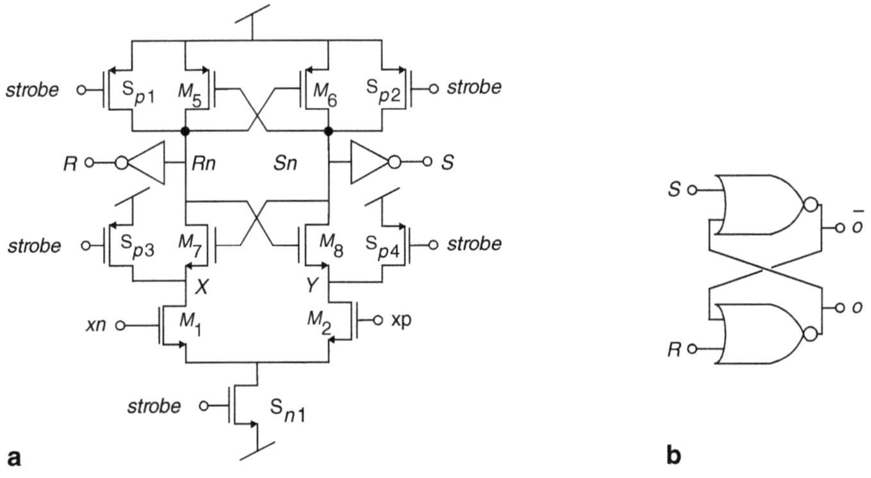

Fig. 5.22 Comparator schematic: **a** regenerative latch, **b** set-reset latch

purpose, transistors M_4 and M_5 are included. These transistors are used as resistors—biased, thus, in linear region—with a high mean resistance of 100 kΩ, given by a very low aspect ratio of 0.2.

The outputs of the preamplifier are connected to the regenerative latch and later to the set-rest latch to keep the previously compared value stable. Figure 5.22 shows the schematic of both the regenerative latch [Wang00] and the set-reset latch. The regenerative latch works as follows: let us suppose that *strobe* is initially at ground voltage, so that switch S_{n1} is off, switches S_{p1}, S_{p2}, S_{p3} and S_{p4} are on, and the internal nodes R_n, S_n, X, and Y are pre-charged at the supply voltage. When *strobe* triggers to a logic one, S_{n1} turns on and transistors M_1 and M_2 will process the differential output of the preamplifier. After a short time, one of the input differential pair transistors will turn off depending on the sign of the input voltage imbalance. Given that transistors M_5-M_8 are initially off, the resulting differential current will

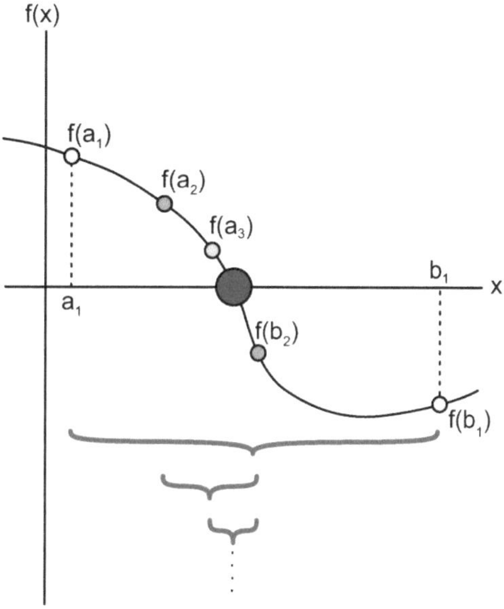

Fig. 5.23 Illustration of the bisectional method algorithm

first flow through the total capacitance present at nodes *X* and *Y*, and will therefore create a differential voltage between both nodes. Meanwhile, M_7 and M_8 will turn on and, later, the cross-coupled scheme will start working, leading to a positive feedback effect and, thus, to the regeneration of the initial imbalance between *xp* and *xn*.

This comparator presents very appealing features for high-speed, low-power and low-offset performance. On the one hand, there is no static power consumption for the regenerative latch due to the use of switch S_{n1}. On the other, only one strobe signal is needed. In addition, the offset of the regenerative latch is dominated by that of its input differential pair. However, its impact is reduced by the preamplifying stage of the comparator, which will also help to lower kick-back noise [Domi03, Figu09].

5.3.4.1 Offset Computation

The method used in Chap. 4 to extract the comparator offset is based on an input ramp waveform. A more efficient method in terms of simulation time is presented here. The bisection method is employed to characterize the input offset of the comparator. This method is a root-finding algorithm which works by repeatedly halving an interval and, then, selecting the subinterval in which the root exists [Pres92]. Figure 5.23 shows the algorithm graphically. Let us suppose that the equation to be solved is $f(x)=0$. Given two points, *a* and *b*, such that f(*a*) and f(*b*) have opposite signs, the intermediate value theorem says that f(*x*) must have at least one root in the interval [*a*, *b*] as long as f(*x*) is continuous [Bolz17]. The bisection method divides the interval into two by computing its midpoint $c = (a + b)/2$. At this point, there are two possibilities: either f(*a*) and f(*c*) have opposite signs, or f(*c*) and f(*b*) have opposite signs. The bisection algorithm is then applied to the sub-interval where the sign change occurs until a solution under a defined tolerance is obtained.

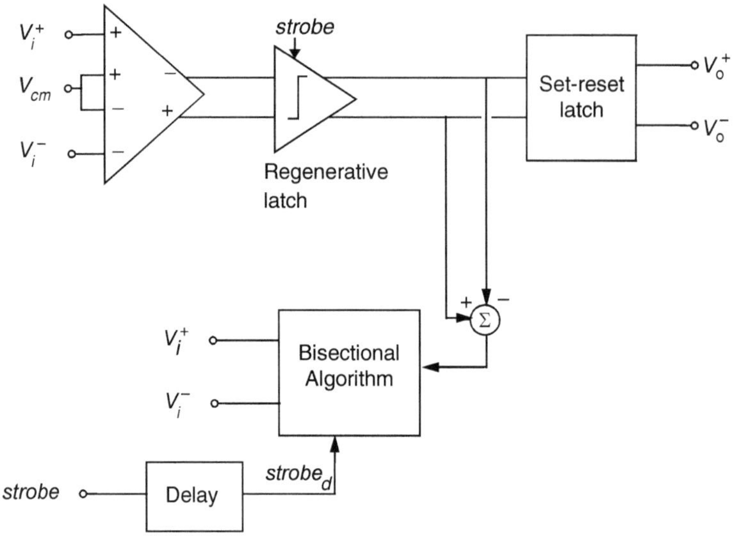

Fig. 5.24 Test-bench for the comparator offset based on the bisectional algorithm

Therefore, if f(x) is a continuous function on the interval $[a, b]$ and $f(a) \cdot f(b) < 0$, then the bisection method converges to a root of f(x). However, the bisection method gives only a range where the root exists, rather than a single solution. The midpoint is the best estimated root location in the smallest interval found. In that case, the absolute error after n steps is at most

$$\frac{|b-a|}{2^{n+1}} \tag{5.13}$$

If, on the contrary, an extreme point of the interval is used, the maximum absolute error is

$$\frac{|b-a|}{2^{n}} \tag{5.14}$$

These formulae can be used to determine in advance the number of iterations that the bisection method would need to converge to a root within a certain tolerance ε. Taking into account (5.14)—the worst case of the two approaches—, the number of iterations n can be defined as

$$n = \log_2\left(\frac{b-a}{\varepsilon}\right) \tag{5.15}$$

to ensure that the error is smaller than the tolerance ε.

Figure 5.24 shows the test-bench for the comparator offset computation. The preamplifier and the regenerative latch are placed inside a feedback loop, in which the preamplifier input signal is controlled by a block that implements the bisectional algorithm described above. A verilogA [Kund04] code is used for this purpose. The input of this block is the differential output of the regenerative latch, so the algorithm

Table 5.13 Results for the comparator performance obtained by electrical simulation

		Gain (dB)		Offset (mV)	
		mean	sigma	mean	sigma
		10.12	0.14	0.38	3.25
t_{lh} (ps)			t_{hl} (ps)		
mean	sigma	worst case	mean	sigma	worst case
234.89	5.55	315.00	183.26	5.95	262.00

basically selects the next input voltage based on the comparator output, following the procedure depicted in Fig. 5.23. Note that a delay is needed for the comparator strobe. The delayed strobe signal ($strobe_d$) triggers the block corresponding to the bisectional algorithm. This time delay must be longer than the response time of the regenerative latch but lower than half of the cycle of the comparator strobe. To sum up, the block that implements the bisectional algorithm imposes an input to the preamplifier that gets closer to the comparator offset with each iteration cycle.

This test-bench was included in a 200-run Monte Carlo simulation in order to take into account transistor mismatches and process variations with a capacitance load of 25 fF[16]. Besides, a corner analysis considering fast and slow device models, $\pm 5\%$ variations in the 1.2-V nominal supply and temperatures in the range of [-40°C, 85°C] was also performed. Table 5.13 shows the values obtained for the mean and standard deviation (sigma in the table) values of the preamplifier gain, the input comparator offset and the low-to-high and high-to-low comparator response times—t_{lh} and t_{hl}, respectively. The worst-case values resulting from the corner analysis are also included. Note that the comparator operates very fast; indeed, the worst-case maximum time response of the comparator is roughly 13 times lower than the time period of the WiMAX operation clock speed (240 MHz).

5.3.5 *A/D/A Conversion*

1.5-bit quantizers are employed in both stages of the cascade to digitize the corresponding analog adder outputs and convert them back to the analog domain in the DACs. The thermometer code quantizer output is then translated into a 1-of-3 code (d_{0-2}) as per the conversion scheme in Fig. 5.25. Former signals control the DAC and decide which of the 3 levels (+1.2 V, 0 V or -1.2 V) will be fed back and integrated. The output of every quantizer in the modulator is binary encoded and stored in a ROM.

5.3.5.1 ADC

Figure 5.26 shows the flash architecture of the 3-level quantizer that compares the analog adder output ($V_i^+ - V_i^-$) with a differential voltage reference given by

[16] This value corresponds to the input parasitic capacitances of the DAC digital gates driven by the quantizer outputs.

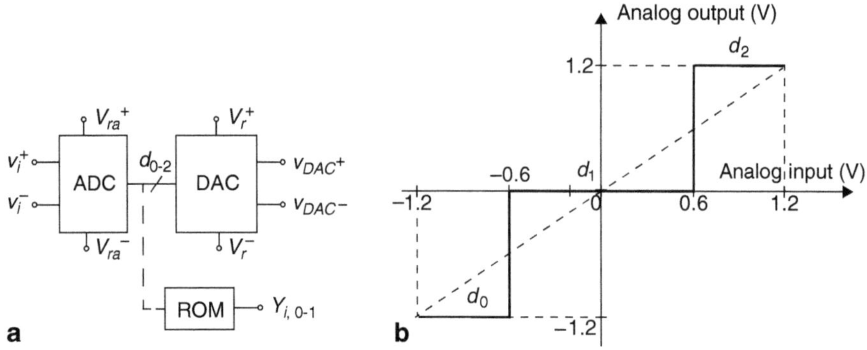

Fig. 5.25 A/D/A conversion: **a** block diagram, **b** I/O characteristic

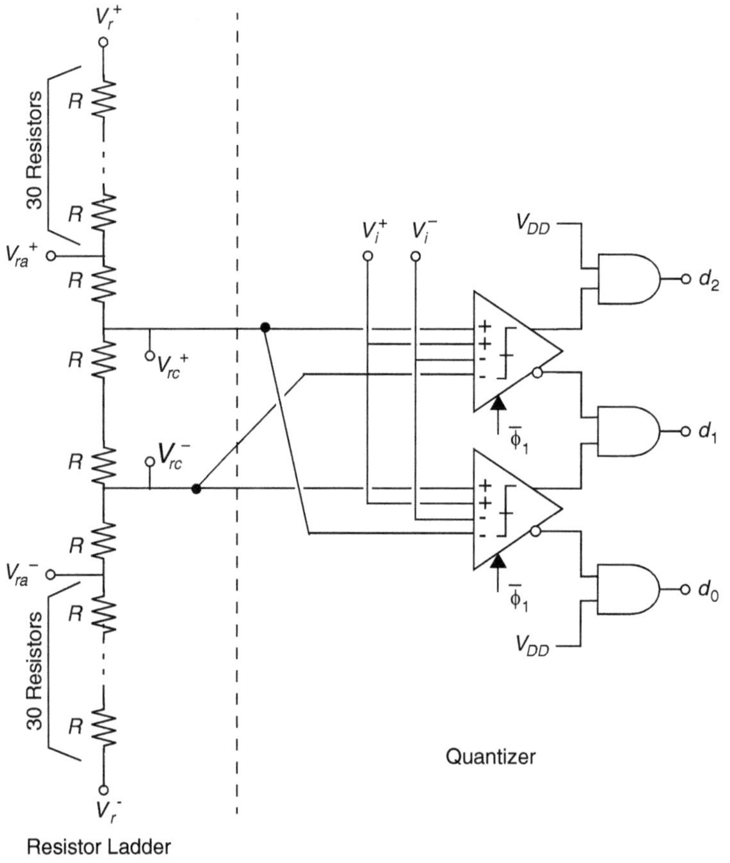

Fig. 5.26 1.5-bit quantizer

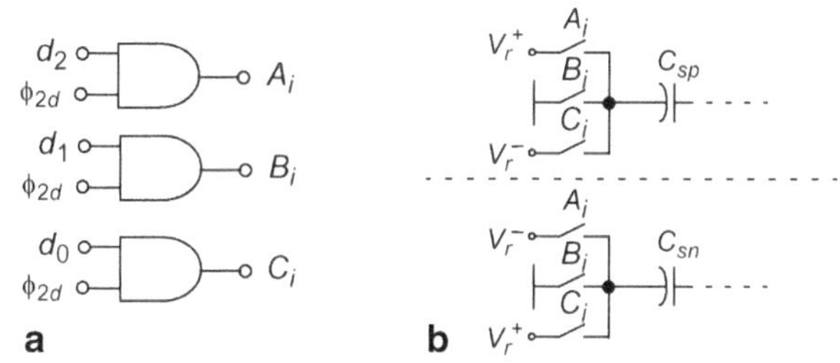

Fig. 5.27 1.5-bit DAC: **a** control gates, **b** D/A feedback operation in the first integrator of the corresponding modulator stage

the differential tabs of the resistor ladder—$(V_{rc}^{+} - V_{rc}^{-})$ and $(V_{rc}^{-} - V_{rc}^{+})$ for the first and second comparators, respectively. Note that there is compression over the references $(V_{ra} = V_{ra}^{+} - V_{ra}^{-})$ given by a resistive ratio of 2-to-32, as mentioned in Sect. 5.2.1.

Comparators in the quantizer are similar to those described in Sect. 5.3.4. Two input differential pairs are used at the pre-amplifying stage in order to operate with two fully-differential inputs; namely, the quantizer input and the compressed references coming from the resistor ladder. The resistor ladder uses 64 segments of 95.9-Ω unit resistors laid out with unsalicided p^{+} poly and connected between the differential reference voltages ($V_r^{+} = 1.2V \text{and} V_r^{-} = 0\text{V}$) thus providing a differential full scale of 2.4 V with a current consumption of 195.5 μA.

5.3.5.2 DAC

Figure 5.27a shows the scheme of the DAC control gates that enable the feedback operation in the first integrator of each modulator stage as depicted in Fig. 5.27b. The feedback voltage to be sampled by the corresponding capacitor at the top of the fully differential implementation (C_{sp} in Fig. 5.27b) is V_r^{+}, V_{cm} or V_r^{-} when the digital code activated by the ADC is d_2, d_1 or d_0, respectively. On the contrary, the capacitor at the bottom (C_{sn} in Fig. 5.27b) samples V_r^{-}, V_{cm}, or V_r^{+} when the digital code activated by the ADC is d_2, d_1 or d_0, respectively. These sampling operations are performed at phase ϕ_{2d} as shown in Fig. 5.27b and described in Sect. 5.2.2.

5.3.6 Clock Phase Generator

Figure 5.28 shows the schematic of the clock phase generator used to produce the non-overlapped and delayed phases illustrated in Fig. 5.4. The schematic of this generator includes a PMOS transistor M_1 inside two of the inverters in order to eliminate the delay at the rising edge between ϕ_i and ϕ_{id} phases, while the delay at the falling edge remains [Park01, Silv04b]. The integration and sampling operations in the ΣΔM are controlled by these phases (and their delayed and complementary versions) which are generated from the schematic in Fig. 5.28 with an input clock signal provided by an external off-chip equipment.

A U-shaped bus is also employed in this prototype to distribute the generated clock phases, their complementary signals, the DAC control signals (A_i, B_i and C_i in Fig. 5.4) and the modified phases that allow the modulator last stage to switch off

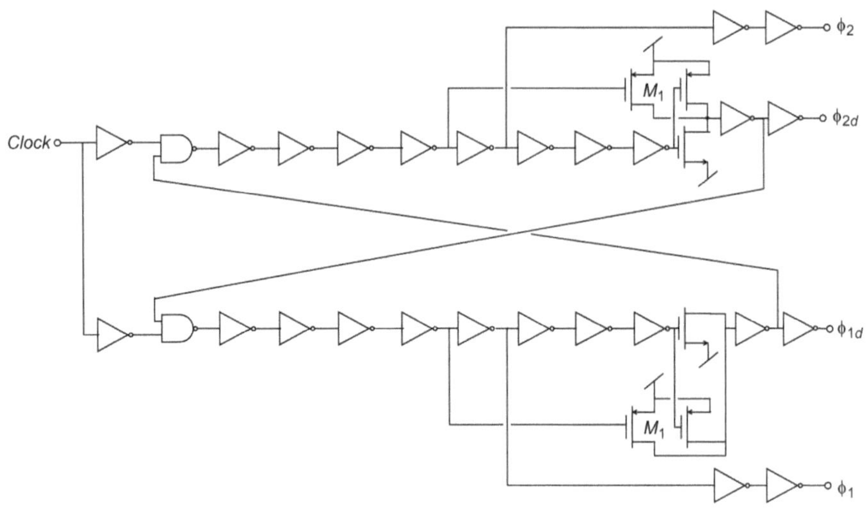

Fig. 5.28 Clock phase generator

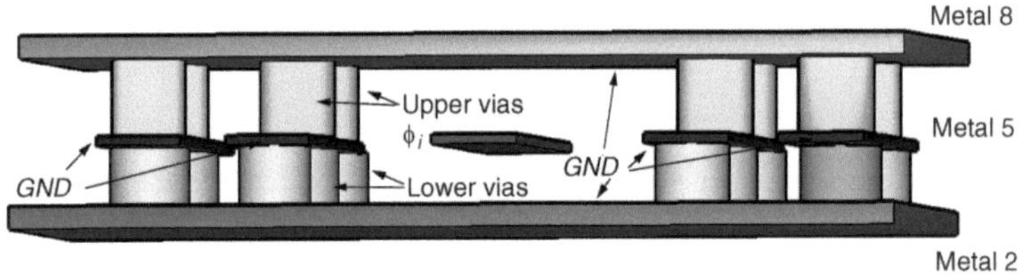

Fig. 5.29 3D illustration of the physical distribution of one isolated clock phase

in order to save power (ϕ_i', ϕ_{id}' and ϕ_i'' in Fig. 5.4). However, the isolation of each signal in this prototype is based on the connection of two parallel ground walls on each side of the signal metal strip, being closed at the top and the bottom by two long horizontal ground planes. In theory, only one metal strip running in parallel to the signal line on each side and connected to the horizontal planes with the necessary vias is enough for this kind of isolation. Unfortunately, this is not compatible with the technological DRC rules because of practical density problems with the metal vias. Two parallel metal strips on each side are therefore necessary to perform the isolation. Altogether, Fig. 5.29 shows a bus section for the distribution and isolation of one signal (ϕ_i in the figure). This signal line and the parallel dedicated ground strips (*GND* in the figure) are laid out in metal 5, while two isolation planes are implemented in metal 8 and 2 at the top and bottom of the structure, respectively. Two parallel ground lines on each side of the signal—thus making four ground lines in total for isolation—are used to alternate vias from the bottom ground plane in metal 2 to the ground strip in metal 5, on the one hand, and from metal 5 to the top plane in metal 8, on the other. If scattering effects are neglected between vias slots in

Table 5.14 Parasitics for the lumped LCR model in Fig. 4.19d

Modulator stage	Worst-case bus length	L_T (pH)	C_T (fF)	R_T (Ω)
First	1mm	440	76	110
Second	750μm	350	60	82.5

Table 5.15 Estimated loads for each clock phase

	$\overline{\phi}_{1d}$	ϕ_{1d}	$\overline{\phi}_{2d}$	ϕ_{2d}	$\overline{\phi}_1, \overline{\phi}_2$	ϕ_1, ϕ_2		
C_L (fF)	149.1	45.7	87.2	31.3	91.8	47.1		
	$\overline{\phi}'_{1d}$	ϕ'_{1d}	$\overline{\phi}'_{2d}$	ϕ'_{2d}	$\overline{\phi}'_1, \overline{\phi}'_2$	ϕ'_1, ϕ'_2	$\overline{\phi}''_1, \overline{\phi}''_2$	?"$_1$, ϕ''_2
C_L (fF)	184.7	52.3	67.6	66.0	100.0	39.1	31.7	11.8

different ground lines in metal 5, in practice a complete Faraday box will be obtained for each signal ϕ_i despite the density issues of the metal vias. However, metal 4 is used for making the connections of the modulator circuitry to the clock phases. In these regions, therefore, vias from metal 3 to metal 5 need to be removed.

A spacing of 0.49 μm between adjacent dedicated ground lines is also maintained to avoid DRC errors due to proximity between vias. In addition, there is a distance of 1.25 μm between the main signal line and the surrounding ground metal strips in order to reduce coupling capacitance. However, 31 signals are routed together in the bus plus the isolating ground lines. This entails a very large area consumption as will be discussed in Sect. 6.4.4.

To extract the equivalent LCR model, a similar approach to the procedure followed in Chap. 4 is applied here as well, the only difference being the inclusion of two different LCR lumped models. One is used for the last modulator stage, which will be laid physically close to the generation of the phases[17], and another is considered for the first stage, which will be farther along. The estimated routing distances of both models are 750 μm and 1 mm, respectively, in which a safety margin of 150 μm is included in both cases to allow for future larger layouts in practice. The resulting parameters of both LRC models are shown in Table 5.14.

The estimated capacitance loads driven by each phase are given in Table 5.15. Note that the corresponding sums of $C_L + C_T$ result in low capacitance values, so the selected phase buffers are library-based. Buffers with fan-outs of 40 and 32 are used for the first- and second-stage clock phases, respectively.

Post-layout simulations have been performed which include not only the clock phase generation but also the buffers and the LCR lumped model in Fig. 4.19d. The effective operation times for the sampling phase are 97.5 and 84.4% of half the clock period for clock frequencies of 40 and 240 MHz, respectively. The respective values

[17] The placement of the modulator last stage close to the clock phase generation helps to reduce the influence of the switching noise coupling on the ΣΔM performance. This will be addressed in Sect. 5.4.

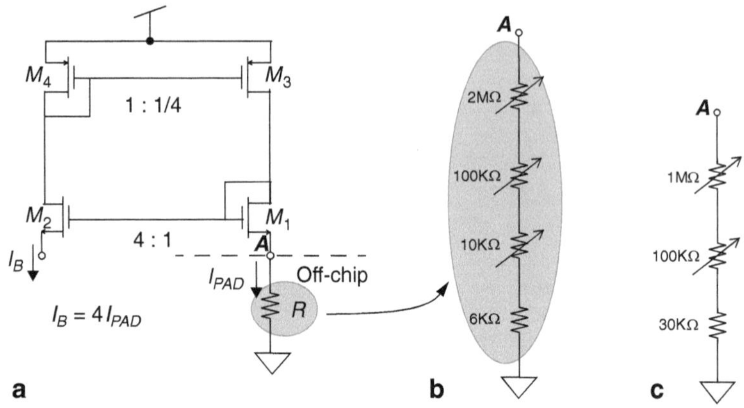

Fig. 5.30 Adaptable bias current generation. **a** schematic, **b** opamp-biasing off-chip resistor, **c** comparator preamplifier-biasing resistor

for the integration phase are 98.2 and 89.0%. The resulting non-overlapping time and the falling phase delays are approximately 80 and 78 ps, respectively.

5.3.7 *Bias Current Adaptation*

The performance of every opamp and all the comparator preamplifiers of each quantizer can be adapted by adjusting the corresponding bias current. To this purpose, the master bias current generator conceptually shown in Fig. 5.30a is used. A set of off-chip resistors is necessary to externally generate the desired currents that are later mirrored to bias all the basic building blocks of the analog core. To adapt the bias current, and, consequently, the power consumption of both the amplifiers and the comparator preamplifiers, an array of off-chip potentiometers is used, as shown in Fig. 5.30b, c. As illustrated in Fig. 5.31, the range of the generated bias current (I_B) goes from 1.8 to 180.0 μA corresponding to off-chip resistors (R) of 2 MΩ and 6 kΩ, respectively. Figure 5.30b suggests that the whole range can be used for the opamps, while the range of the preamplifiers I_B goes from 3.4 to 70.0 μA—see Fig. 5.30c. Note from Fig. 5.13 that a lower range of I_Bs is used for the amplifiers due, for instance, to the appreciable degradation of the opamp DC gain and its non-linearities as shown in this figure.

5.3.8 *Common-Mode Generation*

A folded-cascode amplifier is used in a voltage follower configuration to isolate the generated resistor ladder tabs from all the switching activity at the common-mode voltage that is extracted from the middle node of that same resistor ladder. In

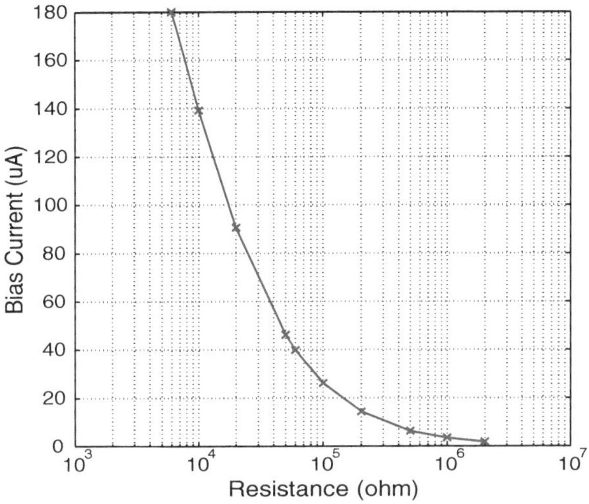

Fig. 5.31 Illustration of bias current (I_B) versus off-chip resistance

addition, a set of decoupling capacitors at different nodes are also necessary. It has been determined that 28.4-, 14.6- and 78-pF capacitors are needed at the middle node of the resistor ladder, the output of the buffer and the output pad of the common-mode signal respectively for correct common-mode generation. This has been verified by transient simulations using CADENCE-SPECTRE [Cade08] for the operation modes of WiMAX and GSM, representing, respectively, the most demanding operation speed and resolution in the modulator. Large off-chip decoupling capacitors of 22-μF are also used between the reference voltages and between the common-mode pad and ground. An additional on-chip decoupling capacitor between the references of 60 pF is employed as well.

5.4 Layout and Prototyping

The modulator has been implemented in a 90-nm 1P9M digital CMOS technology. Figure 5.32a shows the complete layout and highlights its main parts, while Fig. 5.32b shows a microphotograph of the chip. The modulator core occupies an active area of 0.66 mm^2. Figure 5.33 illustrates the layout floorplan with separate analog, mixed and digital supplies, and guard rings surrounding each section of the circuit. The area of the analog part is devoted to the amplifiers, capacitors, preamplifiers and the bias current generators. Analog adders are embedded inside the integrators area since they are implemented only with capacitors and switches. The mixed-mode part has two different supplies: one for the comparator regenerative latches and for the switches at integrators and analog adders ($V_{DDADMix}$), and another for the clock phase buffers ($V_{DDADBuff}$). The digital part encloses all the logic gates, the DAC and the clock phases generator. The latter also includes two blocks: one intended to perform the control logic that creates the controlled phases with *Ctrl* signal (see the bottom of

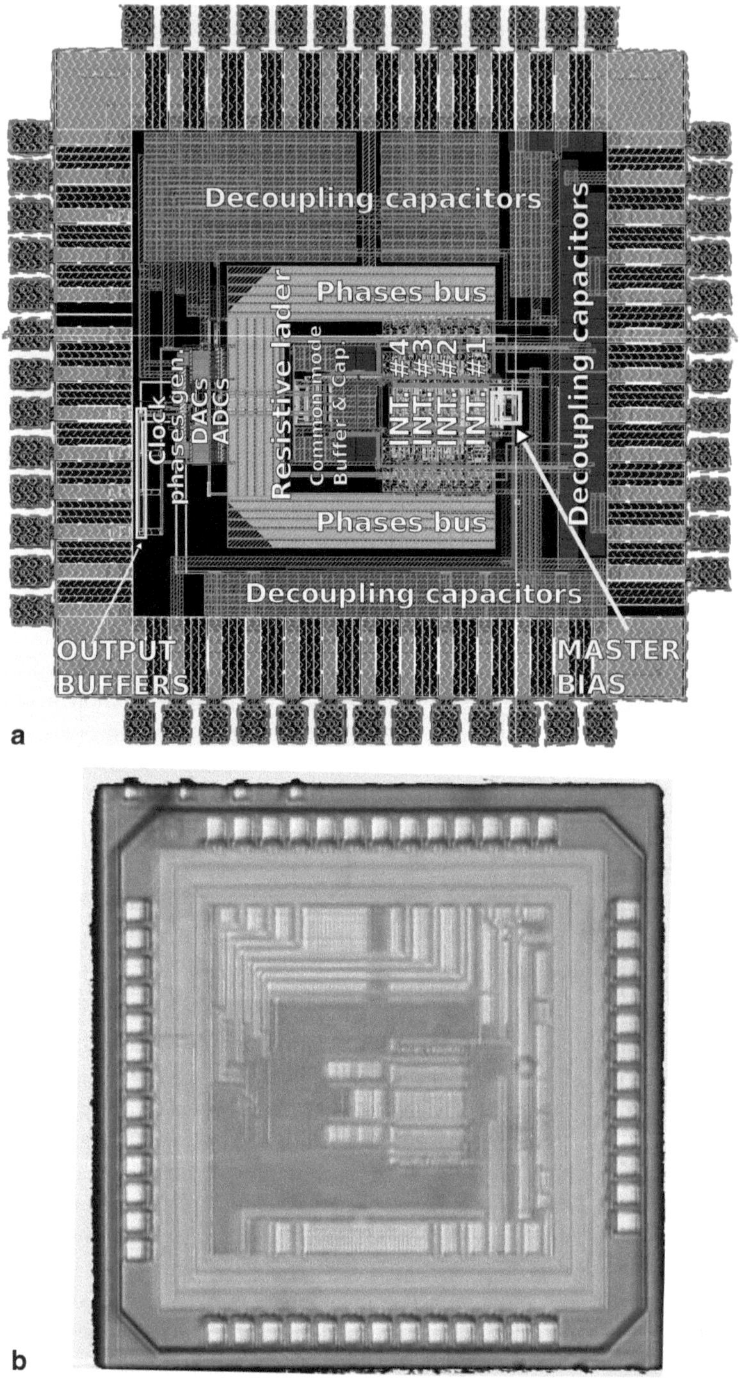

Fig. 5.32 Chip implementation: **a** layout, **b** die microphotograph

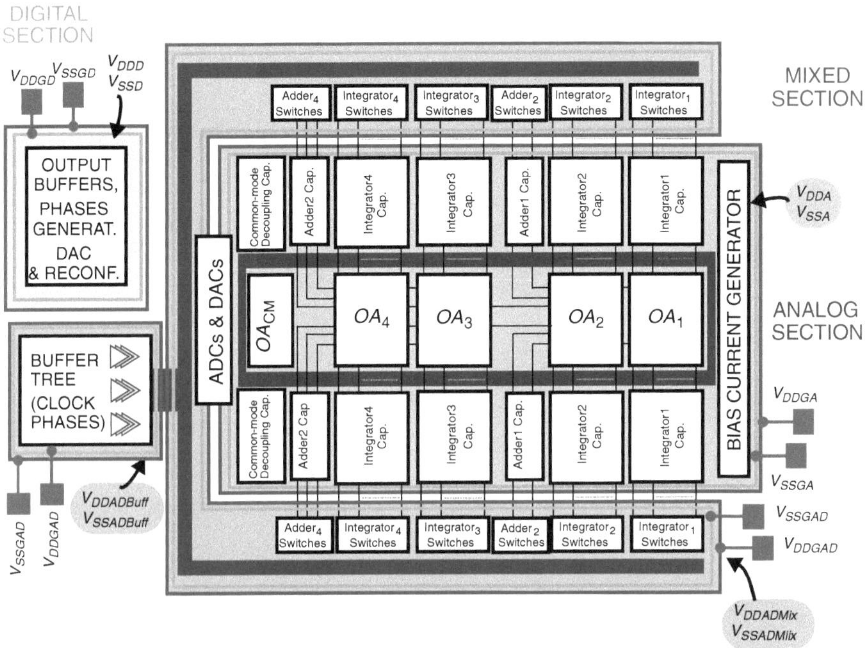

Fig. 5.33 Illustration of the prototype floorplan

Fig. 5.4) and a replica of the former to introduce the same delay in the original phases (using a control signal which is always set to a logic zero).

Apart from this division, the digital part of the pad ring is split into two parts: one devoted to the clock and another to the I/O buffers. This is done to reduce the coupling between both groups of high-swinging signals. In addition, library-based I/O cells with ESD protection are used for the implementation of the pad ring.

The chip is directly bonded over the PCB to reduce package-related parasitics. Double bonding is used for the references in order to halve their bonding parasitic inductances. Three pads—and off-chip pins—are used for the analog supply and ground voltages to shorten the parasitic bonding inductance. Moreover, since their currents are equal but with different signs, their pads and pins are placed in parallel to reduce the total parasitic inductance thanks to the compensation given by the complementary mutual inductances of this configuration [Raza00].

5.4.1 PCB

Figure 5.34 displays the PCB schematic. The strategies employed in its design can be summarized as follows:

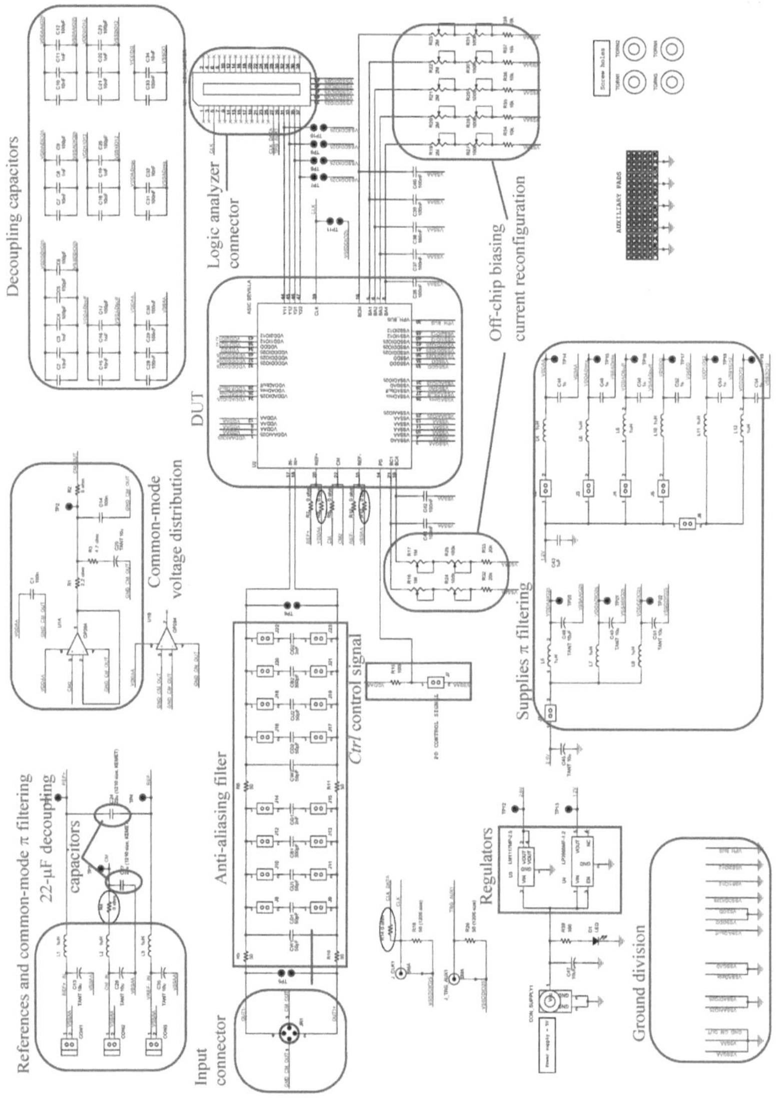

Fig. 5.34 PCB schematic

- The input signal is generated in a fully differential low-distortion external source (Tektronix SG5010 audio oscillator) and, then, filtered by a second-order RC anti-aliasing filter on PCB. There is a whole set of capacitors to be connected depending on the operation mode and fixed 50-Ω resistors. The capacitor values are selected according to the required filtering rejection frequency for each operation mode BW. The input equivalent thermal noise added by this filter has been computed to be well below specifications.
- There are two regulators on the PCB to keep a stable value for the supply voltages. Both devices are connected to an external 5-V voltage generator and provide a 2.5-V (for the chip I/O cells) and 1.2-V supply voltages (for both the chip I/O cells and the on-chip circuitry).
- All the on-chip supplies and voltage references have an independent π filtering at PCB before their connection to the chip. Two capacitors and one inductor are used for each filter. There is a filtering for each supply—after distribution by the PCB regulator—from the DC source to the chip and vice versa. Each supply on-chip is therefore filtered from the others as well. The same applies to the voltage references. Note that jumpers are used to enable the current flowing through the corresponding supply and ground to be measured with an amperimeter.
- The on-chip generated common-mode voltage must be provided to the fully differential input signal generator. To this purpose, a buffer based on a unity feedback opamp is used at PCB. If necessary, the common-mode voltage can be imposed externally[18] with a voltage DC source as an alternative testing approach. To do this, the footprint of a 0-Ω resistor was included.
- 22-μF capacitors are employed as off-chip decoupling capacitors for the common-mode and voltage references. These capacitive values help to reduce ringing and to flatten their temporal responses. This point has been checked in transient simulations, including a macromodel for both the bonding wires and the off-chip capacitors.
- In general, capacitors are employed throughout the PCB to keep a flat frequency response of the impedance when considering PCB parasitics [Alex05]. Note that the smaller the capacitance value, the closer the capacitor is placed to the chip in the PCB layout.
- As mentioned above, the bias currents for opamps and comparator preamplifiers in the modulator are controlled with off-chip potentiometers. A set of on-chip and off-chip capacitors are also used for decoupling these nodes.
- The ground—and supply—layer on the PCB is divided into 9 regions; namely: analog, 2.5-V analog for I/O ring ground, mixed-mode, one for the phase buffers, digital, 2.5-V digital for I/O ring ground, two 1.2-V I/O ring grounds and a dedicated ground (*VPH_BUS* in the figure) for the phase bus isolation. There are 9 main grounds divided through the PCB up to a final point that connects all of them to that of the corresponding voltage source.

---

[18] A π filter is employed for the common-mode voltage distribution on the PCB when this strategy is applied.

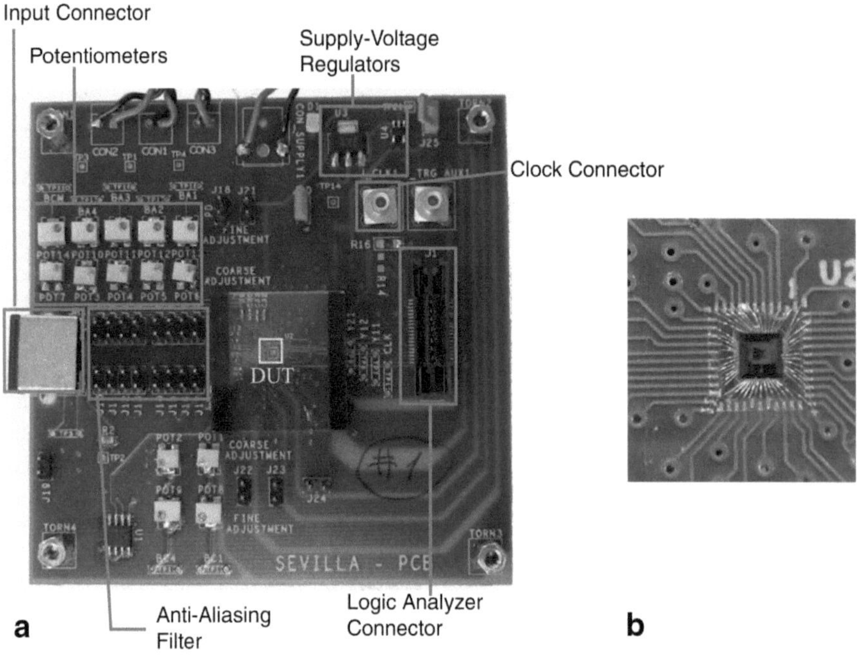

Fig. 5.35 Test PCB: **a** photograph, **b** bonding over PCB detail

Figure 5.35 shows a photograph of the 4-layer PCB. Note that the chip is directly bonded over it, as also shown in the figure.

5.5 Experimental Results

5.5.1 Timing Capturing Issue

Two options are traditionally used for taking the modulator output bit streams in transient simulations. First, a virtual logic analyzer—that can be implemented in the form of a verilogA code [Kund04]—captures this data into a file for later processing. Alternatively, CADENCE-SPECTRE allows the user to capture this same data by introducing a numerical temporal reference[19] [Cade08]. This second option was the one selected although the first approach is the one closer to the procedure used when testing in a laboratory. Unfortunately, the triggering signal—used for an appropriate timing in the capture—was not routed to a pad and then bonded to a pin, so it cannot be taken off-chip. Indeed, the only timing reference outside the chip is the clock

[19] The ideal capturing time is the point at which the output of the comparator is stable. In other words, the falling edge of phase ϕ_{2d} or, alternatively, the rising edge of ϕ_{1d} (see Figs. 5.4 and 5.8).

signal used for the logic analyzer capture operation. Unfortunately, there is no way to know how the clock signal evolves inside the chip and, thus, how far its rising edges are from the ones of ϕ_{id} (the signal that should have been used for triggering the logic analyzer). This is a potential problem when testing high-speed operation modes such as WiMAX. Note that this issue is solved by taking the appropriate triggering signals off-chip for the prototype presented in Chap. 6.

5.5.2 Test Set-Up and Experimental Results

A similar test set-up to the second one used in Chap. 4 (see Sect. 4.6) was employed here. The performance of the chip was evaluated using a logic analyzer to capture the bit streams of the different stages of the modulator. After acquisition, captured data were transferred to a workstation for digital post-processing with MATLAB [Math08a].

Figure 5.36a, b shows the measured output spectra corresponding to GSM and WiMAX operation modes, considering an input sinewave with an amplitude of -12.2 dBFS. The reconfiguration of the noise-shaping order is illustrated in Fig. 5.36c, which shows the output spectra for GPS, on the one hand, and for UMTS and DVB-H modes, on the other; corresponding to the use of the same sampling frequency (120 MHz) with $L=2$ and $L=4$, respectively. Moreover, the measured spectra for all operation modes are compared in Fig. 5.37. It can be observed that thermal noise dominates IBE for GSM, whereas quantization noise dominates in WiMAX. Furthermore, the contribution of thermal and quantization noise is roughly the same in BT and GPS modes. Finally, thermal noise power is slightly greater than quantization noise power for the UMTS mode. On the contrary, quantization noise is marginally greater for DVB-H. Note from Fig. 5.36b that third- and fifth-order harmonic distortion is present for WiMAX. However, although distortion is observable, it does not limit the modulator resolution for WiMAX due to a greater contribution of quantization noise, as can be observed in the figure. Figure 5.38 also depicts the modulator SNDR versus the input signal level for the operation modes under consideration.

Figure 5.39 provides several details of the chip power consumption. First, a budget of the chip consumption is shown for the different supplies. It can be observed that the contribution of the analog circuitry predominates in all modes. In most of the modes, however, the faster the chip operation, the greater the power contribution of the clock phase buffers and the digital blocks[20]. In addition, Fig. 5.39b shows the normalized power consumption to the peak SNDR.

Table 5.16 summarizes the experimental results of the modulator. Aside from the high degree of modulator flexibility for each mode (different OSRs, modulator order

[20] It is well known that the dynamic power of a digital gate is proportional to $f \cdot C \cdot V_{DDD}^2$ [Sedr82], where f, C and V_{DDD} stand for the operation frequency, the load capacitance and the digital supply voltage, respectively.

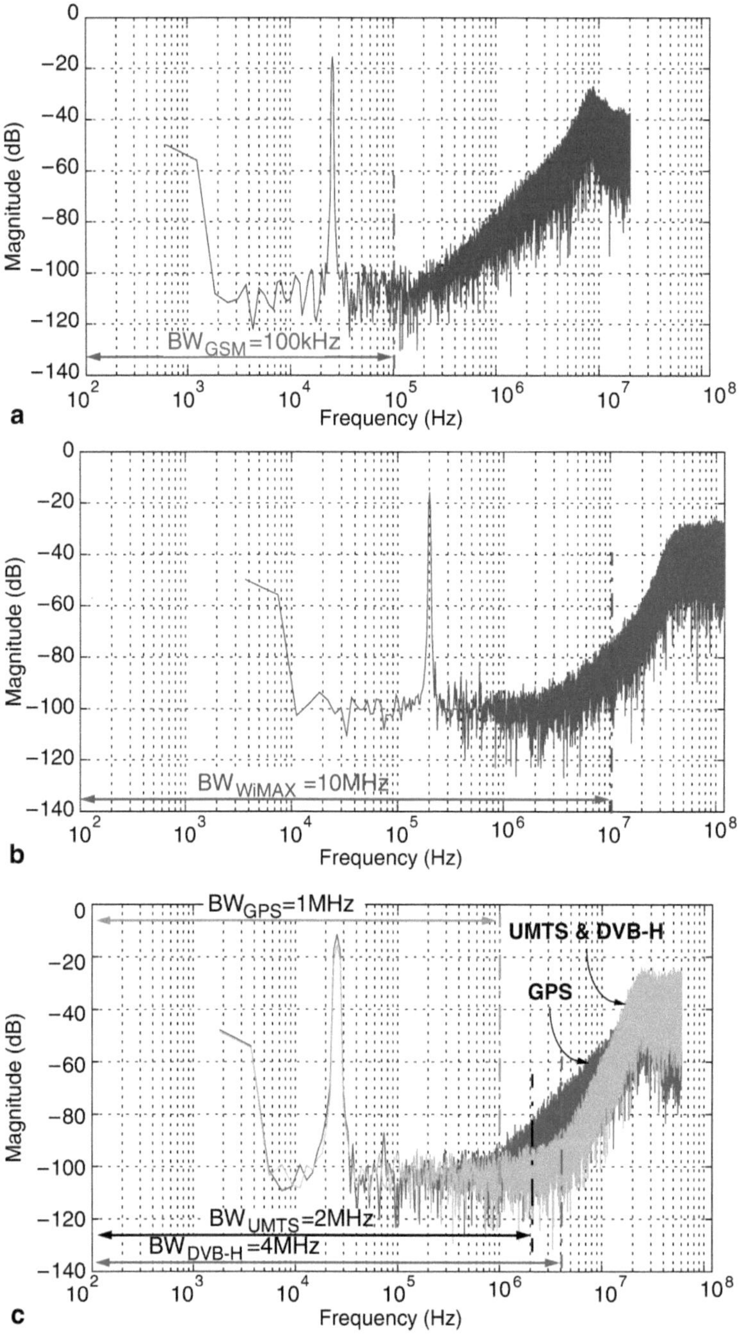

Fig. 5.36 Measured output spectra for **a** GSM, **b** WiMAX, **c** GSP and UMTS

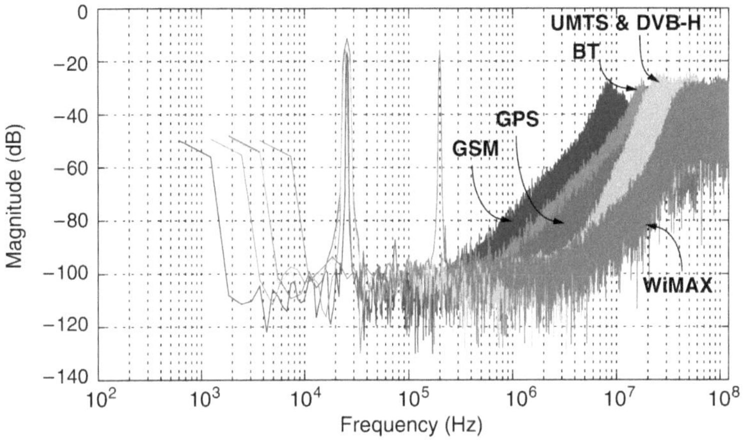

Fig. 5.37 Measured output spectra for all operation modes

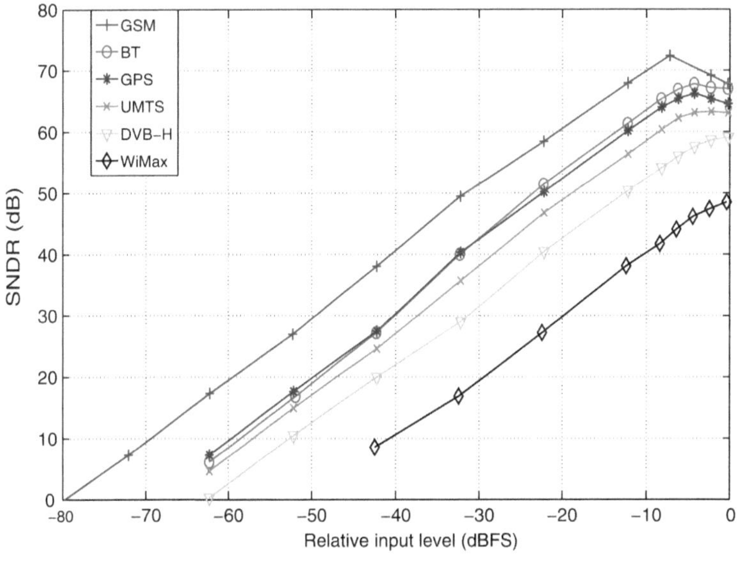

Fig. 5.38 SNDR versus input level for all operation modes

and variable biasing selection for opamps and preamplifiers), the best experimental results are shown in the table.

This prototype considerably improves the performance of the one presented in Chap. 4 (see Table 5.15). In spite of the supply voltage reduction—from 3.3 to 1.2 V, the flexibility capabilities and architectural improvements achieved allow the efficient conversion of modes with larger bandwidths, which allows the range of the modulator bandwidth operation to be widened as well. Furthermore, the power dissipated in the

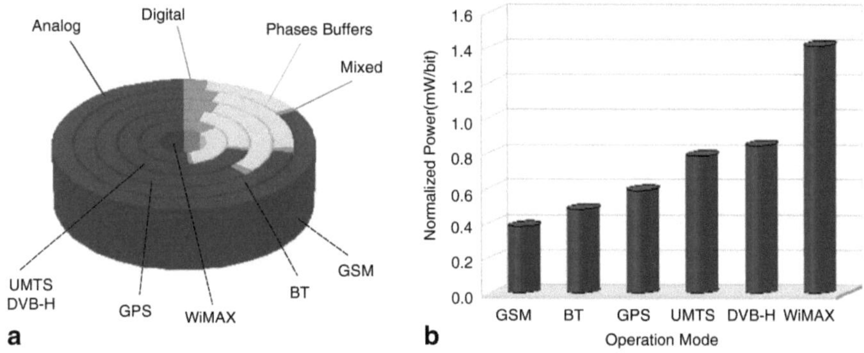

Fig. 5.39 Power consumption for the different operation modes: **a** power budget, **b** normalized power

Table 5.16 Summary of measured performance

	GSM	BT	GPS	UMTS	DVB-H	WiMAX
Modulator architecture	2			2-2		
OSR	200	80	60	30	15	12
BW (MHz)	0.1	0.5	1	2	4	10
Sampling frequency (MHz)	40	80	120			240
DR (dB)	78.0	70.0	71.5	66.0	62.0	52.0
Peak SNDR (dB)	72.3	68.0	65.4	63.3	59.1	48.7
Process	90-nm 1P9M CMOS					
Core area	0.66mm^2 (pads excluded)					
Power consumption (mW)	4.6	5.3	6.2	8.0		11.0
Analog core	3.8			5.7		
Phases Buffer	0.6	1.1	1.8			3.6
Mixed section	0.0	0.1				0.2
Digital section	0.1	0.3	0.4			1.4

modulator is largely reduced—for instance, to only 18% of the power for UMTS—in comparison with the design in Chap. 4. This results in a significant improvement in the values of FoM_1 and FoM_2, as will be shown in Chap. 7.

5.6 Summary

This chapter describes the design, implementation and experimental characterization of a multi-mode ΣΔM capable of converting a large number of standards with good performance. The chosen architecture is a 2-2 cascade ΣΔM based on unity-STF stages, in which the back-end stage can be switched off to save power. The modulator is therefore capable of varying the order of the loop filter, the sampling frequency and the bias currents of its amplifiers and comparator preamplifiers.

The chapter discusses the selection of the modulator topology and the architectural coefficients, addressing practical architectural issues and their solutions. Details on the design of the modulator building blocks are also presented at transistor level.

The prototype was fabricated in a 90-nm CMOS technology and features dynamic ranges of 78.0/70.0/71.5/66.0/62.0/52.0 dB and peak SNDRs of 72.3/68.0/65.4/63.3/59.1/48.7 dB within bandwidths of 100 kHz/500 kHz/1 MHz/2 MHz/4 MHz/10 MHz, while consuming 4.6/5.3/6.2/8.0/8.0/11.0 mW, respectively. These results show a performance competitive with state-of-the-art multi-standard $\Sigma\Delta$ modulators, covering one of the widest regions in the resolution-bandwidth plane as will be shown in Chap. 7.

The architectural advantages of this prototype (mainly provided by the unity-STF stages in the cascade) enable low-power high-speed A/D conversions for a multi-mode hand-held device. This can easily be observed when its performance is compared with that obtained by the prototype in Chap. 4.

Nevertheless, if the device functionality requires N standards to be converted in parallel, N ADCs will be placed on the chipset despite all their individual multi-standard capabilities. Concurrency is therefore the next step in the roadmap of these kinds of devices [Hash03]. Next chapter shows the design, implementation and results of a prototype that gives a solution to the aforementioned problem while also introducing much more flexibility in its reconfiguration capabilities.

Chapter 6
A 1.2-V 90-nm CMOS Adaptive Concurrent Resonation-Based 2-2-2 Cascade $\Sigma\Delta$M for SDR

Flexibility in baseband processing is becoming one of the most important necessities in future terminals for mobile communications and wireless networks [Blue03]. Concurrent reception of wireless standards will help to boost the needed flexibility, while making more efficient use of the available hardware. The $\Sigma\Delta$M introduced in this chapter implements several flexible strategies. Apart from all the reconfiguration strategies used in Chaps. 4 and 5, concurrency and resonation are also implemented. The latter allows an optimum selection of the resonation zeros in order to reduce the quantization noise for a given bandwidth, while the former increases the parallel processing capabilities of the $\Sigma\Delta$M by re-using its own hardware. There is also a set of reconfiguration strategies that helps to trade power consumption for performance for a set of bandwidths in a reconfigurable $\Sigma\Delta$M. The sampling frequency and, consequently, the oversampling ratio, and the order of the in-loop filter can be adjusted to the requirements of each operation mode. The biasing of the $\Sigma\Delta$M building blocks and the resolution of the embedded quantizers can be also modified so that the modulator too achieves the required resolution with minimum power. All these strategies for flexibility are incorporated in the $\Sigma\Delta$M presented in this chapter. Moreover, this $\Sigma\Delta$M IC adopts and improves the design of some of the building blocks of the $\Sigma\Delta$M described in Chap. 5. This chapter also examines the design, implementation, layout and experimental results of a flexible $\Sigma\Delta$M, intended to fuel the A/D conversion of 4G wireless telecommunication transceivers. This chip is capable of converting up to 3 standards in parallel operation. The $\Sigma\Delta$M architecture consists of a 2-2-2 cascade with adaptable order and concurrent capabilities for each stage. The experimental characterization of the chip shows a proper operation of modulator order reconfiguration, adaptive biasing, variation of the internal quantization and concurrency.

This chapter is organized as follows: Sect. 6.1 shows the architecture and the flexible strategies embedded in the $\Sigma\Delta$M. Section 6.2 illustrates the SC schematic and the timing operation in the modulator. Section 6.3 explains how the aforementioned strategies are implemented in practice. Afterwards, more details of all the $\Sigma\Delta$M building blocks at transistor level are provided in Sect. 6.4. The layout of the chip and the PCB design are illustrated in Sect. 6.5. Finally, the experimental performance of the chip is described in Sect. 6.6.

A. Morgado et al., *Nanometer CMOS Sigma-Delta Modulators for Software Defined Radio*, 201
DOI 10.1007/978-1-4614-0037-0_6, © Springer Science+Business Media, LLC 2011

6.1 Architecture and Strategies for Flexibility

The main purpose of this design is the implementation of a flexible $\Sigma\Delta$M with parallel conversion capabilities. Performance improvements for every wireless standard—as usually pursued in mono-standard designs—is a minor objective, with the priority being a more flexible solution that can, eventually, lead to concurrent processing. Furthermore, additional flexibility is typically accompanied by a better performance, because the system circuitry and its architecture can be adjusted to each operation mode more efficiently. Several building blocks of the design presented in Chap. 5 have also been re-designed here to boost their performances.

Apart from the inclusion of several reconfiguration strategies, the principal novelty in this prototype is the parallel processing of more than one operation mode; in other words, concurrency to enable the parallel conversion of a number of inputs.

This section shows the employed architecture and its main parameters. Special attention is paid to the different reconfiguration strategies at the architectural level, such as the adaptive modulator order, resonation and concurrency. The drawback is an increase in the complexity in the modulator design, because additional SC branches and modulator building blocks are required for the extra functionality.

6.1.1 Architecture

Figure 6.1 shows the block diagram of the proposed architecture. It consists of a sixth-order 3-stage reconfigurable 2-2-2 cascade $\Sigma\Delta$M. Unity STF is obtained through the use of feed-forward paths in each single-loop stage. Local resonation is incorporated in the second and third modulator stages following the scheme in Fig. 3.10b. This modulator can also adapt its order to the operation mode. The scheme in Fig. 6.1 presents three possible inputs (X_1, X_2 and X_3) that can be used for concurrency as indicated below. The digital cancellation logic (DCL), shown in Fig. 6.2, needs to be adjusted to match the architectural configuration in the modulator.

6.1.1.1 Architectural Coefficients

The in-loop coefficients employed in the architecture are different from those in Chap. 5. Table 6.1 shows the values of all the in-loop coefficients except those used to implement resonation, which will be given later. The values of the coefficients are selected to keep output swing opamp requirements low[1], while the number of unit

[1] The output swing requirements will be roughly doubled—45% of the reference voltage—for the first and third integrators if the modulator's internal quantization is of 3 levels (see Table 5.4). By contrast, if all embedded quantizers use 5 levels, the output swing requirements will be approximately 22% for the first and third integrators, and 6% for the second and fourth integrators.

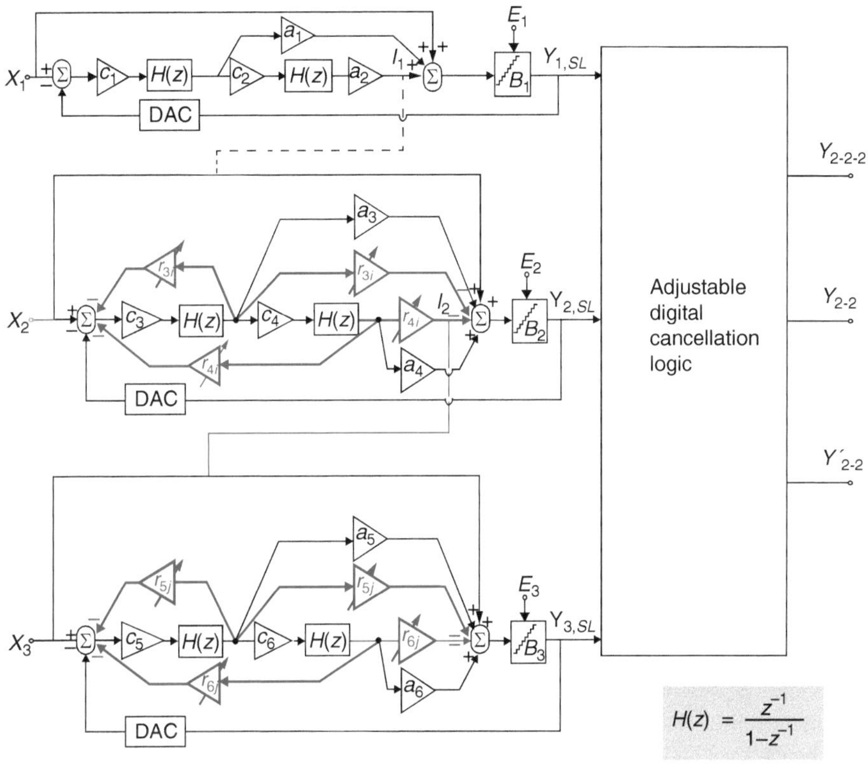

Fig. 6.1 Block diagram of the proposed reconfigurable $\Sigma\Delta$M

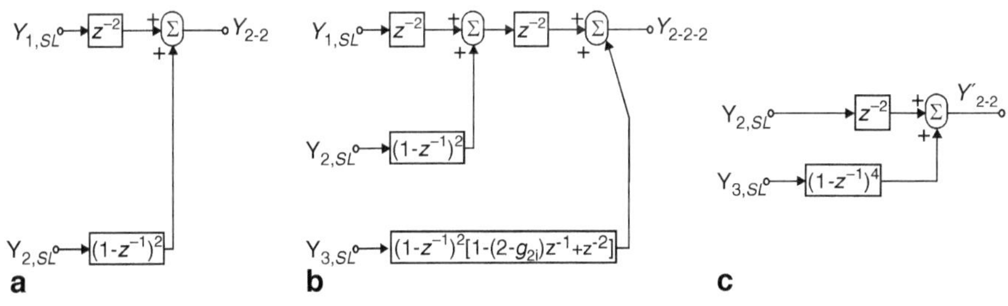

Fig. 6.2 Adjustable DCLs in Fig. 6.1: **a** DCL 1, **b** DCL 2, **c** DCL 3

capacitors per stage is reduced compared to the prototype presented in Chap. 5[2]. This technique is convenient for area reduction. Note that there is an additional $\Sigma\Delta$M stage in this prototype, so larger area consumption is expected.

---

[2] Only 2×28 unit capacitors are needed for a 2-2 cascade configuration in this modulator, while the in-loop coefficients employed in Chap. 5 demand the use of 2×35 units.

Table 6.1 Coefficients in Fig. 6.1

c_1	c_2	a_1	a_2	c_3	c_4	a_3	a_4	c_5	c_6	a_5	a_6
0.5	0.5	4	4	0.5	0.5	4	4	0.5	0.5	4	4

During resonation, coefficients r_{3i}, r_{4i}, r_{5j} and r_{6j} change according to the required resonation factors; otherwise, they are always 0. The corresponding values of these coefficients, together with the allowable resonation factors, are given below.

6.1.2 Modulator Order Reconfigurability and Concurrency

Every stage in Fig. 6.1 can work either separately or as part of a cascade. The different actual architectural configurations are the following:

- A 2-2-2 cascade $\Sigma\Delta$M that processes the input signal X_1. In this case, a 6th-order high-pass filtering is obtained for the NTF. The modulator output is $Y_{2\text{-}2\text{-}2}$ and the DCL 2 in Fig. 6.2 is used. In other words, DCL 1 and 2 provide a 4th- and 6th-order shaping of the last-stage quantization error for a cascade made up of the first two stages and of the three stages, respectively.
- A 2-2 cascade $\Sigma\Delta$M. In this configuration, the last stage can be switched off to save power or can work in a single-loop configuration, processing input X_3 concurrently. The outputs for the two options are $Y_{2\text{-}2}$ and $Y_{3,\,SL}$, respectively. The DCLs 2 and 3 in Fig. 6.2 are therefore not used.
- First stage operating in a 2nd-order single-loop configuration, with respective input and output X_1 and $Y_{1,\,SL}$. There are several possibilities in this case. On the one hand, the back-end stages can work as an additional 2-2 cascade or can be switched off when not used. Alternatively, one or both of the back-end stages can operate concurrently. The DCL 3 in Fig. 6.2 is only used when the second and the third stages work as a 2-2 cascade processing input X_2. DCLs 1 and 2 are never used in this case.

NTF changes according to the configuration selected. If a single-loop configuration is selected for one of the stages, a second-order in-loop filtering will be obtained as follows:

$$\mathrm{NTF}_{SL}(z) = (1 - z^{-1})^2 \tag{6.1}$$

On the contrary, if any of the two subsequent stages are working in a 2-2 cascade configuration (i.e., first and second stages, or second and third stages), the corresponding NTF will be given by:

$$\mathrm{NTF}_{2\text{-}2}(z) = (1 - z^{-1})^4 \tag{6.2}$$

Finally, when all stages are connected together, the overall $\Sigma\Delta$M works as a 2-2-2 cascade with the following NTF:

$$\mathrm{NTF}_{2\text{-}2\text{-}2}(z) = (1 - z^{-1})^6 \tag{6.3}$$

The use of local resonation that would include zeros in the NTF is not yet considered. The implementation of this strategy will be discussed below. The resolution of each of the embedded quantizers can be also adapted to 3 or 5 levels.

In general, there is only one master clock in a cascade $\Sigma\Delta$M and the clock phases are generated from this signal. Let us suppose that more than one stage in the scheme in Fig. 6.1 is working concurrently. If there is only one clock governing the generation of all the clock phases in the modulator, the speed of all concurrent stages will be dominated by the stage that requires the fastest operation. This leads to a waste of power for the slowest concurrent stages. Different clocks are therefore needed. The solution to this problem in this modulator is given by incorporating an on-chip clock divider for each stage, so that the speed of each concurrent stage can be adjusted as required.

6.1.3 Resonation Strategies

Both the second and the third stage in Fig. 6.1 make use of tunable coefficients to allow programable resonation operation. Table 3.1 is adapted to provide the relationships between the resonation coefficients r_{3i}, r_{4i}, r_{5j} and r_{6j}, the resonation factors g_{2i} and g_{3j} and the in-loop coefficients in Table 6.1 as follows:

$$\left.\begin{aligned} r_{3i} &= 2g_{2i} \\ r_{4i} &= 4g_{2i} \\ a'_{3i} &= a_{3i} - 2g_{2i} \\ a'_{4i} &= a_{4i} - 4g_{2i} \end{aligned}\right) \qquad (6.4)$$

$$\left.\begin{aligned} r_{5j} &= 2g_{3j} \\ r_{6j} &= 4g_{3j} \\ a'_{5j} &= a_{5j} - 2g_{3j} \\ a'_{6j} &= a_{6j} - 4g_{3j} \end{aligned}\right) \qquad (6.5)$$

where the effective feed-forward coefficients in Fig. 6.1—denoted with a'_{ij}—are implemented as a subtraction of the feed-forward coefficients in Table 6.1 (a_{ij}) and those used to implement resonation as defined in (6.4) and (6.5); i.e., $2g_{2i}$ or $4g_{2i}$ for (6.4), and $2g_{3i}$ or $4g_{3i}$ for (6.5). Each of these latter resonation coefficients (g_{2i} and g_{3j}) can have three possible values and either of these values—which are included in Table 6.2 together with the corresponding resonation coefficients—can be combined[3] with the other two. The resulting permutations are given in Table 6.3.

Resonation is also compatible with the above-mentioned architectural strategies. A 2-2 MASH $\Sigma\Delta$M with a zero notch in the NTF will therefore be obtained if two stages are configured in a cascade configuration and their in-loop coefficients are

[3] The actual way to implement this combination of coefficients is merely to use three different SC branches with different control signals corresponding to three different values of g_{ji}.

Table 6.2 Resonation coefficients of the proposed ΣΔM

	Second Stage			Third Stage			
	g_{21}	g_{22}	g_{23}	g_{31}	g_{32}	g_{33}	
g_{2i}	0.250	0.125	0.0625	0.35	0.25	0.10	g_{3j}
r_3	0.500	0.250	0.125	0.7	0.5	0.2	r_5
r_4	1.00	0.50	0.25	1.4	1.0	0.4	r_6
a'_3	3.500	3.750	3.875	3.3	3.5	3.8	a'_5
a'_4	3.00	3.50	3.75	2.6	3.0	3.6	a'_6

Table 6.3 Other resonation coefficients of the proposed ΣΔM

	Second Stage				Third Stage				
	$g_{21}+g_{22}+g_{23}$	$g_{21}+g_{22}$	$g_{21}+g_{23}$	$g_{22}+g_{23}$	$g_{31}+g_{32}+g_{33}$	$g_{31}+g_{32}$	$g_{31}+g_{33}$	$g_{32}+g_{33}$	
g_{2i}	0.4375	0.375	0.3125	0.1875	0.70	0.60	0.45	0.35	g_{3j}
r_3	0.875	0.750	0.625	0.375	1.4	1.2	0.9	0.7	r_5
r_4	1.75	1.50	1.25	0.75	2.8	2.4	1.8	1.4	r_6
a'_3	3.125	3.250	3.375	3.625	2.6	2.8	3.1	3.3	a'_5
a'_4	2.25	2.50	2.75	3.25	1.2	1.6	2.2	2.6	a'_6

given by (6.4) and (6.5). In this case, the resulting NTF would be:

$$\mathrm{NTF}_{2\text{-}2}(z) = (1 - z^{-1})^2 \cdot \left[1 - (2 - g_{ij}) \cdot z^{-1} + z^{-2}\right] \tag{6.6}$$

Another possible configuration is that in which all stages are connected to produce a 2-2-2 cascade configuration and there is a resonation coefficient g_{3j} implemented in the last stage. The corresponding NTF would then be given by:

$$\mathrm{NTF}_{2\text{-}2\text{-}2}(z) = (1 - z^{-1})^4 \cdot \left[1 - (2 - g_{3j}) \cdot z^{-1} + z^{-2}\right] \tag{6.7}$$

Finally, if both back-end stages have g_{2i} and g_{3j} as their respective resonation factors, the NTF would change to:

$$\mathrm{NTF}_{2\text{-}2\text{-}2}(z) = (1 - z^{-1})^2 \cdot \left[1 - (2 - g_{2i}) \cdot z^{-1} + z^{-2}\right] \cdot \left[1 - (2 - g_{3j}) \cdot z^{-1} + z^{-2}\right] \tag{6.8}$$

Note that if the resonation factors, g_{2i}, is not power of 2, it will result in a complex, costly implementation of DCL 2 [Sanc06].

Note that resonation can work together with concurrency. This way, the second and the third stages can implement resonation with respective resonating factors g_{2i} and g_{3j}, while operating concurrently. In this case, the resulting NTFs are:

$$\mathrm{NTF}_{SL}(z) = 1 - (2 - g_{ij}) \cdot z^{-1} + z^{-2} \tag{6.9}$$

where g_{ij} stands for the corresponding resonating factor; namely, either g_{2i} or g_{3j}.

6.1.4 Summary of All Flexible Strategies

All reconfiguration and flexible techniques presented in this section can be summarized as follows:

- The $\Sigma\Delta$M order can be adjusted to the operation mode. It can therefore be configured as either a 2nd-order single-loop; a 2-2 cascade or a 2-2-2 cascade $\Sigma\Delta$M. In addition, stages which are not used can be switched off to save power.
- Concurrency is also implemented, so each stage can work separately by processing different input signals. Up to three modes can operate in parallel. If the $\Sigma\Delta$M is processing two modes concurrently, there will be two possible configurations. On the one hand, a 2-2 cascade $\Sigma\Delta$M and a single-loop $\Sigma\Delta$M will be working in parallel. On the other, two single-loop $\Sigma\Delta$Ms will be able to process two inputs concurrently, while another stage might be switched off to save power.
  Multi-rating is beneficial for concurrency because it enables the use of different sampling frequencies for each stage.
- The resolution of each embedded quantizers can be set to 3 or 5 levels.
- The last two stages can resonate to reduce quantization noise. This appealing feature can be implemented when the stage is operating in a concurrent configuration or when it is part of a cascade $\Sigma\Delta$M.

The SC implementation of the scheme in Fig. 6.1 and, later, the deployment of all these flexible features at circuit level will be described in the next two sections.

6.2 SC Implementation

Figure 6.3 shows the single-ended version of the proposed $\Sigma\Delta$M SC schematic. The first stage is depicted in Fig. 6.3a. Both the second and the third stages are identical. Fig. 6.3b illustrates the implementation of only the second stage. There is a block named DAC in both Fig. 6.3a and b. This block is responsible for the generation of the control signals that will govern the operation of the feedback sampling capacitors of the front-end integrator in the corresponding stage. There are three SC branches (and thus three sampling capacitors) for the DAC operation in the aforementioned integrators. One of the SC branches is used for the 3-level configuration of the quantizer and, consequently, of the DAC. Only one sampling capacitor—C_{s11}, C_{s13} and C_{s15} for the first[4], second and third stages, respectively—is employed for this purpose in each stage. The other two SC branches are needed for the 5-level configuration of the DAC. The corresponding sampling capacitors in this case are C_{s21} and C_{s31} for the first stage, C_{s23} and C_{s33} for the second, and C_{s25} and C_{s35} for the third. Although there is a 5-level configuration in the quantizers, no DEM is implemented on-chip.This might prevent the use of this configuration in either the

[4] Note that capacitor C_{s11} is shared for the sampling of the 3-level DAC voltage and the input voltage X_1.

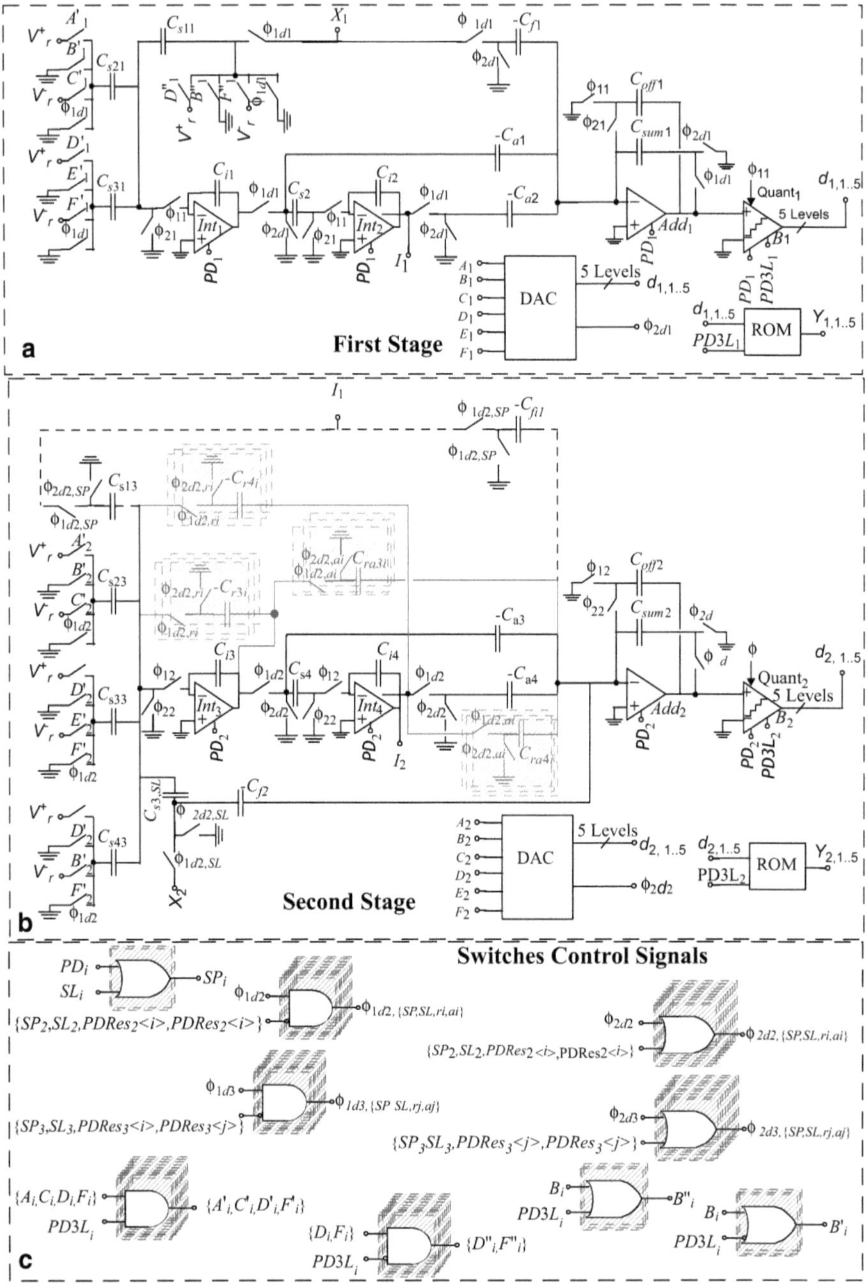

Fig. 6.3 SC implementation of the proposed flexible ΣΔM: **a** first stage, **b** second stage, **c** reconfiguration digital logic

first stage of one of the cascaded configurations or in the modulator stages when operating as single-loop $\Sigma\Delta$Ms, because of the non-linearities introduced by the mismatches of the DAC elements. However, this 5-level configuration of the quantizers is completely compatible with dual-quantization in a cascade [Bran91] and can help to increase resolution, specially in wideband modes, and to reduce opamp output swings.

The output digital streams of each stage are stored in a ROM and then captured off-chip. The DCL blocks are shown in Fig. 6.2 but not in Fig. 6.3c. Fig. 6.3c depicts the digital logic blocks necessary to implement the main reconfiguration strategies in this $\Sigma\Delta$M. The functionality of these digital logic gates will be explained in Sect. 6.3.

The analog adders are active [John97]; i.e., based on opamps. This leads to an implementation with a virtual ground and that is free of the inherent capacitance division of a passive approach. At first, no scaling factor is therefore imposed on the quantizer reference voltages. Nevertheless, the opamp of the first-stage adder needs an output swing as large as the maximum voltage swing at the input of the embedded quantizer of the first stage[5] [Ghar06]. In this design, the reference voltage is equal to the supply voltage to maximize SNDR. If the full scale of the quantizer is reached (or a close value), the output swing of that opamp will need to be as high as this voltage. This will eventually result in low headroom for the output branch of the opamp and, as a consequence, in a power hungry opamp. A scaling factor of 1/4 is implemented on the input signal together with the addition operation in order to avoid this situation. This factor will later be compensated by scaling the reference voltages of the quantizers as was done in Chap. 5. The operation of the active analog adder is as follows. First, the signals to be added are sampled by the capacitors C_{fi}, C_{fij} and C_{ai}. Then, C_{sumi} performs the summation, while C_{offi} is employed to hold and subtract the opamp offset. The opamp-based adder performs the addition together with an inversion in sign [John97]. Negative capacitors compensate that negative sign when the corresponding architectural coefficient is positive. In the practical fully-differential implementation, the negative capacitors in the SC schematic in Fig. 6.3 are implemented by swapping the differential nodes that are used to connect them.

The 1/4 scaling factor over the quantizers reference voltages imposes a specific requirement on the comparator offset equal to:

$$\text{Offset} < \frac{V_{ref}}{K} \cdot \alpha \tag{6.10}$$

as defined in Sect. 5.2.1. Note that V_{ref} (defined as $V_r^+ - V_r^-$), K and α stand for the reference voltage[6] (1.2 V), the number of levels in the quantizer and the scaling factor 1/4, respectively. The maximum offset requirements are equal to 100 and 60 mV for the respective quantizer resolutions of 3 and 5 levels.

Table 6.4 gives the capacitors values of the SC schematic in Fig. 6.3. These values are selected to have low equivalent capacitive loads for the opamps—their values will be shown in Sect. 6.4.1—while kT/C noise contributions are also kept

[5] This also applies to all stages in the modulator because they can operate concurrently.

[6] The differential reference voltage is $2 \cdot V_{ref} = 2.4$ V.

Table 6.4 Capacitor values (in fFs)

C_{s11}	C_{s21}	C_{s31}	C_{i1}			C_{s2}	C_{i2}	C_{f1}	C_{a1}	C_{a2}	C_{sum1}	C_{off1}	
200	100		400			75	150	50	200	200	200	50	
C_{s13}	C_{s23}	C_{s33}	C_{s43}	$C_{s3,SL}$	C_{i3}	C_{s4}	C_{i4}	C_{fi1}	C_{a3}	C_{a4}	C_{f2}	C_{sum2}	C_{off2}
400	50		100	100	200	50	100	200			50	200	50
C_{s15}	C_{s25}	C_{s35}	C_{s45}	$C_{s5,SL}$	C_{i5}	C_{s6}	C_{i6}	C_{fi2}	C_{a5}	C_{a6}	C_{f3}	C_{sum3}	C_{off3}
400	50		100	100	200	50	100	200			50	200	50

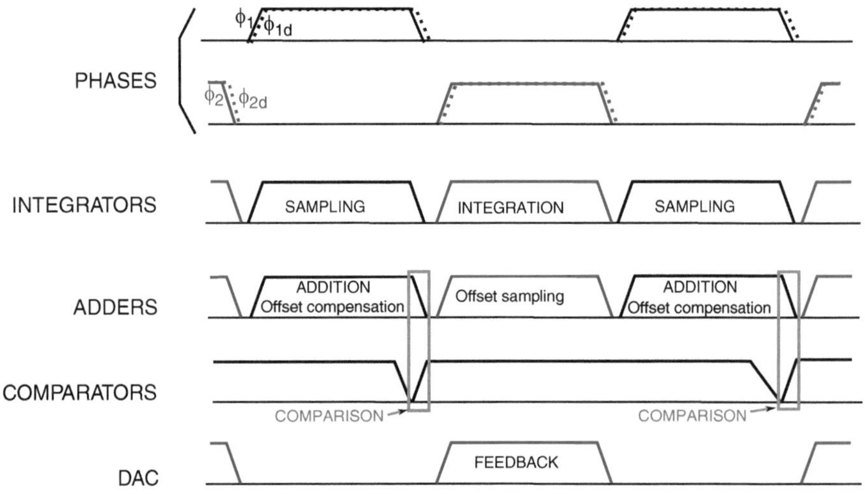

Fig. 6.4 Timing operation of the building-blocks of the SC schematic in Fig. 6.3

low for the given OSRs. The minimum unit capacitor is selected as 50 fF for matching requirements[7].

Figure 6.4 illustrates the timing operation of the main building-blocks of the schematic in Fig. 6.3. The timing scheme depicted is similar to that in Fig. 5.8, the only difference being the analog addition operation. In this design, capacitor C_{offi} samples the offset of the opamp at ϕ_2 and compensates it in the next phase ϕ_1 in which the addition is performed by capacitor C_{sumi}.

6.3 Implementation of Reconfiguration Strategies at Circuit Level

This section details the building blocks used to implement the reconfiguration capabilities in this prototype.

[7] This limit will not be considered when implementing low resonation coefficients.

6.3.1 Adaptability and Concurrency at Circuit Level

Figure 6.3c illustrates how reconfigurability is performed in the SC schematic by means of a set of control signals and the associated digital logic shown in the figure. The main signals and their corresponding functionalities are summarized as follows:

- PD_i is used to switch off the stage number i and power down its amplifiers and comparator preamplifiers in the corresponding embedded quantizers.
- SL_i is responsible for the concurrent operation of the corresponding stage, i. If SL_i is active, the concurrent path that processes input signal X_i will be enabled and the one that samples the inter-stage path coming from the previous stage is deactivated. Note that this only works for the back-end stages since the first stage has no inter-stage path.
  In practice, SL_2 and SL_3 activate the SC branches of capacitors $C_{s3,SL}$, C_{f2}, and $C_{s5,SL}$ and C_{f3}, respectively. In contrast, when the aforementioned control signals are off, the last SC branches are disconnected and the inter-stage paths are enabled by activating the SC branches of capacitors C_{s13}, C_{fi1}, and C_{s15} and C_{fi2}, respectively.
- Control signals $PDRes_i$<j> enable the resonation operation in stage i. If the last signal is activated, the SC branches of capacitors C_{r3i}, C_{r4i}, C_{ra3i} and C_{ra4i}, and C_{r5j}, C_{r6j}, C_{ra5j} and C_{ra6j} will perform resonation for the second and third stages, respectively.
- Finally, $PD3L_i$ sets the quantization resolution of the ADC and the DAC in stage i. The digital logic that controls the D/A feedback operation (DAC in Fig. 6.3) generates the control signals A_i, B_i, C_i, D_i, E_i and F_i (see Sect. 6.4.5). A logic operation between the last signals and $PD3L_i$ is then performed to obtain A'_i, B'_i, B''_i, C'_i, D'_i, D''_i, E'_i, F'_i and F''_i. These signals control the switches that perform the above-mentioned in-loop feedback operation. Note that the activation of signal $PD3L_i$ leads to a 5-level operation in the embedded quantizer and the feedback DAC, while a logic zero for the last signal establishes a 3-level quantization.

6.3.1.1 Quantizer Resolution Adjustment

Depending on their location in the quantizer structure, there are two types of comparators: type A and B. Their respective schematics are depicted in Fig. 6.5a and b. Two type-A comparators are used for 3-level quantization, whereas two type-B comparators are incorporated to the former ones when 5-level quantization is selected. The control signals governing the quantizer resolution are as follows:

- $PD3L = '1'$ makes the quantizer operate in the 5-level mode.
- $PD3L = '0'$ activates the 3-level quantization. Fig. 6.5c indicates that $PDstg5L = '1'$ in this case. This way, the preamplifiers of the type-B comparators are powered down to save power in this configuration of the quantizer, as suggested in Fig. 6.5b.

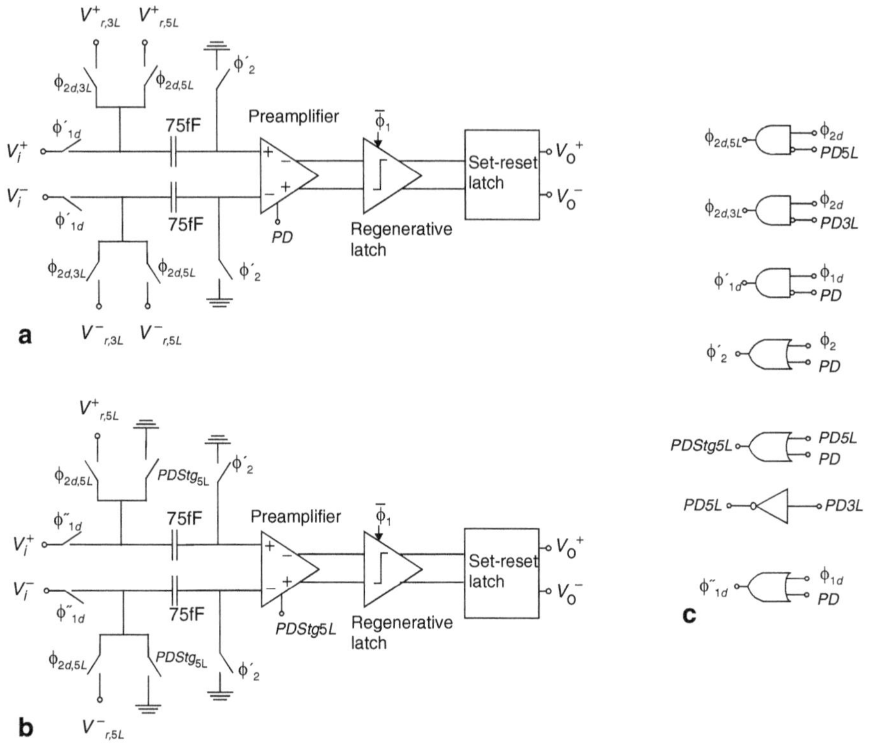

Fig. 6.5 Comparators SC input branch, reconfiguration and types: **a** type A, **b** type B, **c** reconfiguration logic

- $PD = '1'$ means that the whole stage is switched off, so the preamplifiers of all comparators in the quantizer are powered down[8] and their inputs are set to the common-mode voltage.

Note that the comparator type A needs two differential SC branches for its reference voltages and these are selected according to the quantizer resolution mode. This is indicated in Fig. 6.5a for voltages $V^+_{r,\,3L}$, $V^+_{r,\,5L}$, $V^-_{r,\,3L}$ and $V^-_{r,\,5L}$. Fig. 6.5c shows the modification of the phases that control the sampling of these comparator references (or tabs of the resistor ladder in the quantizer) with the aforementioned control signals *PD*3*L* and *PD*.

A single resistor ladder is implemented to generate the 3- and 5-level voltage tabs in order to save both area and power. Figure 6.6 shows how the I/O characteristic of the embedded A/D/A converter changes with the quantizer resolution.

Figure 6.7 shows the schematic of every embedded quantizer in the modulator. This schematic is divided into 3 parts. First, a resistive ladder is needed to provide the

[8] Section 6.3.3 shows that the comparator trigger signal $\overline{\phi_1}$ becomes static when *PD* is on. The regenerative latch is therefore also powered down in this case.

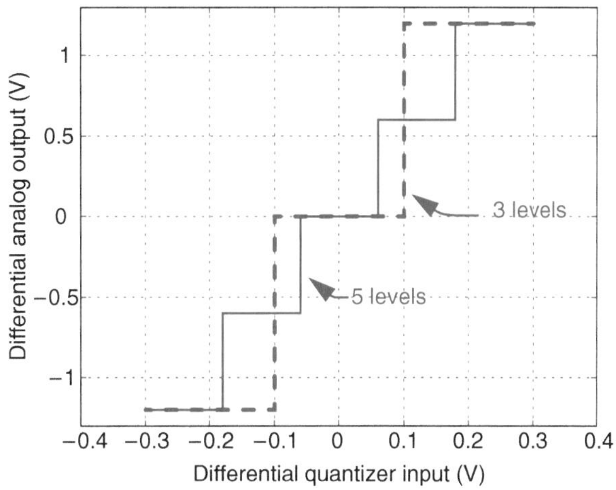

Fig. 6.6 Illustration of the reconfiguration in the A/D/A conversion I/O characteristic

voltage tabs for the appropriate comparisons, performed by the comparators ladder. The use of the different kinds of comparators is illustrated in the figure by a label inside a circle. Comparator type A is therefore used for the extremes of the comparator ladder, while the intermediate comparators are type B. Finally, digital encoding is used to transform the thermometric code given by the comparator ladder into a 1-of-5 code. The delays of all the digital paths in Fig. 6.7 have been compensated with buffers (not shown in the figure).

The resistor ladder compensates the 1/4 factor of the analog adder operation. This ladder is used for the generation of the voltage tabs for both 3 and 5 levels configurations. The resistor ladder is made up of 120 segments[9] of 20-Ω resistors and consumes 417-μW. The values of the voltages generated are shown in Fig. 6.7. The common-mode voltage is also taken from the midpoint of the resistive ladder. An on-chip capacitance of 52.5 pF together with an off-chip 22-μF capacitance are used to decouple this node.

6.3.1.2 Biasing Reconfiguration of the ΣΔM Building Blocks

Adapting the biasing of the main ΣΔM building blocks is a popular way of adjusting their power consumptions to their required performances [Lim06, Ouzo07, Chri07, Morg07b, Crom10, Chri10, Morg10]. This can be done either based on off-chip elements, as in Chap. 5, or by using current mirrors that are switched according to a set of control signals. Indeed, the prototype presented in Chap. 4 follows the latter approach (see Fig. 4.10). In the current prototype, the on-chip current mirroring has been improved to adopt a binary-weighted approach [Bult92].

First, a master current is generated through an external resistor, R, of 620 kΩ connected to a PMOS current mirror as shown in Fig. 6.8. A set of output currents

[9] Each segment is made up of two 40−Ω resistors in parallel configuration to improve matching.

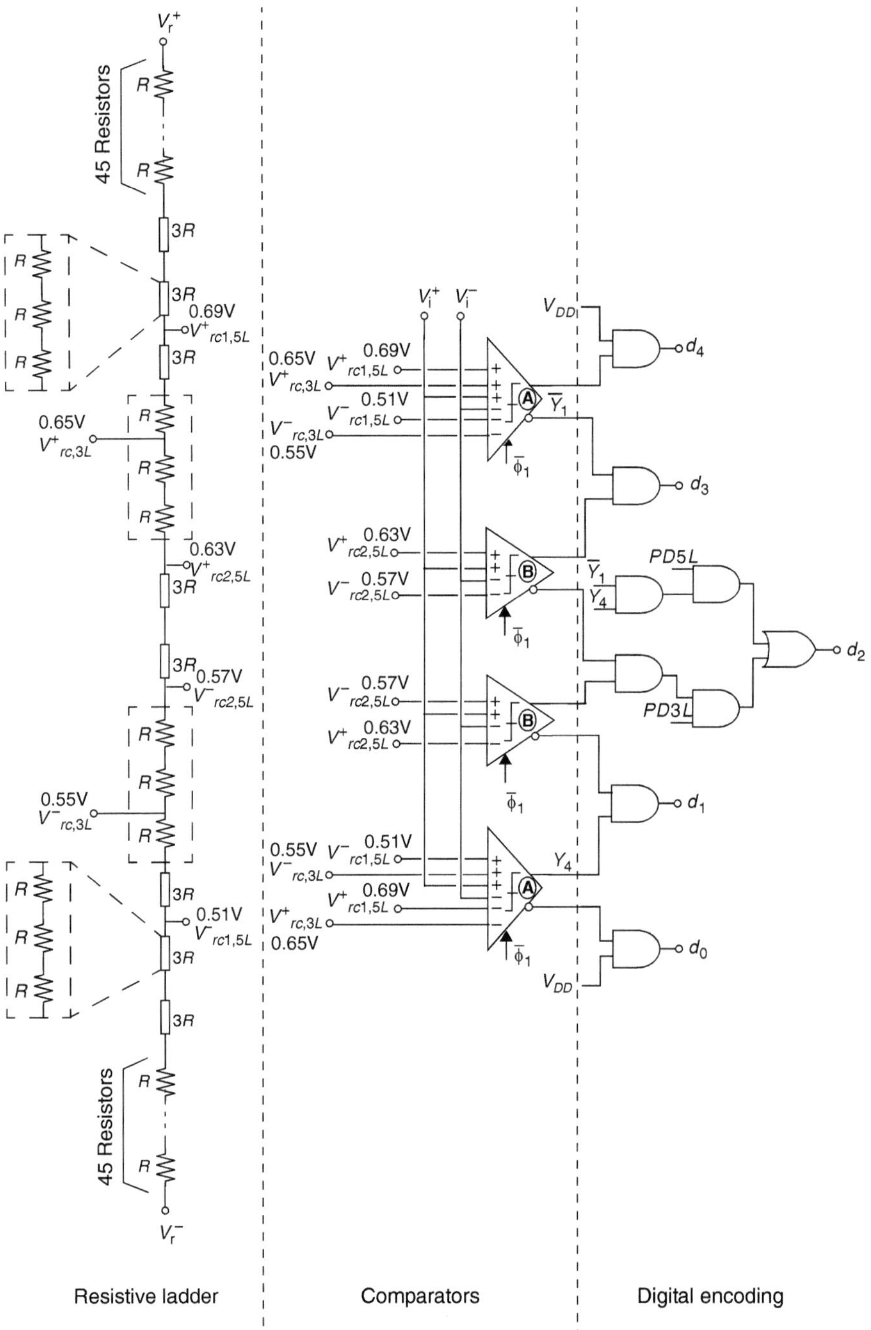

Fig. 6.7 Reconfigurable quantizer schematic

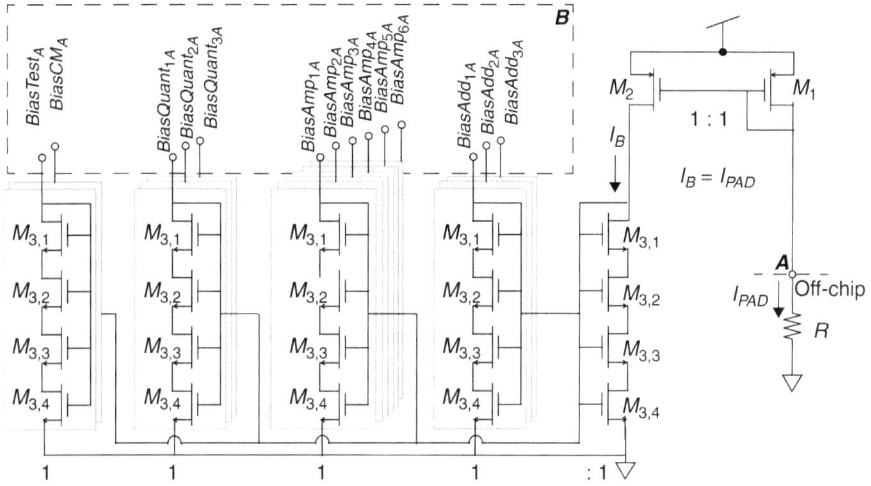

Fig. 6.8 Generation and mirroring of the master currents

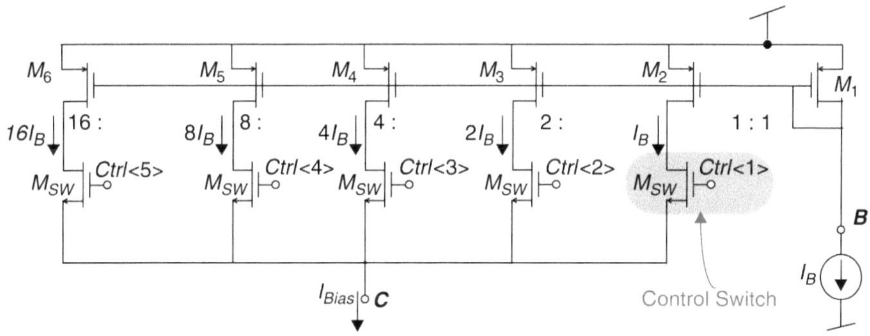

Fig. 6.9 Reconfiguration in the generation of the bias currents

are mirrored with unity factor. The effective transistor lengths in the NMOS mirrors are increased by placing transistors in series. This helps to draw a more squared and, thus, more compact layout of the elements. Once these intermediate currents (I_Bs) are generated, reconfiguration is performed. Fig. 6.9 shows how a 5-bit binary-weighted current mirror based on PMOS transistors is used for this purpose. All mirrored currents are then selected by control signals, *Ctrl<i>*, applied to the gates of NMOS-based control switches. These mirrored currents go directly to all the opamps in the modulator, all of them having a dedicated signal *Ctrl<i>* to control their bias current independently. The same scheme applies to the quantizers, so all the preamplifiers in each quantizer have the same bias current controlled with a 5-bit binary-weighted scheme as indicated in Fig. 6.9. However, this last mirrored bias current for the quantizers (I_{Bias} flowing through node C in Fig. 6.9) needs to be replicated in 4 tails to bias the four preamplifiers that conform a quantizer (see Fig. 6.7). Figure 6.10

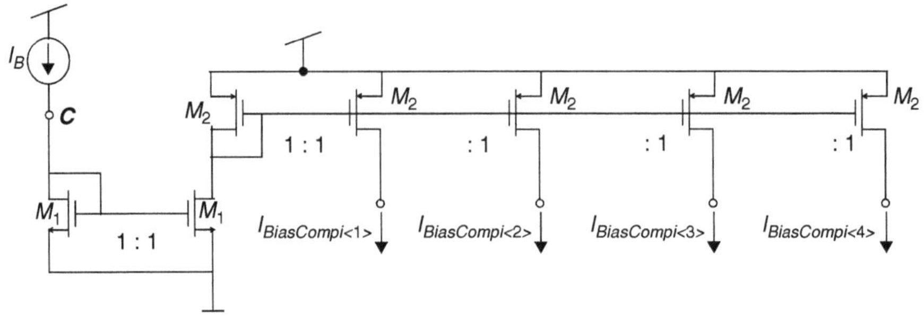

Fig. 6.10 Current mirror for the biasing of the comparator preamplifiers in the quantizers

depicts the current mirror used for the biasing of the comparator preamplifiers in the quantizers. An extra transistor M_2 has been incorporated to the scheme in Fig. 6.10 for the third-stage quantizer. This extra current tail is given to an off-chip resistor, so the variation of the biasing of the comparator preamplifiers can be measured for every possible configuration of the corresponding control word. One 5-bit binary-weighted current mirror, shown in Fig. 6.9, is also connected to an off-chip resistor to measure the biasing of the opamps. This is incorporated to this prototype for testing purposes and can be used to compensate for process, temperature and supply variations, and also for transistor mismatches.

Transistor-level simulations that consider corner and Monte Carlo analyses indicate that approximate maximum currents of 125 μA and 75 μA with an incremental step of 5 μA are obtained for the opamps and comparator preamplifiers, respectively. These values fall well within the desired operation range[10].

6.3.2 Implementation of Resonation Reconfiguration

Figure 6.3b shows the implementation of resonation for the second or the third stages. The signals $PDRes_2$<*i*> and $PDRes_3$<*j*> control the switches operation in the SC branches of capacitors C_{r3i}, C_{r4i}, C_{ra3i} and C_{ra4i}, and C_{r5j}, C_{r6j}, C_{ra5j} and C_{ra6j} for the second and the third stages, respectively. There are three possible values for each of these capacitors and the three corresponding SC branches are activated separately with $PDRes_2$<*i*> and $PDRes_3$<*j*> for any of the three possible *i*s and *j*s. This leads to the practical implementation of all the coefficients in Table 6.2 and 6.3. Table 6.5 shows the values of all capacitors used for resonation. Note that the permutations in Table 6.3 can be obtained by connecting a number of the SC branches corresponding to the capacitors in Table 6.5. This table also includes the value of the unit capacitor

[10] These values roughly double the maximum of the initially desired values, so there is useful room for higher operation in the opamps.

Table 6.5 Resonation capacitances

i	g_{2i}	C_{ra3i}		C_{ra4i}		C_{ui} (fF)	C_{r3i}		C_{r4i}		C_{ui} (fF)
		Value	N	Value	N		Value	N	Value	N	
1	0.25	25fF	2, Series	50fF	1	50	25fF	2, Series	50fF	1	50
2	0.125	12.5fF	4, Series	25fF	2, Series		12.5fF	2, Series	25fF	2, Series	
3	0.0625	6.25fF	2, Series	12.5fF	1	12.5	6.25fF	8, Series	12.5fF	4, Series	
j	g_{3j}	C_{ra5j}		C_{ra6j}		C_{uj} (fF)	C_{r5j}		C_{r6j}		C_{uj} (fF)
		Value	N	Value	N		Value	N	Value	N	
1	0.1	10fF	2, Series	20fF	1	20	10fF	2, Series	20fF	1	20
2	0.25	25fF	2, Series	50fF	1	50	25fF	2, Series	50fF	1	50
3	0.35	35fF	1	70fF	2, Parallel	35	35fF	2, Series	70fF	1	70

(C_u), the employed number of unit elements (N) and their connection configurations, either in series[11] or in parallel.

A number of capacitances are very small. Larger capacitors could have been implemented if larger unit capacitors had been used for the front-end integrator and the analog adder of the corresponding $\Sigma\Delta$M stage. However, this would have meant greater capacitance loads to be driven by the opamps and, therefore, more demanding dynamic requirements. The sizes of the switches of the resonation SC branches are scaled according to the value of the connected capacitor in order to reduce parasitic capacitances. Nevertheless, routing parasitics are unavoidable and their influence may prevent some of the resonation coefficients—especially the smaller ones—from working in practice.

6.3.3 Implementation of Multi-rating

Multi-rating needs different operations for frequencies in the $\Sigma\Delta$M stages depending on the operation mode [Colo02, Bos10]. There are two possible implementations for multi-rating, namely:

- To use as many clock signals as the number of stages in the $\Sigma\Delta$M. However, this solution needs a number of dedicated pins equal to the number of stages and can lead to synchronization problems[12].
- To internally divide a single clock signal by a ratio determined by the operation mode and the $\Sigma\Delta$M stage. In this case, even if there is a division factor of 1, there will be no synchronization problems for a cascade configuration.

[11] Note from Table 6.5 that a number of resonation capacitors are implemented in series. This way, larger unit capacitors can be used to reduce the actual capacitance by the number of elements in series. This helps to reduce capacitance mismatching. However, the influence of the bottom parasitics in the series connections may prevent correct resonation functionality.

[12] The implementation of a standard cascade $\Sigma\Delta$M would need that at least two successive stages to have a clock division factor of 1. A serious synchronization problem may therefore occur if 2 different off-chip clocks are provided to the chip because they are probably not correlated. One solution would be to take the same off-chip clock for this division factor of 1.

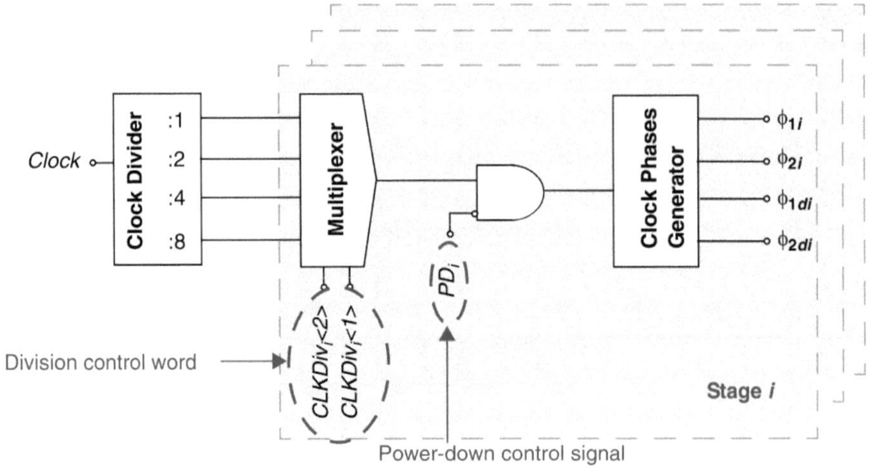

Fig. 6.11 Clock division and phase generation for the ΣΔM stages

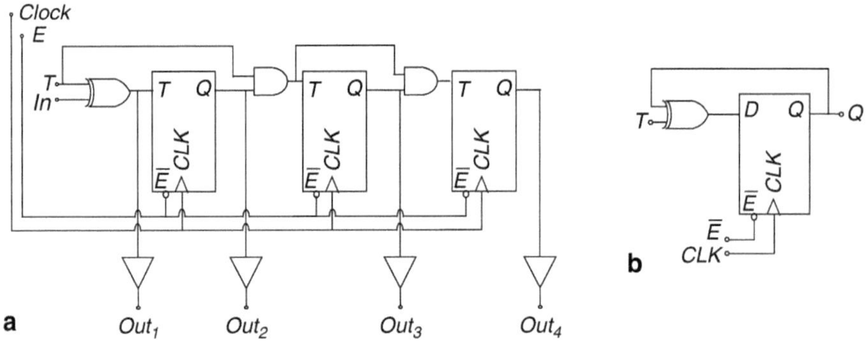

Fig. 6.12 Clock divider: **a** schematic, **b** T flip-flop

The second solution is the one implemented in this chip, so every phase in Fig. 6.3 is associated with the corresponding stage. For instance, ϕ_{1d1}, ϕ_{1d2} or ϕ_{1d3} correspond to ϕ_{1d} for the respective stage numbers 1, 2 or 3. The generation of all the phases is briefly explained below.

Figure 6.11 shows how the clock division and the phase generation of the stages are carried out. The external clock signal is connected to a frequency divider with division factors 1, 2, 4 and 8. The divider outputs are selected via a multiplexer controlled by signals *$CLKDiv_i$<j>* for each stage *i*. An AND gate will then turn off the master clock of each stage if PD_i is active. This prevents any switching activity in the phases of the stage that is switched off with PD_i. This strategy results in less power consumption and less switching noise. Details on the design of the clock phase generator at transistor level will be provided in Sect. 6.4.4.

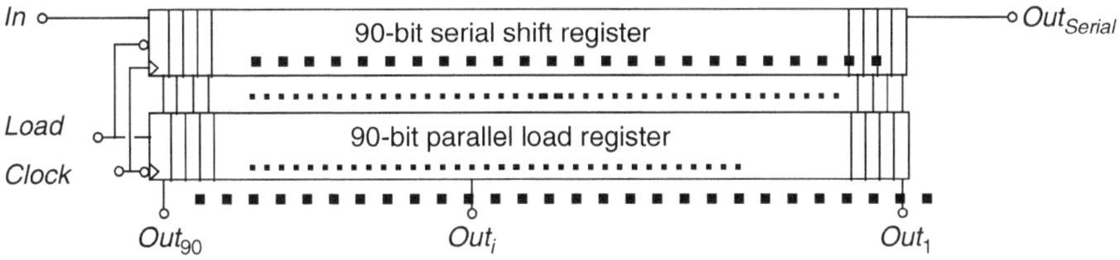

Fig. 6.13 Block diagram of the serial-in parallel-out register

6.3.3.1 Clock Divider

Figure 6.12 shows the clock divider schematic. Four different clock signals ($Out_{1\text{-}4}$) are obtained from the division in frequency of the input clock (*Clock*) by factors 1, 2, 4 and 8, respectively. The division is based on T flip-flops in series. Each of these elements makes use of an XOR gate together with a D flip-flop inside a local feedback loop to perform a division by 2 in frequency.

6.3.4 Control Signals Generation

Altogether the modulator reconfiguration strategies entail the use of 90 control signals. Such a large number requires a serial-to-parallel solution to provide the signals to the chip. To this purpose, a serial-to-parallel register—the block diagram of which is depicted in Fig. 6.13—is used to collect serial input data (*In*) and transform it into 90 parallel control bits (Out_i). The serial output of the shift register (Out_{Serial}) can be captured off-chip to verify that the shifting and loading operations were performed correctly. The figures show how two registers are used. The one at the top is used to shift the data in serial mode, while the one at the bottom places all the signals previously shifted by the register at the top in parallel. A clock input signal (*Clock*)—different from that of the $\Sigma\Delta$M—is needed for the operations of the registers. An additional input signal (*Load*) enables either the shifting operation in the top register or the loading in the bottom register in Fig. 6.13.

Figure 6.14 illustrates the encoding of the serial input word, while Table 6.6 provides a brief description of the functionality of the control signals. There are 5 control signals in the table—denoted as *NoUse*<4:0>. These were initially contemplated but finally not needed.

6.3.4.1 Serial-to-Parallel Control Register

2×90 D flip-flops (DFFs) are needed to implement the two registers in Fig. 6.13. If a flat scheme is directly implemented, fan-out issues [Geig90] will impose the use of power-hungry buffers in order to drive *Load* and *Clock* signals to the 2×90 DFFs. In order to drive this heavy load more efficiently, the DFFs of the 90-bit register are

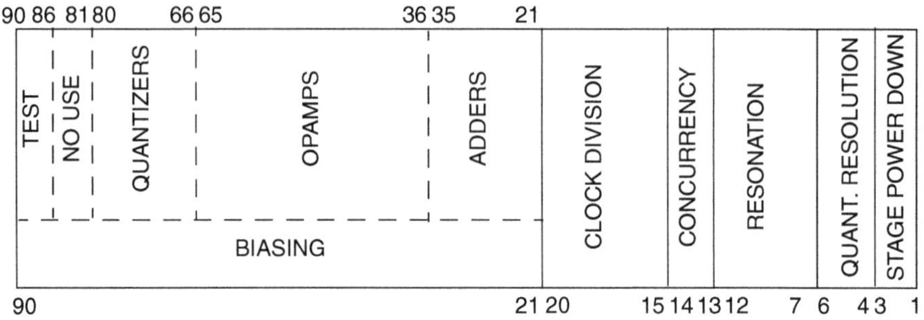

Fig. 6.14 Encoding of the serial input word

Table 6.6 Description of the serial control word

Index	Signal name	Description
1..3	$PD_{1:3}$	Stages power down signals
4..6	$PD3L_{1:3}$	Control of quantizers resolution
7..9	$PDRes_2$<1:3>	Resonation control for 2nd stage
10..12	$PDRes_3$<1:3>	Resonation control for 3rd stage
13	SL_2	Enabling concurrency for 2nd stage
14	SL_3	Enabling concurrency for 3rd stage
15, 16	$CLKDiv_1$<1:2>	Clock division for 1st-stage clock phases
17, 18	$CLKDiv_2$<1:2>	Clock division for 2nd-stage clock phases
19, 20	$CLKDiv_3$<1:2>	Clock division for 3rd-stage clock phases
21..25	$BiasAdd_1$<0:4>	Analog adder biasing
26..30	$BiasAdd_2$<0:4>	
31..35	$BiasAdd_3$<0:4>	
36..40	$BiasAmp_1$<0:4>	Opamp biasing
41..45	$BiasAmp_2$<0:4>	
46..50	$BiasAmp_3$<0:4>	
51..55	$BiasAmp_4$<0:4>	
56..60	$BiasAmp_5$<0:4>	
61..65	$BiasAmp_6$<0:4>	
66..70	$BiasQuant_1$<0:4>	Comparator preamplifiers biasing in the quantizers
71..75	$BiasQuant_2$<0:4>	
76..80	$BiasQuant_3$<0:4>	
81..85	*NoUse*<0:4>	No use
86..90	*BiasTest*<0:4>	Opamp biasing test

grouped into 4-bit sections with small local buffers to internally drive the signals. Fig. 6.15 depicts this basic cell comprising 4 shifting flip-flops and 4 loading ones. Note from Fig. 6.15 that local buffers are also used at each parallel output. These buffers do not need high driving capabilities since the corresponding signals will also be buffered locally when used far away from the driving point.

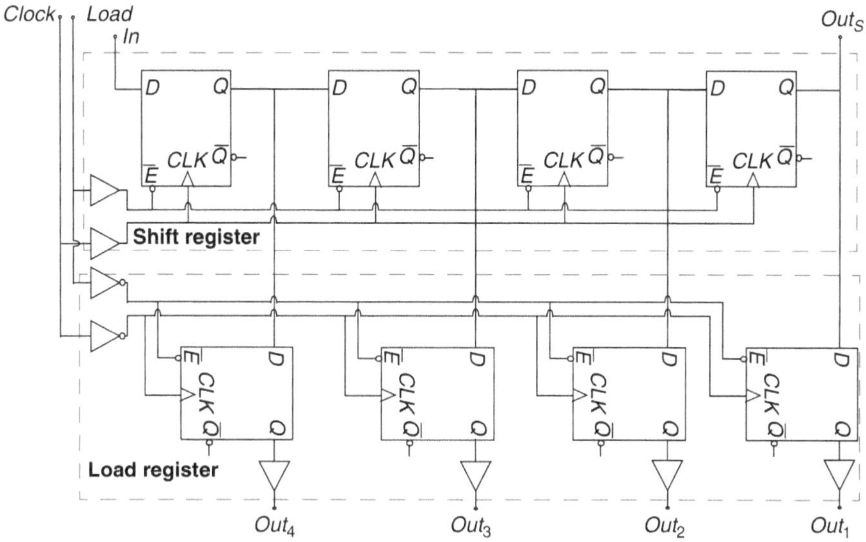

Fig. 6.15 Schematic of the basic register cell

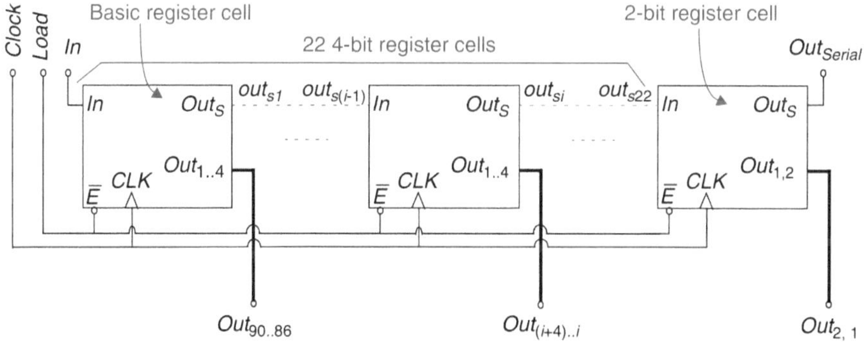

Fig. 6.16 Schematic of the complete serial-to-parallel control register

Figure 6.16 shows the complete schematic of the serial-to-parallel register, in which the cell in Fig. 6.15 is used as the basic element. The last element in the schematic is made up of only 2 shifting and 2 loading flip-flops. Note that this implementation allows a tree structure for the distribution and routing of the register clock (*Clock*) and the signal *Load*, which simplifies the design and the layout as well as reducing buffer power consumption as stated above.

6.3.4.2 Timing Operation

Figure 6.17 illustrates the timing diagram of both the shifting and the loading operations, sequentially. The serial input signal *In* changes at a rate given by the clock

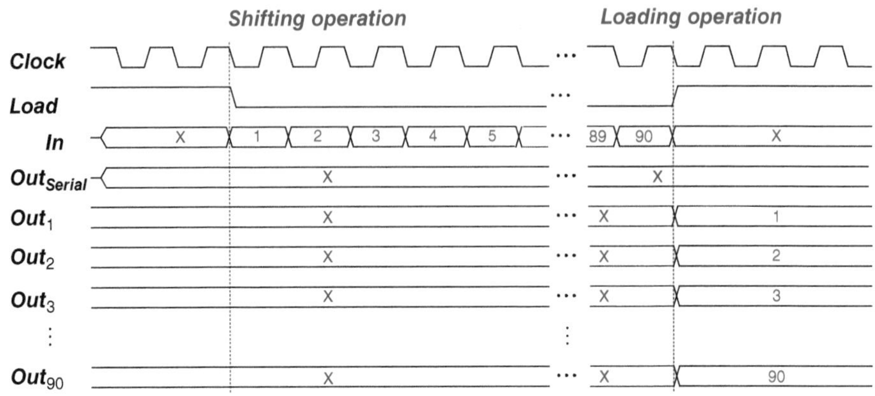

Fig. 6.17 Timing diagram of the serial-to-parallel control register

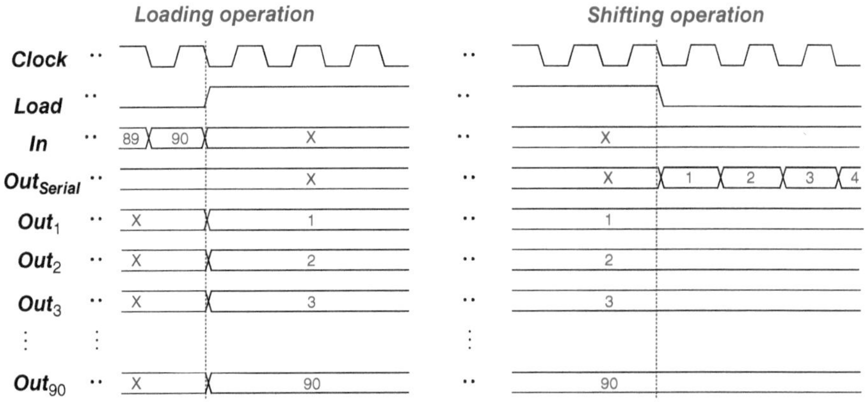

Fig. 6.18 Timing diagram of the output extraction of the data in series

signal *Clock* when *Load* is in shifting mode (logical 0). In this mode, each of the flip-flops in the top row of Fig. 6.15 shifts the serial stream at every rising edge of the *Clock* signal. The loading operation occurs at the rising edges of *Load* signal by capturing the output of the 90 shifting flip-flops at that very moment; i.e., the output of the last 90 cycles in which *Load* was in shifting mode. Following this scheme, exactly 90 cycles are needed in shifting mode to provide the 90 control signals. No reset is included in the serial-to-parallel register, so the values of the loading register outputs remain unknown until the *Load* signal is activated. Note that, after the preloading of the register as indicated in Fig. 6.17, Out_{Serial} provides the output of the shift register in an in-series configuration; i.e., each bit of the serial output shifts sequentially with the falling edges of the *Clock* signal when *Load* is $'0'$ as indicated in Fig. 6.18. This can be used to verify that the serial register is shifting the bits properly when testing.

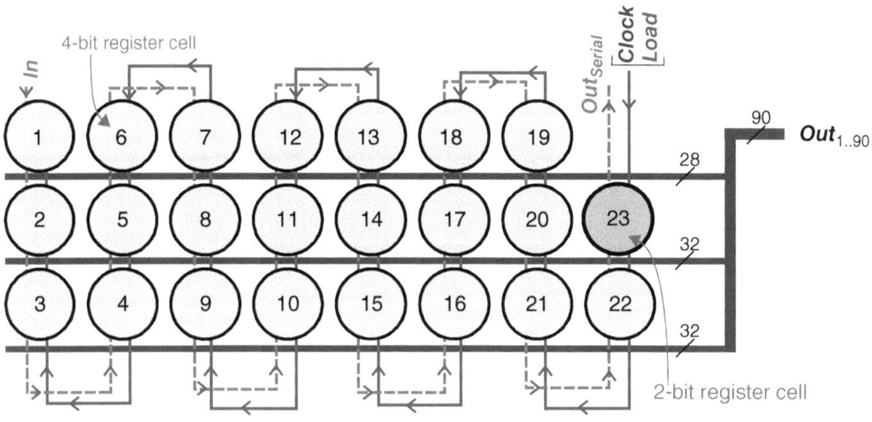

Fig. 6.19 Floorplan of the serial-to-parallel control register

6.3.4.3 Layout

Figure 6.19 shows the layout floorplan of the serial-to-parallel register, together with the distribution of all its signals. The figure indicates that the last cell register is a 2-bit cell, while all the previous ones are 4-bit register cells. The input signal serial data, *In*, is routed from the first to the last cell. In contrast, *Clock* and *Load* signals are provided from the last cell sequentially to the preceding cells. This routing strategy potentially prevents the register from triggering a flip-flop before another positioned later in the serial connection due to routing delays. Otherwise, this might have resulted in an erroneous serial-to-parallel conversion. Furthermore, the speed of this clock can be made as low as desired, so high-frequency issues can easily be avoided.

6.4 Electrical Design: Reuse and Improvements of Previously Designed Building Blocks

The previous section illustrates how the reconfiguration strategies in the $\Sigma\Delta$M have been implemented at the electrical level. This section presents the design of the main modulator blocks (amplifiers, comparators, etc) at the transistor level. As will be seen, most of them have been reused from the prototype presented in Chap. 5. The modifications and improvements of these blocks will be also addressed.

6.4.1 Amplifiers

The amplifier used in this modulator prototype is basically the folded-cascade amplifier presented in Fig. 5.10, with a minor modification; namely, an increase of

Table 6.7 Equivalent capacitive loads of each integrator and adder

L	Eq. load	Int. #1	Int. #2	Adder #1	Int. #3	Int. #4	Adder #2	Int. #5	Int. #6	Adder #3
6	$C_{eq,s}$ (pF)	0.62	2.04	0.35	0.71	2.55	0.35	0.71	0.82	0.35
	$C_{eq,i}$ (pF)	0.42	0.31	1.02	0.76	0.29	1.24	0.76	0.29	1.24
4	$C_{eq,s}$ (pF)	0.62	2.04	0.35	0.71	0.82	0.35			
	$C_{eq,i}$ (pF)	0.42	0.31	1.02	0.76	0.29	1.24			
2	$C_{eq,s}$ (pF)	0.62	0.68	0.35						
	$C_{eq,i}$ (pF)	0.42	0.31	1.02						

Table 6.8 Results for amplifier performance obtained by simulation at the transistor level

I_B (μA)	5	10	20	30	40	50	60
DC gain (dB)	47.09	47.16	47.31	47.13	46.71	46.1	45.34
***gm* (mA/V)**	1.02	1.71	2.77	3.66	4.35	4.87	5.46
GB_i (MHz)	524	878	1422	1879	2233	2500	2803
GB_s (MHz)	79	133	216	285	339	380	426
Phase margin (°)	68.54	70.19	71.91	73.02	73.65	74.09	74.61
***Io* (μA)**	42.77	91.66	193.9	298.4	404.1	510.5	617.4
SR_i (V/μs)	57	125	269	417	571	726	884
SR_s (V/μs)	7	15	33	52	71	90	110
$C_{eq,i}$ (pF)	0.31						
$C_{eq,s}$ (pF)	2.04						
Input cap. (fF)	162.5	172.2	179.6	184.0	187.6	190.8	193.7
Output cap. (fF)	26.72	20.57	14.14	12.14	10.12	9.08	8.46
Output swing (V)	±0.96	±0.91	±0.83	±0.76	±0.7	±0.64	±0.59
Power (mW)	0.28	0.53	1.01	1.48	1.94	2.4	2.85

150% in the transistor sizes at the differential input pair. This results in a larger DC gain and *GB* for the same bias current, I_B. However, the input parasitic capacitance of the opamp is also higher. The equivalent capacitive loads during integration and sampling—defined in (5.9)—may be also greater. This is partially compensated by using small unit capacitors if possible. Table 6.7 shows the equivalent capacitive loads of the amplifiers for the different orders to which the ΣΔM can be configured. Note that their values are comparable to those in the previous prototype, listed in Table 5.8.

Table 6.8 summarizes the amplifier performance for different values of I_B. Note that Table 6.8 considers the equivalent capacitive loads given by Table 6.7 for the second integrator in either a 2-2-2 cascade configuration or 2-2 cascade configuration. Dynamic performance is shown based on the transconductance (*gm*), GB during integration and sampling phases (GB_i and GB_s, respectively), the current at the opamp output branch (I_o) and SR during integration and sampling phases (SR_i and SR_s, respectively). GB_i, GB_s, SR_i and SR_s are used to compare dynamic performance between the opamp in Chap. 5 and this one. Since the values of the equivalent

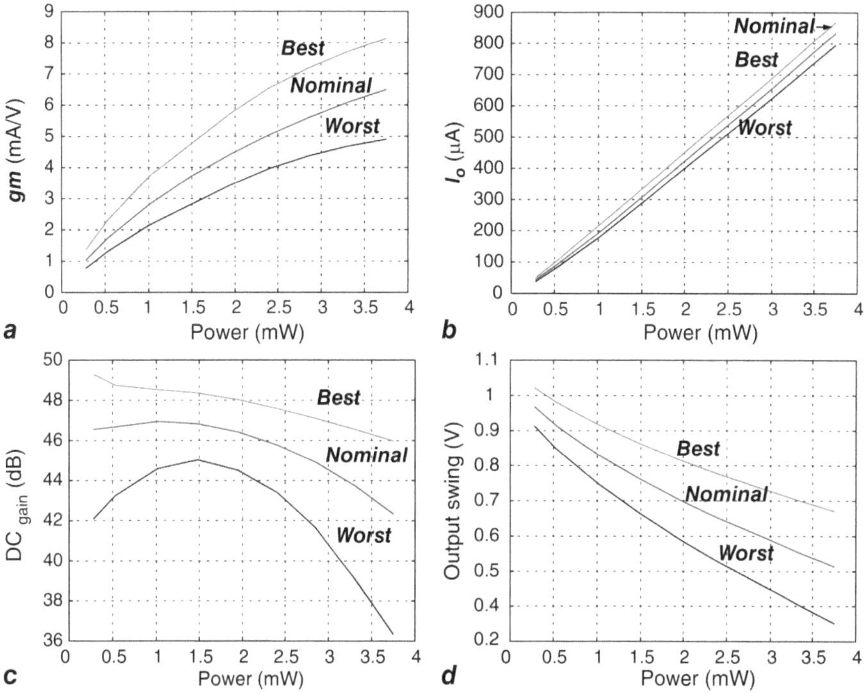

Fig. 6.20 Opamp performance adaptability on: **a** transconductance, **b** output current, **c** DC gain, **d** output swing

capacitive loads are similar in both cases and *gm* is greater in this case, better dynamic performance can be expected. This can easily be verified by a visual comparison of Tables 5.9 and 6.8 for the amplifiers in Chap. 5 and those in this design. For instance, GB_i and GB_s are 194 and 170% greater when this design is compared with that presented in Sect. 5.3.1 for $I_B = 40\ \mu$A, while SR_i and SR_s are 120 and 134% greater. These results coincide with the reduction in the equivalent loads (79 and 90% for the integration and sampling equivalent capacitive loads, respectively) but also vindicate the use of greater sizes of transistors for the input differential pair. This strategy also helps to increase the opamp gain by approximately 3 dB.

The variation in the opamp simulated results versus I_B are considered together with a corner analysis that includes a ± 5% supply variation, slow and fast devices and a temperature range of [−40°C, +85°C]. Figure 6.20 illustrates the variation in the opamp *gm*, I_o, DC gain and output swing with I_B. The best, nominal and worst-case results of the corner simulations are also shown in the figure.

Figure 6.21 illustrates the variation in the opamp gain non-linearities with I_B. Fig. 6.21a depicts the DC transfer characteristic and how it is modified by I_B, while Fig. 6.21b shows the evolution of the opamp non-linear gain versus I_B.

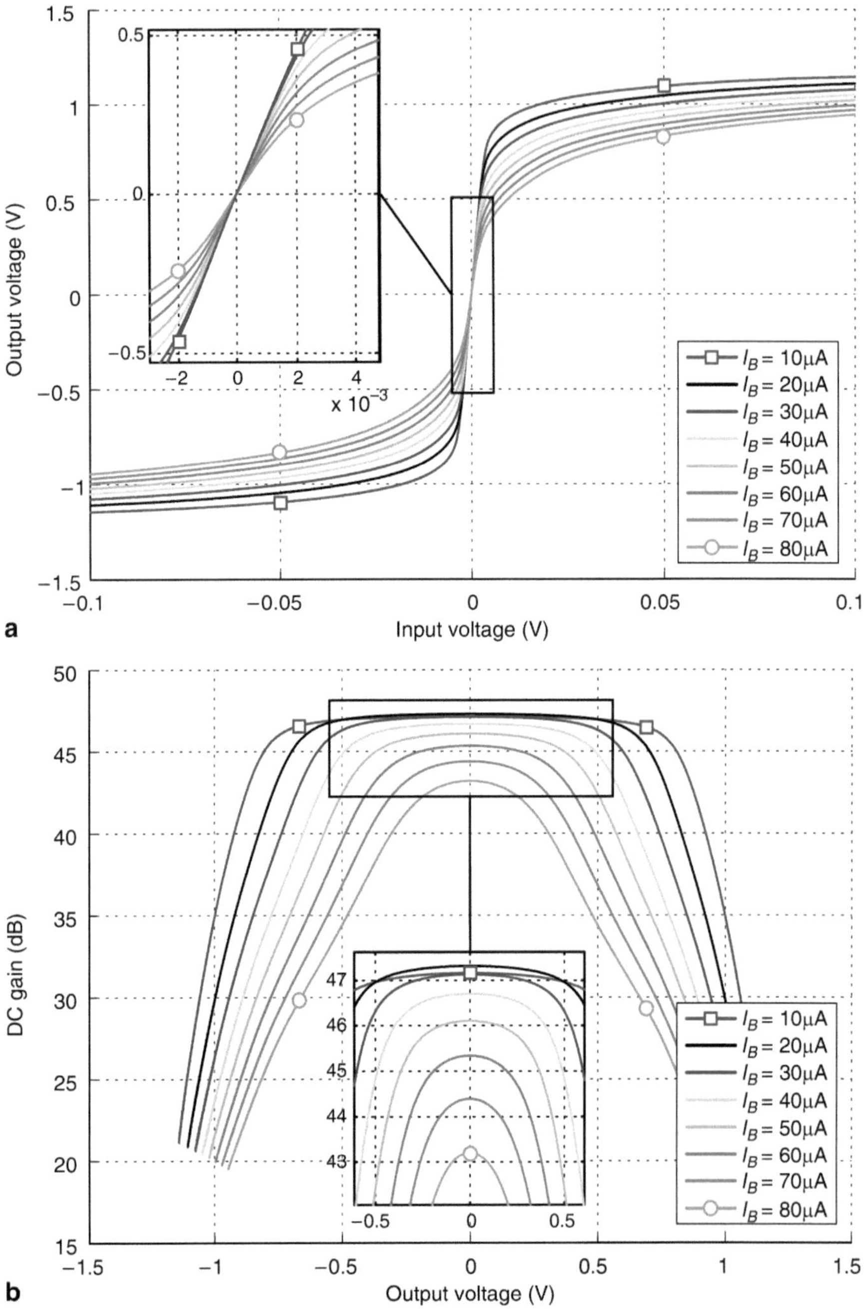

Fig. 6.21 Illustration of the opamp gain non-linearities: **a** I/O characteristic, **b** gain versus output voltage

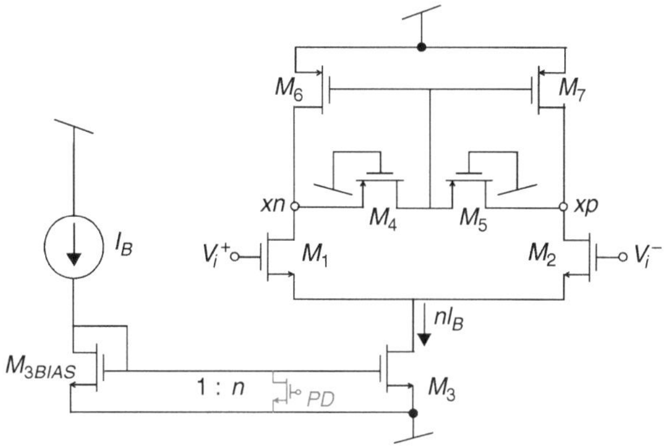

Fig. 6.22 Comparator preamplifier schematic

6.4.2 *Comparators*

The same comparator architecture from previous prototypes—a preamplifier, regenerative and set-reset latches—is again used here. However, an SC network is included at the comparator input to improve preamplifier performance and to enable the implementation of a reconfiguration strategy for the embedded quantizer resolution—either 3 or 5 levels. The regenerative and set-reset latches are the same as in Chap. 5, whereas the sizing of the preamplifier is modified to improve its performance.

Given that the sampling and subtraction of the differential signals and references is performed in this prototype by an SC network at the comparator input, only one input differential pair is now required at the preamplifier compared to previous prototypes.

Figure 6.22 shows the preamplifier schematic, in which the DC current flowing through the output loading transistors is halved for the same bias current nI_B when compared to Fig. 5.21. This leads to greater output resistance of the preamplifier and, consequently, to higher DC gain. The ratio n of the NMOS current mirror is also modified from 4 to 16. This helps to increase SR and GB for a lower I_B that can be adjusted on-chip. This preamplifier is based on that presented in Sect. 5.3.4 but with an increase of 150% in the sizes of the transistors in the input differential pair. This improves the DC gain and the GB and SR performances. M_3 and M_{3BIAS} sizes are also changed to keep transistor M_3 in strong saturation in accordance with the operation range of I_B. A NMOS transistor is included in the preamplifier schematic in Fig. 6.22 to power down the block with either *PD* or $PDStg_{5L}$ signals depending on the comparator type as indicated in Fig. 6.5.

The preamplifier performance is evaluated through transistor-level simulations. Then, the offset is characterized based on a different method, which will later be conceptually compared with the ones used in the previous chapters. Monte Carlo and corner simulations are used for these simulations.

Figure 6.23 illustrates the variation in several preamplifier features as a function of its bias current; namely, GB, DC gain, SR and power consumption.

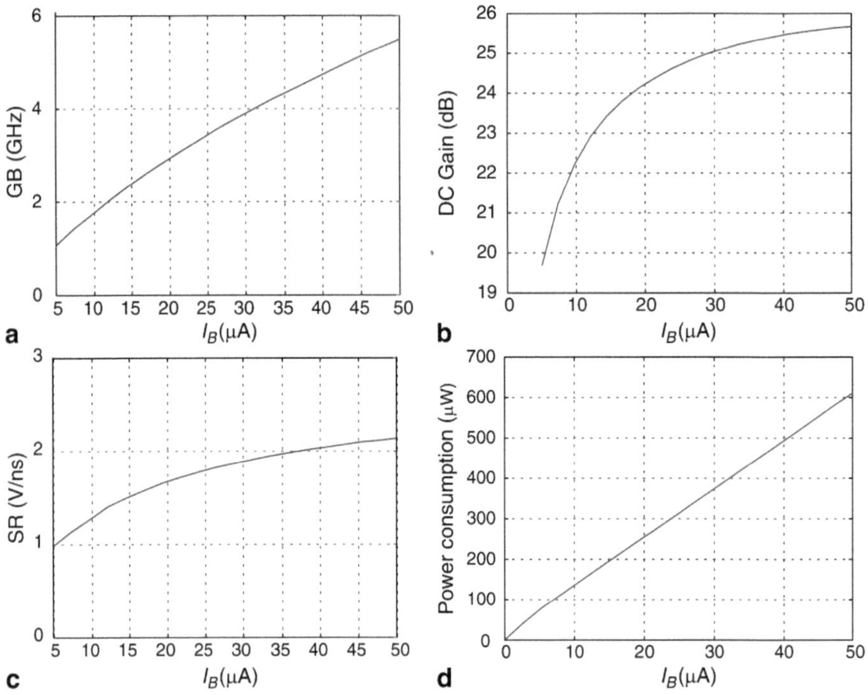

Fig. 6.23 Preamplifier performance versus its bias current: **a** GB, **b** DC gain, **c** SR, **d** power consumption

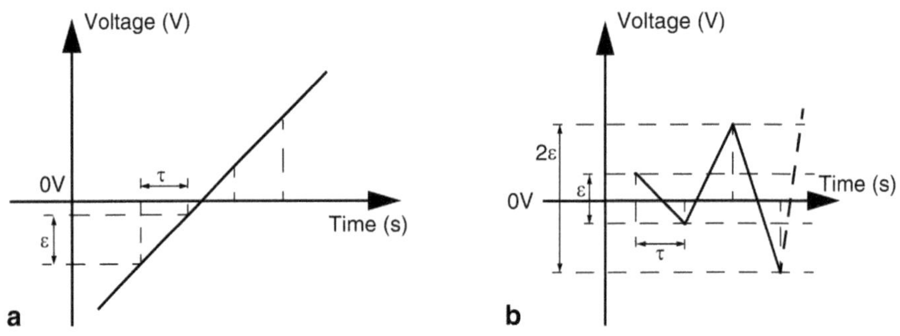

Fig. 6.24 Input signal for offset characterization: **a** ramp signal, **b** triangular approach

Offset can be characterized in a number of ways. For instance, a ramp input signal might be used or the bisectional method employed as presented in Chaps. 4 and 5, respectively. Fig. 6.24 compares the approach based on a ramp input with another based on a triangular input. All the regenerative latches of the comparators are discrete-time because a trigger signal is needed for the regenerative operation. This solution, compared to its continuous-time counterpart, results in faster comparators [Malo01]. However, a time grid is needed for the characterization of these systems

Table 6.9 Results of comparator performance obtained by electrical simulation

			Offset (mV)		
		mean	sigma	worst case	
		0.78	1.56	1.60	
t_{lh} (ps)			t_{hl} (ps)		
mean	sigma	worst case	mean	sigma	worst case
280.7	20.9	325.0	174.4	3.9	270.0

given by the trigger signal periodicity (τ). This entails a certain offset grid in voltage, ε, when either a ramp or a triangular input signal are employed[13].

A method based on an input ramp does not allow for memory or hysteresis issues in the comparator operation because the voltage value of the input signal is always growing. The ramp method can therefore be considered an optimistic approximation. A more realistic solution that does allow for the aforementioned problems in the comparator can be implemented with a sinusoidal or triangular input signal. If a sinusoidal input is used, there must not be any correlation in frequency between its periodic frequency and that of the comparator trigger; otherwise, the same input signal will be taken reiteratively. However, more conservative results are obtained with a triangular input signal because this signal is forced to take alternative signs in order to allow for memory effects. In this case, the values of the input signal change at the same time as the comparator trigger. The absolute value of the input amplitude is increased every 2 cycles by the chosen comparator grid ε. Note that the slope of the input signal changes its sign every cycle to produce the triangular shape.

Table 6.9 shows the offset results for the triangular input signal approach. Low-to-high and high-to-low resolution times (t_{lh} and t_{hl}, respectively) are also included in the table. 200-run Monte Carlo simulations, together with the worst case of a corner analysis considering fast and slow device models, $\pm 5\%$ variation in the 1.2-V supply and temperatures in the range of [−40 °C, +85 °C], are considered.

6.4.3 Switches

As shown in Sect. 5.3.3, a ratio of $W_{\mathrm{PMOS}}/W_{\mathrm{NMOS}} = 3.125$ provides an equalized DC response of the switch on-resistance. This ratio is maintained for all the switches in this $\Sigma\Delta$M. There is also a trade-off between the mean value of the switch on-resistance and the drain and source parasitic capacitances of its CMOS transistors. Fig. 6.25 illustrates how these two features change with the PMOS transistor width, while maintaining the relation $W_{\mathrm{PMOS}}/W_{\mathrm{NMOS}} = 3.125$. The criterion for the sizing of the switch transistors is to use large transistors in the switches in which high linearity and high speed are required, and low sizes for low values of capacitive parasitics in

[13] Note that the bisectional approach also presents a certain offset resolution or grid that is dependent on the number of interval divisions as described in Sect 5.3.4.

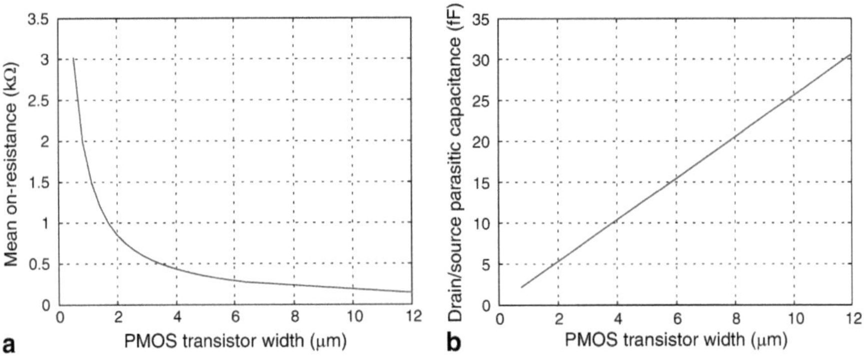

Fig. 6.25 Electrical simulation results of the switches versus transistors sizes: **a** on-resistance, **b** parasitic capacitances

the SC branches with small capacitors. This way, the influence of capacitive parasitics is diminished. The sizes employed are those with PMOS widths of 12, 6, 3, 1.5 and 0.75 μm.

6.4.4 Clock Phase Generators

As stated in Sect. 6.3.3, each stage of the ΣΔM can operate at different frequencies. To that purpose, the external clock signal is initially divided in frequency by factors 1, 2, 4 and 8 and the corresponding set of non-overlapped phases is the generated for each ΣΔM stage (see Fig. 6.11).

Details on the clock phase generators and the distribution and routing of the clock phases are given below.

6.4.4.1 Clock Phase Generation

This circuitry transforms its input clock into two non-overlapped phases and their delayed versions. Fig. 6.26 shows its schematic, which is very similar to the one presented in Chap. 5. The only difference is the inclusion of 2 additional inverters (highlighted in the figure) after the two-input inverter; i.e., just before driving the output phases. This leads to an increase in the non-overlapping time between the phases [Park01, Silv04b, Ruiz09] as will be indicated below in the post-layout simulations. This helps the ΣΔM to operate faster.

6.4.4.2 Phase Distribution

The chip presented in Chap. 5 makes use of a very wide bus for phases routing and distribution. Modification of this scheme is unavoidable in this case since each stage

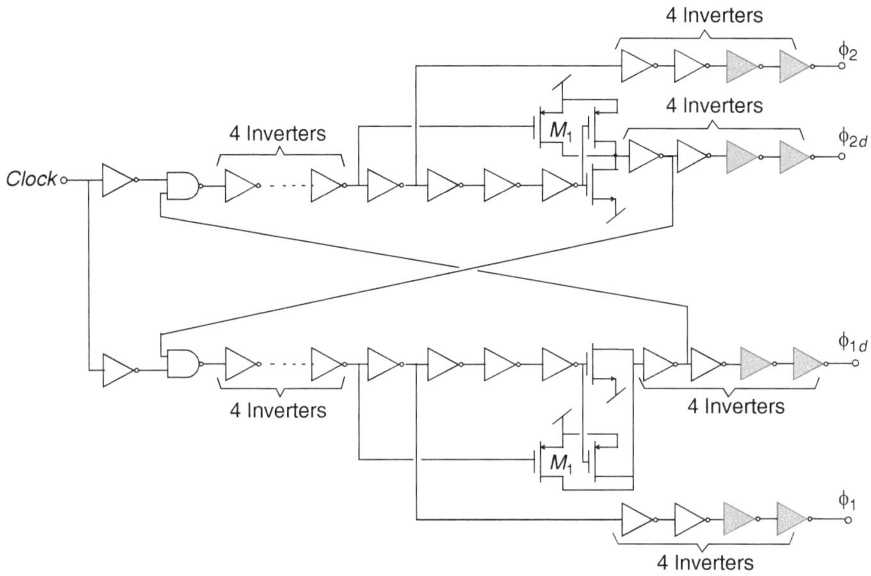

Fig. 6.26 Clock phases generator

must be able to operate at different frequencies. This drastically increases the number of phases to be routed, which—together with a larger number of DAC levels (from 3 to 5) and of $\Sigma\Delta$M stages (from 2 to 3)—makes it necessary to reduce the width of the bus.

The phase distribution is similar to that described in Sect. 5.3.6. A U-shaped bus is therefore used for this purpose. Each phase is isolated using a Faraday cage [Krau92] with 2 ground walls at each side of the signal. A number of modifications are carried out in this design to reduce the bus width, namely:

- Only the phases, their delayed versions and the DAC outputs are routed along the chip. Unlike the phases bus in Chap. 5, the inverted versions of all previous signals are not globally routed but locally generated by an inverter instead. This results in the distribution of 30 signals, whereas 31 were needed for the phases in Chap. 5.
- The spacing among the 2 parallel isolating ground paths and the distance from the signal path to the surrounding isolating grounds are both reduced.

Altogether, these strategies reduce the bus width by 45%.

The LCR lumped model of the bus routed along the chip has been obtained following the procedure described in Sect. 4.4.6—see Fig. 4.19d—so it has been considered in the electrical simulations of this prototype. The model is divided into three sections, each of them corresponding to a stage in the modulator. The lengths of the bus for the first, second and third stages are estimated in a worst case as 1.5, 1.4 and 1.2 mm, respectively.

Table 6.10 Parameters of the LCR lumped model for the bus routing

ΣΔM stage	Worst-case bus length	L_T (pH)	C_T (fF)	R_T (Ω)
First	1.5 mm	600	150	315
Second	1.4 mm	560	140	294
Third	1.2 mm	480	120	252

Table 6.10 indicates the parameters of the LCR lumped model. Note that, although the bus width is considerably reduced compared to the prototype in Chap. 5, the value of parasitics, specially capacitive parasitics, has increased.

All clock phases are locally buffered to the transistors gates of the switches, using one buffer for the NMOS transistor gate and an inverter for the PMOS gate. This relaxes the driving requirements of the phase buffers that are routed along most of the layout in this ΣΔM. Their fan-outs are therefore reduced by 50% and, consequently, the power consumption is also diminished. Furthermore, most of the reconfiguration digital control gates—see Fig. 6.3c—are also implemented locally to each switch when needed. This also helps to reduce the number of routing signals in the phases bus and, thus, to diminish both power—because there are less phases buffers—and also area consumed by the phases bus.

Post-layout simulations contemplating the buffers and the LCR lumped models mentioned above have been performed. The effective operation times for the sampling and the integration phases are 77 and 79% of half of the clock period respectively for a clock frequency of 320 MHz. The resulting non-overlapping time is 2.3 times greater than that in Chap. 5 (see Sect. 5.3.6), while the falling phase delay is approximately 75 ps.

6.4.5 DAC and ROM

Figure 6.27 shows the digital gates used to binary encode the quantizer digital outputs d_j that control the D/A operation of the feedback switches. A number of buffers—not shown in the figure—are used, and their fan-out and that of the DAC digital gates are adjusted to obtain the same delay for every output of the DAC control logic in Fig. 6.27.

Figure 6.28 depicts the ROM schematic, the outputs of which are directly captured off-chip. Note that the control signals *PD3L* and *PD5L* are needed to allow for the dual resolution mode of the quantizers.

For a given active code in the quantizer d_j, there is a certain codification of the modulator output bit streams in the ROMs. There is also a relationship between the latter codes (active code in the quantizer output and modulator output code) and the in-loop fed back voltage, which is indicated in Table 6.11. Note from this table that the encoding of the modulator's output (i.e., the binary codification of the ROM) is not implemented as a conventional increasing binary code. In contrast, this code allows the transistors in Fig. 6.28 to be shared for both types of configurations (3 and 5 levels) and, as a result, simplifies the circuitry in the ROM. As described in

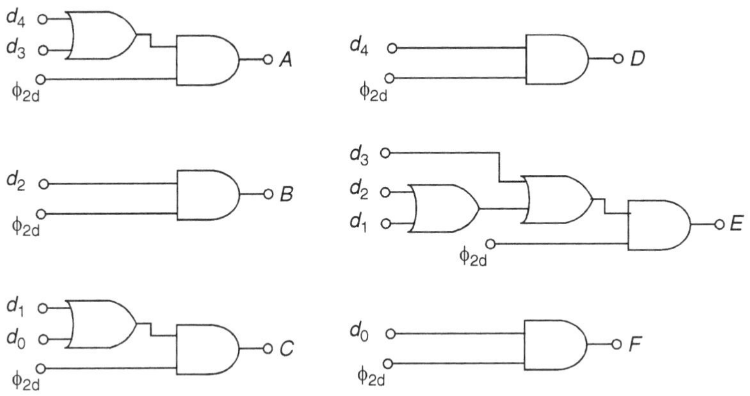

Fig. 6.27 DAC digital logic implementation

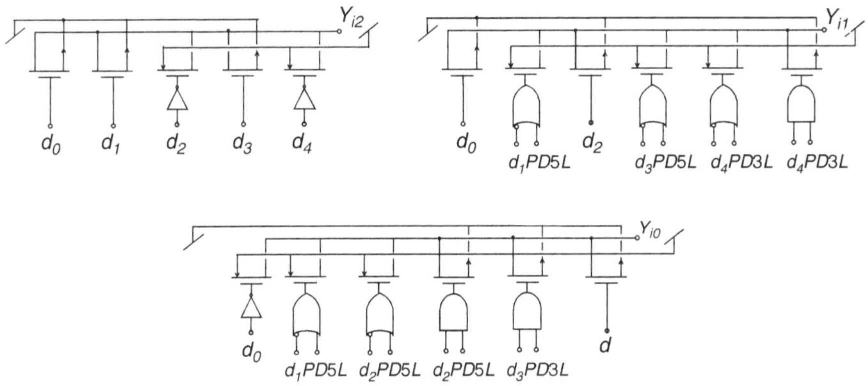

Fig. 6.28 ROM schematic

Sect. 6.2 (see Fig. 6.3), two sampling capacitors are used for the in-loop feedback operation in a 5-level configuration of the DAC and only one sampling capacitor in the 3-level configuration. The logic governing the connection of the corresponding SC branches to V_r^+, V_{cm} and V_r^- in phase ϕ_{2d} is also included in Table 6.11.

6.5 Layout, PCB and Test Set-up

This section describes the layout of the chip, putting special emphasis on the pad I/O ring. The schematic and the layout of the PCB are shown and the main differences with that in Chap. 5 are addressed. Finally, the test set-up, that will be the starting point of the experimental characterization in next section, is explained.

Table 6.11 Truth table for the ROM encoding and the corresponding DAC output voltages

Active code of the quantizer output	Encoding of the modulator output			DAC output voltage					
				1st SC branch			2nd SC branch		
	Y_{i2}	Y_{i1}	Y_{i0}	V_r^+	V_{cm}	V_r^-	V_r^+	V_{cm}	V_r^-
5 levels									
d_4	1	0	0	1	0	0	1	0	0
d_3	0	1	0	0	1	0	1	0	0
d_2	1	0	1	0	1	0	0	1	0
d_1	0	1	1	0	0	1	0	1	0
d_0	0	0	1	0	0	1	0	0	1
3 levels									
d_4		1	0	1	0	0			
d_2	---	0	0	0	1	0			
d_0		0	1	0	0	1			

6.5.1 *Layout*

Figure 6.29 shows the layout of the chip, in which the main $\Sigma\Delta$M building blocks are highlighted, while Fig. 6.30 shows its microphotograph. This layout makes use of all the techniques reported in Chaps. 4 and 5 to improve performance, such as the extensive use of common-centroid techniques, capacitive decoupling, guard rings, supplies and grounds isolation of the different layout regions and fully differential implementation.

The prototype in Chap. 5 (see Fig. 5.33) uses a dedicated region for the buffers of the clock phases but the corresponding guard ring isolating ground is shared with that of the chip's main mixed-mode region. The mixed-mode region in this prototype is divided into 2 regions with different guard rings and corresponding isolating grounds. The main mixed-mode part is devoted to the switches of integrators and analog adders, and to the regenerative latches of comparators, while the buffers of the clock phases are included in a different region with its own ground, supply and guard ring isolating ground. This helps to reduce coupling of the switching noise generated in the phase buffers into the other mixed-mode sections of the chip.

6.5.1.1 Pad I/O Ring

Figure 6.31 depicts the pad I/O ring, which makes use of several techniques—explained below—to improve isolation of the different layout regions. Digital I/O cells type A are implemented based on I/O library cells, while the others—including the type B for digital supplies—are full custom. The main layout regions are isolated on the I/O ring as well. Also, the fastest switching digital part is placed as far as possible from the most sensitive analog part in both the chip layout and the I/O ring.

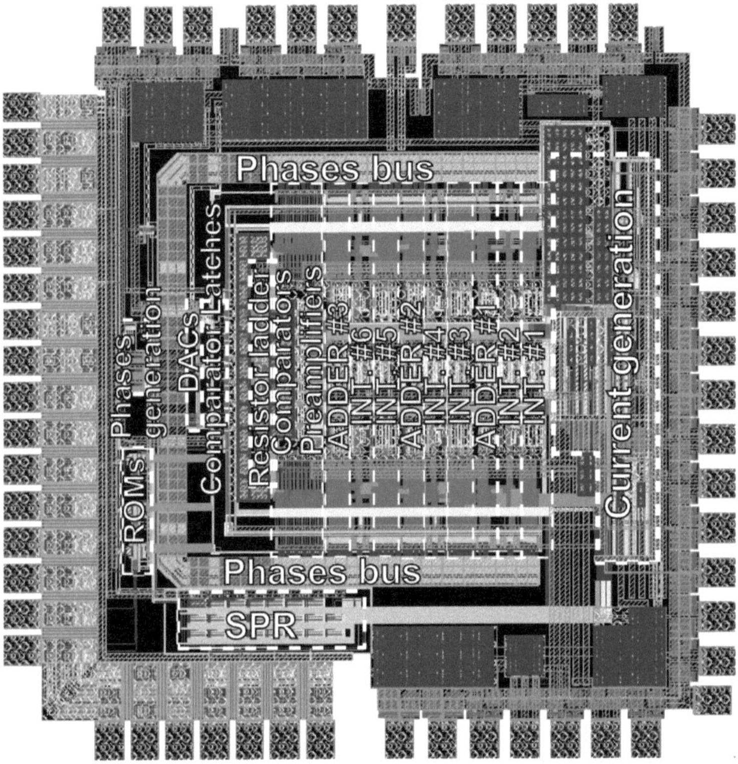

Fig. 6.29 Layout of the $\Sigma\Delta$M prototype (SPR stands for serial-to-parallel register)

The I/O ring is physically cut at 2 points to avoid switching noise coupling. This is done between the digital portion of the I/O ring type A—which includes the clock, triggers, $\Sigma\Delta$M outputs and signals devoted to the SPR, and the I/O supplies—, and that of the I/O ring type B and the analog region. Furthermore, the supplies of each region are not connected in the I/O ring. The grounds, however, are connected through 2 diodes in anti-parallel configuration to provide not only a virtual cut but also ESD protection [Chun06]. The supplies and grounds of each region need to be routed inside their associated I/O ring portion. Wide strips and stacked metals are used to reduce resistance parasitics. These parasitics reduction techniques are specially exploited for the analog supplies and grounds. Note that the term "supplies" in Fig. 6.31 refers to the supply, ground and isolating ground employed for the corresponding guard ring.

Three pads are used for biasing purposes. One of them is used to generate the on-chip master current via an off-chip resistor directly connected to the pad as shown in Fig. 6.8. The other two are used for testing purposes. One is devoted to the opamp biasing while the other is devoted to the comparator preamplifier biasing. This allows the on-chip generated currents to be the measured as described in Sect. 6.3.1.

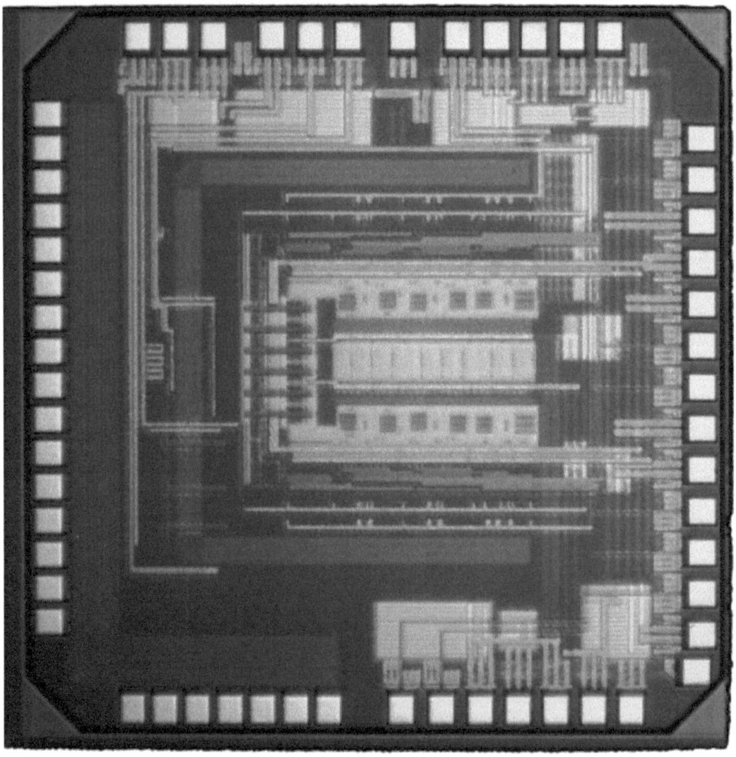

Fig. 6.30 Microphotograph of the chip

There are 4 dedicated pads for the serial-to-parallel register, as shown in Fig. 6.31. The input pins are the serial data signal, the input clock and the loading/shifting activation signal. The serial output can be captured for testing purposes.

A dedicated ground called V_{PH_BUS} is used to take the noisy isolating ground of the clock phases bus out of the chip. V_{PH_BUS} is not connected to the I/O ring in any way, so as to prevent coupling with other grounds or signals. Indeed, wide metals in stack are used to route this signal out of the chip through a very low impedance path.

Figure 6.31 also shows how double bonding is used for the $\Sigma\Delta$M reference voltages (V_r^+ and V_r^-), to halve their bonding parasitic inductances. The analog ground and supply also use 3 and 2 pads, respectively. These pads are placed in an interleaved parallel configuration to reduce their parasitic inductances as explained in Sect. 5.4.

The lack of reference in the capturing time of the chip in Chap. 5—as described in Sect. 5.5.1—has been solved in this prototype. To this purpose, a set of triggering signals—denoted as triggers in Figs. 6.31 and 6.33—are extracted off-chip and, later, used in the logic analyzer to enable the capture. Three triggering signals are provided off-chip because of the possible multi-rating operation in each stage, so a triggering

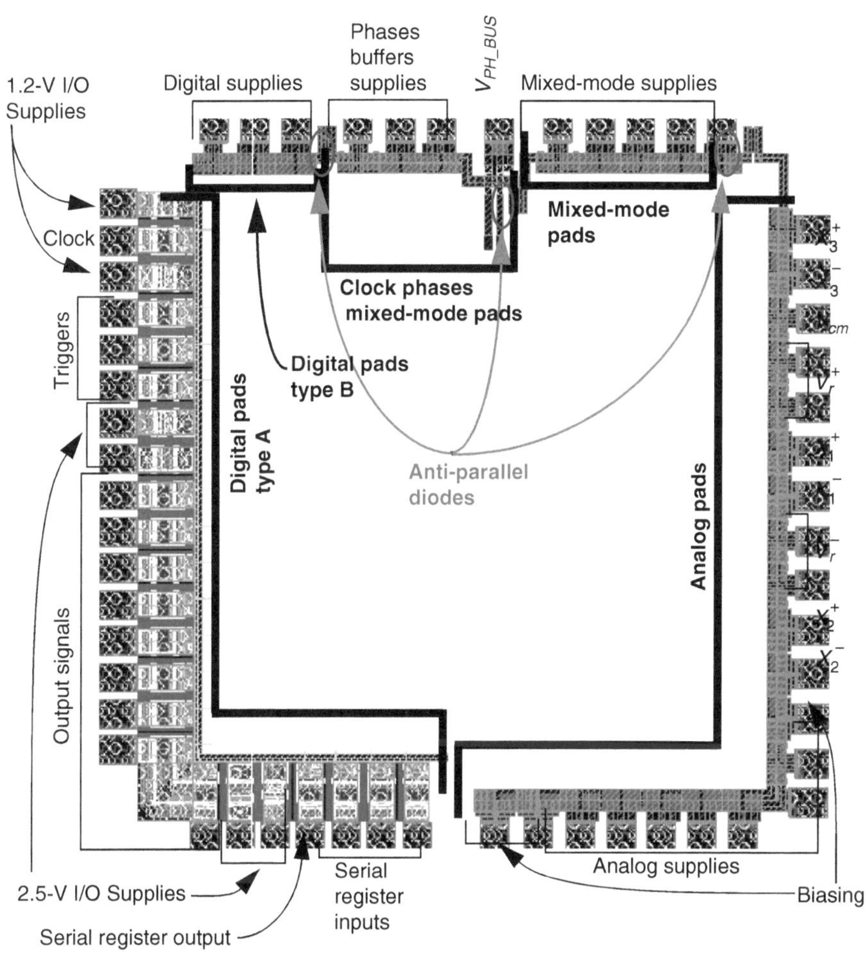

Fig. 6.31 Layout of the I/O pad ring and its corresponding region divisions

signal for each stage is needed when concurrency is enabled in all stages. In these signals the outputs of the comparators are as stable as possible; i.e., the farthest in time from the comparator strobe. Note from Fig. 6.4 that the triggering operation therefore corresponds to the falling edge of signals ϕ_{2j}.

Figure 6.32 shows the bonding diagram of the chip, which is depicted with the same orientation as the layout in Fig. 6.29. The employed package is a 64-pin ceramic QFP [Phil00]. The crosses indicate the double bonding pins in the figure ($\Sigma\Delta$M reference voltages). A number of pins are unconnected. Two of them are placed between the analog and the digital portion of the I/O ring to reduce parasitics due to mutual inductance coupling. Also, the V_{PH_BUS} pin has an unconnected pin on both sides. The other unconnected pins are placed in the corners of the cavity because these pins have the longest bonding wires and, thus, the largest parasitic inductances.

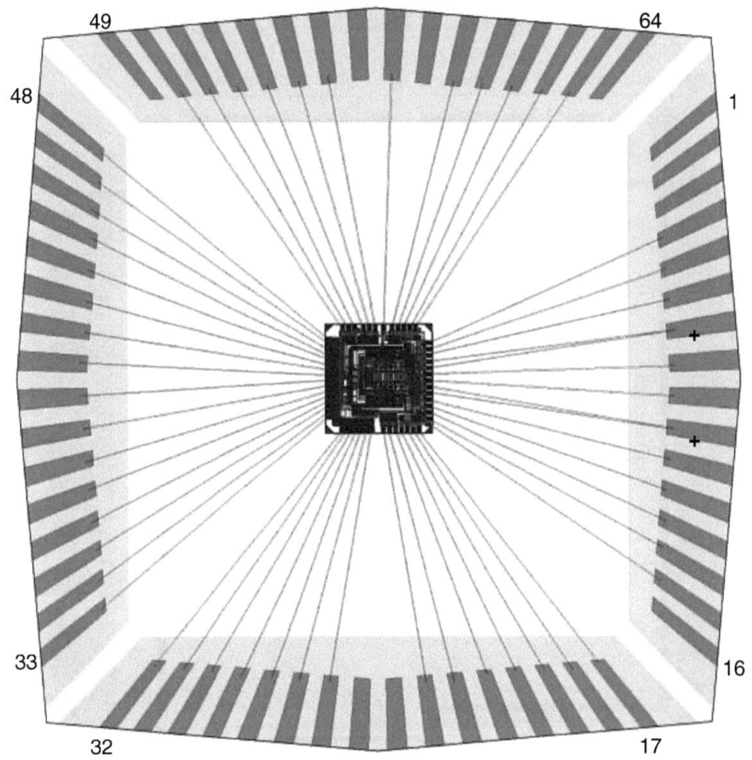

Fig. 6.32 Bonding diagram of the chip prototype

6.5.2 PCB

Figure 6.33 shows the schematic of the PCB that has been designed to characterize the chip. The main differences with the PCB presented in Sect. 5.4.1 are listed below:

- There are 3 different input signals which can be provided off-chip in order to convert them in parallel when the 3 stages work concurrently. Three anti-aliasing filters are therefore needed. There are also two different kinds of connectors. One of them is for slow and medium speed. This connector is prepared for a differential signal generator with a common-mode voltage generated on-chip. The other input connector is prepared for high speed. It is based on a single-ended signal generator that, in general, is more common in test equipment than its differential counterpart[14]. However, an extra small PCB with a fully-differential opamp in follower configuration is needed to transform the single-ended input signal into a fully-differential one with the common-mode voltage imposed by the chip.

[14] This is specially the case here given the number of input signals that can be processed in parallel.

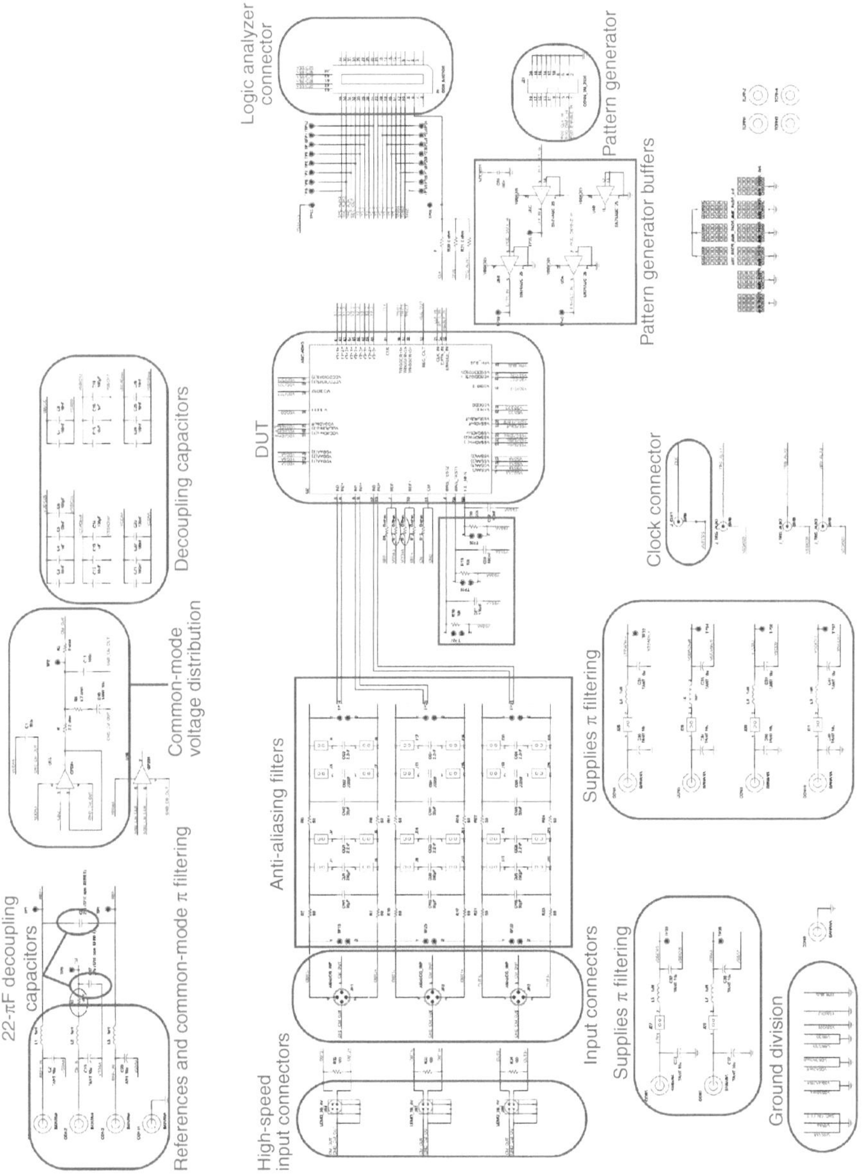

Fig. 6.33 PCB schematic

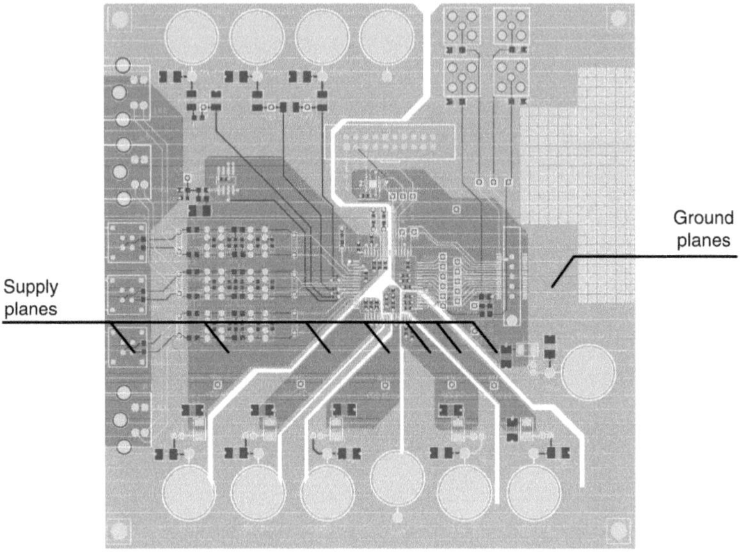

Fig. 6.34 Layout of the PCB

- There is no regulator for the supplies. This allows more flexibility for testing since each supply can be independently varied off-chip. However, a large number of connectors are needed for all these supplies, thus leading to an area-hungry solution. π filtering and capacitive decoupling are also employed here for the supplies and the reference voltages.
- The serial input control word is provided by an off-chip pattern generator. The pattern generator provides not only the serial data but also the associated clock and the loading/shifting activation signal[15]. This allows an efficient implementation of the timing diagram in Fig. 6.17 and avoids synchronization problems among the 3 signals. A set of buffers is mounted on the PCB to adapt the voltage levels of the pattern generator (3.3 V) to that of the on-chip I/O pad ring cells (2.5 V).
- The grounds and supplies separations on the PCB follows a similar procedure to the one explained in Sect. 5.4.1. Fig. 6.34 illustrates this on the PCB layout, where the physical separations between the supplies and grounds PCB planes can be clearly seen. Fig. 6.35 shows a photograph of the PCB, in which the divisions of the planes used for supplies and grounds are visible. As visible from the photo, the use of so many supply connectors makes the PCB very large.

[15] The generation of the aforementioned signals has been automated via some MATLAB [Math08a] scripts and a Visual Basic [Bale99] function that has been included in the pattern generator.

Fig. 6.35 Photograph of the PCB

6.5.3 *Test Set-up*

The test set-up employed is the same used for the prototype in Chap. 5 (see Sect. 5.5.2) and the second test set-up in Chap. 4 (see Sect. 4.6). The experimental characterization of the chip was therefore based on a logic analyzer that captures the bit streams of the different stages in the modulator. After acquisition, MATLAB [Math08a] is used to perform the digital post-processing on the captured data.

6.6 Experimental Results

Measurement results on the chip performance are presented in this section. These results indicate a correct operation of most of the flexible strategies incorporated to the prototype.

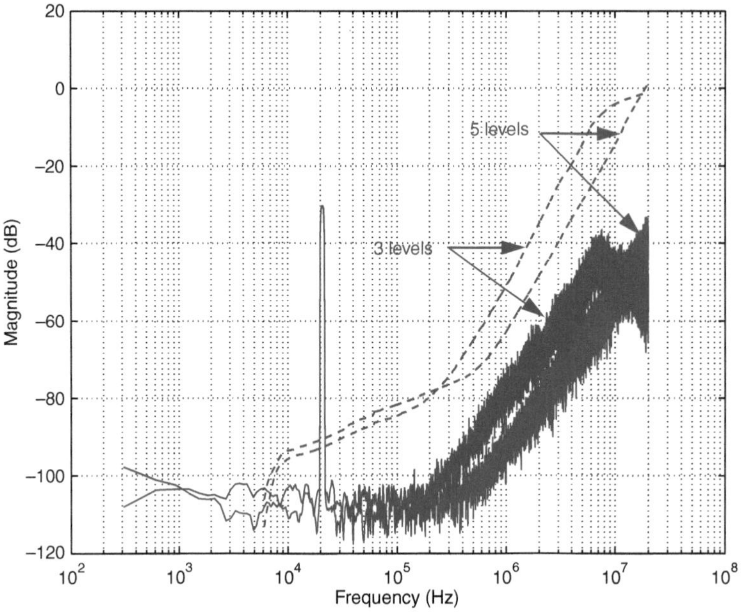

Fig. 6.36 Illustration of internal quantization reconfiguration

6.6.1 Reconfiguration of the Internal Quantization

Figure 6.36 shows two spectra obtained from a 256 k-point FFT of the first-stage modulator output and the corresponding cumulative error power. These measurements have been obtained for a modulator clock frequency of 40 MHz and a –30dBFS input tone. The number of levels in the first-stage embedded quantizer is reconfigured from 3 to 5. Note from the high-frequency region in Fig. 6.36 that the quantization error power considerably reduces when 5 levels are used in comparison to 3 levels. White noise in the low-frequency range however increases, but the in-band error power is the same for both configurations at 240 kHz approximately.

Note that there is no visible offset component in Fig. 6.36 (and in the remaining spectra in this section). Although the offset power is similar to that obtained for the prototype in Chap. 5, mean value of the codified output bit stream and, consequently, the offset component in the spectra have been removed by software for the sake of clarity.

6.6.2 Adjustable Modulator Order

The variation of the modulator order has been also experimentally checked. The last two stages have been switched off and their building blocks powered down. The

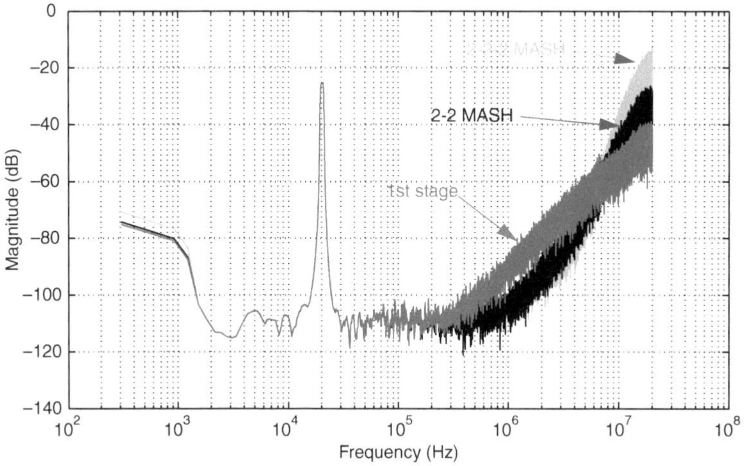

Fig. 6.37 Illustration of the reconfiguration of the modulator order

output bit streams corresponding to the second and the third stage in the modulator remained static, as checked with an oscilloscope. The triggering outputs (phases ϕ_{2i} in Fig. 6.3) used to synchronize the capture operation of the logic analyzer also remain quiet for the second and third modulator stages.

Figure 6.37 illustrates the reconfiguration of the modulator order. Three different configurations can be obtained as shown in the figure; namely, a 2nd-order single-loop, a 4th-order 2-2 cascade and a 6th-order 2-2-2 cascade. In this measurement, all embedded quantizers employ 5 levels and the clock frequency is 40 MHz. The input signal is a -30-dBFS single tone. Fig. 6.37 indicates that, from 300 kHz to 7.5 MHz approximately, the 2-2 cascade configuration reduces quantization noise when compared to the 2nd-order single-loop configuration due to the increased noise shaping. Note from the figure that a reduction of approximately 5 dB in the IBE power is only obtained for the 2-2-2 cascade when compared to the 2-2 cascade. This is due to noise leakages of the first modulator stages as expected. The third stage can however work concurrently as will be shown next.

6.6.3 Concurrency

Figure 6.38 shows two experimental examples of concurrent operation. Fig. 6.38a depicts two modulator stages operating in parallel. To this purpose, a clock frequency of 40 MHz is used and a −37dBFS@20 kHz input tone is provided to the first stage, while a −42dBFS@200 kHz to the second stage at the same time. Fig. 6.38b illustrates the concurrent operation of the first and third stages with respective inputs of −12-dBFS@200 kHz and −37dBFS@20 kHz. Note from Fig. 6.38 that concurrency is functional in both cases, although the noise floor and non-linearities increase.

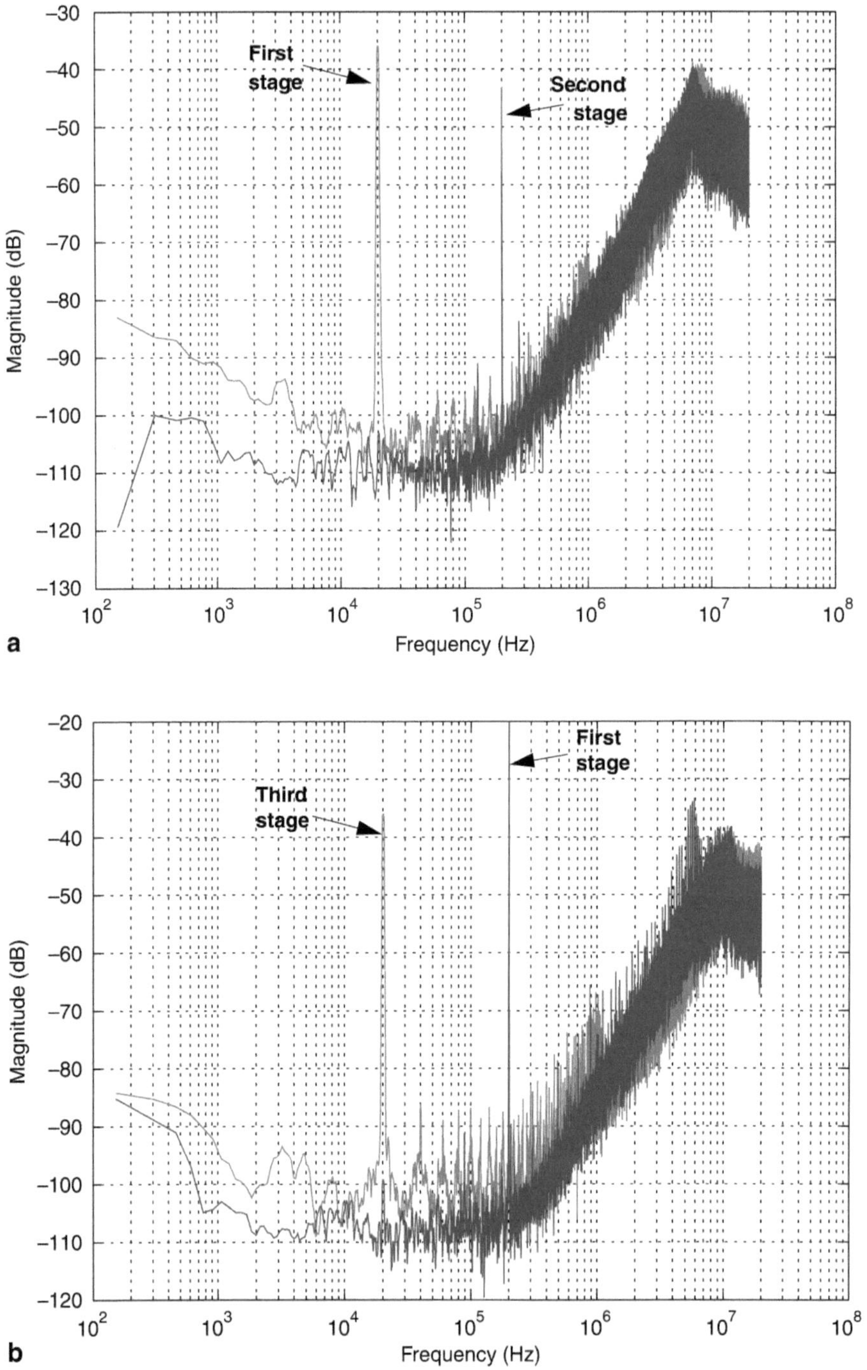

Fig. 6.38 Illustration of concurrency: **a** first and second stages in parallel, **b** first and third stages operating concurrently

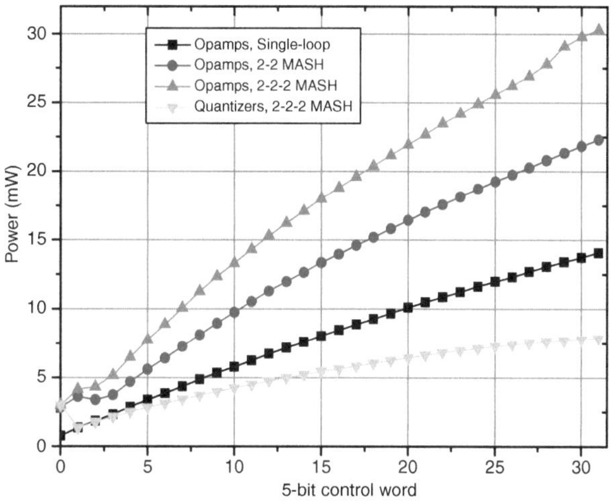

Fig. 6.39 Reconfiguration of the analog power consumption

6.6.4 *Adaptive Biasing*

As stated in Sect. 6.3.1, the modulator incorporates on-chip adaptation of the bias currents for the opamps and comparator preamplifiers in the embedded quantizers. There are two test points that can be used to measure the reconfigurable currents for both opamps and also comparator preamplifiers. Experimental measurements have been performed for the opamps, indicating that the corresponding bias currents can be successfully adapted on-chip. The opamp bias currents can be reconfigured from 4 μA to 75 μA.

Figure 6.39 shows the reconfiguration of the power dissipation of the analog part of the modulator by varying the control word of the on-chip binary-weighted current mirrors. Two cases are contemplated in the figure; namely:

- The variation of the bias currents in opamps, all to the same value. In this case, the comparator preamplifiers in the quantizers are not biased, so they do not contribute to the overall analog power. To this purpose, none of the control switches in the binary-weighted current mirror are activated for the comparator preamplifiers (see Fig. 6.9). In addition, three modulator configurations are included in the figure: 2nd-order single-loop, 2-2 MASH and 2-2-2 MASH configurations.
- The variation of the bias currents of comparator preamplifiers in the embedded quantizers. Opamps are not biased for these measurements and the modulator is configured as a 2-2-2 cascade $\Sigma\Delta$M with 3-level internal quantization.

As shown in Fig. 6.39, biasing reconfiguration operates correctly and provides a monotonic control of the bias current value. The dissipation reduction due to the power down of non-used stages can also be observed by simply comparing the three modulator configurations for opamp biasing.

6.7 Summary

This chapter shows the design and experimental characterization of a very flexible $\Sigma\Delta$M intended to enable the A/D conversion in 4G and concurrent transceivers. Many reconfiguration strategies have been employed to increase chip functionality and adaptability to each operation mode. Resonation, quantizer resolution adaptation and multi-rating are novel strategies incorporated to those used in Chaps. 4 and 5. This prototype is also capable of processing a number of wireless standards in parallel. To the authors' knowledge this is the first chip that implements this appealing feature, making it the first of a possible series of devices with boosted flexibility and concurrent capabilities for an SDR scenario.

Experimental results validate the use of most of the flexible strategies implemented, including concurrency or parallel capabilities.

Chapter 7
Performance of the Prototypes and Comparison with the State of the Art

This chapter briefly sums up the measured performance of the three prototypes presented in this book. The two first designed ICs are compared with the state of the art in reconfigurable ADCs intended for multi-standard wireless applications.

7.1 Performance of the Designed Prototypes

Table 7.1 shows the measured performance of the first two designed $\Sigma\Delta$ modulators in Chaps. 4 and 5. This comparison contemplates the figures of merit (FoM_1 and FoM_2), defined in Sect. 2.7 and respectively proposed by [Good96] and [Rabi97].

Table 7.2 summarizes the flexible strategies implemented in the prototypes presented in this book. They are classified as strategies at the architectural level—such as adaptive modulator loop-filter order, adjustable resolution of the internal quantization, use of resonation and concurrency—and at the circuit level—which contemplates adaptive biasing of the modulator building blocks and programmable sampling frequency. The main features, reconfiguration capabilities and performance of the three prototypes will be also discussed below.

Prototype labelled 'A' refers to that presented in Chap. 4. This reconfigurable SC $\Sigma\Delta$M uses a 4th-order 3-stage cascade topology (2-1-1 $\Sigma\Delta$M). The prototype is implemented in a 130-nm 1P8M CMOS process, with supply voltages of 1.2 V (for digital IOs) and 3.3 V (for the rest of the circuitry). A differential reference voltage of 2.4-V is used. The last stage of the modulator is switchable and its building blocks can be powered down when the stage is switched off to save power. This way, the order of the cascade and, consequently, also of the NTF filtering are varied from 4 to 3. The internal resolution of this last stage can be modified from 1 to 2 bits, so a dual-quantization scheme can be employed in the modulator for further reducing in-band quantization noise. The modulator also incorporates circuit-level reconfigurable techniques, such as adaptive opamp biasing and variation of the sampling frequency. The use of these strategies allows the modulator to feature dynamic ranges of 83.8/75.9/58.7 dB within signal bandwidths of 200 kHz/1 MHz/4 MHz with a power consumption of 23.9/24.5/44.5 mW.

A. Morgado et al., *Nanometer CMOS Sigma-Delta Modulators for Software Defined Radio*,
DOI 10.1007/978-1-4614-0037-0_7, © Springer Science+Business Media, LLC 2011

Table 7.1 Measured performance summary of the first two designed prototypes

PROTOTYPE A			
Standard	**GSM**	**BT**	**UMTS**
Modulator architecture	2-1	2-1-1(2bits)	
Differential reference voltage	2.4 V		
OSR	100	20	10
BW (MHz)	0.2	1	4
Sampling frequency (MHz)	40		80
DR (dB)	83.8	75.9	58.7
Peak SNDR (dB)	78.7	71.3	53.7
FoM_1	8.5	4.1	14.1
$FoM_2 \times 10^5$	2.1	1.8	0.1
Core area	0.65/0.76[a] mm^2		
Power consumption[b] (mW)	23.9	24.5	44.5
Supply voltage	3.3 V/1.2V		
Process	130-nm 1P8M CMOS		

PROTOTYPE B						
Standard	**GSM**	**BT**	**GPS**	**UMTS**	**DVB-H**	**WiMAX**
Modulator architecture	2(3-level)			2(3-level)-2(3-level)		
Differential reference voltage	2.4 V					
OSR	200	80	60	30	15	12
BW (MHz)	0.1	0.5	1	2	4	10
Sampling frequency (MHz)	40	80	120			240
DR (dB)	78.0	70.0	71.5	66.0	62.0	52.0
Peak SNDR (dB)	72.3	68.0	65.4	63.3	59.1	48.7
FoM_1	5.3	2.4	2.0	1.7	1.4	2.5
$FoM_2 \times 10^5$	2.0	2.3	1.9	1.8	1.3	0.2
Core area	0.66 mm^2 (pads excluded)					
Power consumption (mW)	4.6	5.3	6.2	8.0		11.0
Supply voltage	1.2 V					
Process	90-nm 1P9M CMOS					

a This value includes the area occupied by the bandgap and the reference buffer.
b Digital output buffers and reference buffer are not included.

Prototype labelled 'B' refers to that presented in Chap. 5. It consists of a 2-2 cascade with unity-STF stages and 3-level embedded quantizers. This chip is implemented in a 90-nm digital CMOS technology and operates with a single supply voltage of 1.2 V. A 2.4-V differential reference voltage is used to maximize resolution. The order of the modulator can be varied from 4 to 2. In the latter case, the back-end stage is switched off and its building blocks are also powered down to save power. This prototype allows the variation of the biasing in opamps and comparator preamplifiers. To this purpose, off-chip potentiometers are used, so that the number of biasing configurations—i.e., the number of possible generated bias currents—is

Table 7.2 Reconfiguration techniques implemented in the prototypes of this book

<table>
<tr><th colspan="11">ARCHITECTURAL LEVEL</th></tr>
<tr><th rowspan="2">Prototype</th><th colspan="3" rowspan="2">NTF filtering order</th><th colspan="2" rowspan="2">Internal quantization</th><th colspan="2" rowspan="2">Resonation</th><th colspan="3">Concurrency</th></tr>
<tr><th>1st stage</th><th>2nd stage</th><th>3rd stage</th></tr>
<tr><td rowspan="2">A</td><td rowspan="2">3rd</td><td colspan="2" rowspan="2">4th</td><td colspan="2">Back-end quantizer</td><td colspan="5" rowspan="3">---</td></tr>
<tr><td>1bit</td><td>2bit</td></tr>
<tr><td>B</td><td>2nd</td><td colspan="2">4th</td><td colspan="2">---</td></tr>
<tr><td rowspan="4">C</td><td rowspan="4">2nd</td><td rowspan="4">4th</td><td rowspan="4">6th</td><td colspan="2">All quantizers</td><td colspan="2" rowspan="3">Stage</td><td colspan="3">All stages</td></tr>
<tr><td rowspan="3">3levels</td><td rowspan="3">5levels</td><td>2SL[a]</td><td>2SL</td><td>2SL</td></tr>
<tr><td>2SL</td><td colspan="2">2-2 MASH</td></tr>
<tr><td>2nd</td><td>3rd</td><td colspan="2">2-2 MASH</td><td>2SL</td></tr>
</table>

<table>
<tr><th colspan="6">CIRCUIT LEVEL</th></tr>
<tr><th rowspan="2">Prototype</th><th colspan="3">Biasing</th><th colspan="2">Sampling frequency</th></tr>
<tr><th>Control</th><th>Controlled blocks</th><th># of biasing modes</th><th>Control</th><th>Range (MHz)</th></tr>
<tr><td>A</td><td>On-chip</td><td>Opamps</td><td>3</td><td rowspan="2">Modulator</td><td>40-80</td></tr>
<tr><td>B</td><td>Off-chip</td><td>Opamps and preamplifiers</td><td>---</td><td rowspan="2">40-240</td></tr>
<tr><td>C</td><td>On-chip</td><td>Opamps and preamplifiers</td><td>2^5(for each opamp and quantizer)</td><td>Modulator and each stage (multi-rating)</td></tr>
</table>

a This term (2SL) refers to the use of the stage as a stand-alone 2nd-order ΣΔM.

determined by the control steps in the potentiometers. The sampling frequencies can be varied from 40 to 240 MHz in order to adjust the modulator OSRs. Experimental measurements feature a dynamic range of 78/70/71.5/66/ 62/52 dB within a signal bandwidth of 100 kHz/500 kHz/1 MHz/2 MHz/4 MHz/10 MHz and with a power dissipation of 4.6/5.3/6.2/8.0/8.0/11.0 mW.

Prototype denoted as 'C' alludes to the modulator described in Chap. 6. This chip is a 2-2-2 cascade with unity-STF stages. Similarly to prototype 'B', a 1.2-V 90-nm CMOS technology and a 2.4-V differential reference voltage are also employed. The chip can reconfigure its loop-filter order from 6 to 4 or 2, by switching off one or two stages, respectively. The number of levels in each quantizer can be independently set to either 3 or 5. Every stage can work either concurrently or as part of a cascade, so that there are three possible configurations when performing concurrent operation (see Table 7.2); namely, either the three stages working as part of a 2-2-2 cascade, or the first or last two stages as a 2-2 cascade and the remaining stage as a stand-alone 2nd-order single-loop. The architecture incorporates resonation in the last two stages and it can work together with concurrency. This way, the last two stages can resonate in one mode, while operating concurrently or, alternatively, a back-end stage can incorporate resonation in a cascade configuration for other mode. There are 7 possible resonation factors in each stage. A set of 5-bit binary-weighted current mirrors are used on-chip to adjust the bias current of each opamp and of comparator preamplifiers in every quantizer. The sampling frequency can be set off-chip—as performed in prototype 'B'—or independently adjusted for each stage via an on-chip clock divider.

Let us compare the prototypes proposed in this book. First, prototype 'A' will be compared with 'B'. Prototype 'C' will not be considered for a comparison in terms of performance but only based on its reconfiguration capabilities.

Note from Table 7.1 that prototype 'B' considerably improves the performance of prototype 'A'. In fact, both the bandwidth and the number of considered standards are expanded. Modulator 'A' is capable to operate on 1 decade and 1 octave—from 200 kHz to 4 MHz, while modulator 'B' processes 2 decades of bandwidths in frequency—from 100 kHz to 10 MHz. Furthermore, power consumption is greatly reduced in prototype 'B'. This difference is specially appreciable for a signal bandwidth of 4 MHz (UMTS and DVB-H in prototype 'A' and 'B', respectively), in which 1 bit more is obtained for prototype 'B'. In this case, power consumption is reduced to 18% (from 44.5 to only 8.0 mW). This reduction is also high for GSM (reduction of 19%) but the achieved resolution decreases in approximately 1 bit for prototype 'B'. The reduction in power dissipation are mainly due to:

- A lower supply voltage, which helps to reduce power consumption, although it complicates the transistor-level design because of the low voltage headroom.
- The use of USTF, which helps to reduce the requirements of the opamps with low supply voltage. As a result, prototype 'B'—that uses a 2-2 cascade with USTF stages—can operate at higher sampling frequencies with noteworthy lower power consumption.
- More granularity in the reconfiguration strategies, specially those used at circuit level. Note from Table 7.2 that sampling frequencies rise up to 240 MHz in prototype 'B', so larger OSRs can be used—specially for broadband standards such as DVB-H or WiMAX. The modulator resolution is therefore increased but the amplifiers speed must be increased as well. Prototype 'B' benefits from a faster transistor operation because of technology scaling. In addition, opamp SR requirements are greatly reduced by the use of USTF in stages.
  There is a fine adjustment of the opamp biasing in comparison to prototype 'A' and the comparator preamplifiers are controlled as well. This feature is further improved in prototype 'C' that incorporates an on-chip control.

Prototypes 'B' and 'C' use similar topologies but prototype 'C' also incorporates additional architectural improvements, such as a larger modulator order, resonation, adaptive quantization resolution in each stage, concurrency and multi-rating.

7.2 Comparison with the State of the Art

Table 7.3 and 7.4 sum up the state of the art in multi-mode multi-standard CMOS ADCs, extracted from ICs reported in open literature [Burg01, Dezz03, Salo03, Aria06, Chri07, Ouzo07, Bos10, Crom10, Chri10]. The performance is

Table 7.3 State of the art in CMOS multi-standard ΣΔ ADCs

Ref.	Standards	SNDR (bit)	BW	Power (mW)	Technology process	Reconfiguration strategy	FoM_1	FoM_2 $x10^5$
DT ΣΔ Converters								
[Bos10]	GSM	12.5	100kHz	3.4	90nm	Sampling/Multi-rating/ # stages	3.0	4.9
	Bluetooth	12.3	500kHz	3.7			0.7	18.0
	UMTS	10.5	2MHz	6.8			1.2	3.1
[Burg01]	GSM	11.7	200kHz	11.5	0.25 μm	Sampling/Reconfigurable filtering/Adaptive biasing	8.8	0.9
	UMTS	8.3	3.84MHz	13.5			5.4	0.15
[Chri07]	EDGE	13.8	100kHz	2.9	0.13 μm	Sampling/ # stages/Programmable notches	1.0	36.4
	UMTS	12.5	1.92MHz	7.4			0.3	6.6
	WLAN	10.2	10MHz	20.5			0.9	5.3
[Chri10]	EDGE	13.4	100kHz	2.0	0.13 μm	Sampling/Programmable notches/# bits/Adaptive biasing	0.9	29.7
	UMTS	12.2	1.92MHz	7.2			0.3	40.8
	LTE	11.5	5MHz	13.6			0.5	15.1
		10.7	10MHz	20.2			0.6	6.6
		10.4	20MHz	34.7				5.3
[Dezz03]	GPRS	13.2	100kHz	2.4	0.13 μm	# stages/# bits	1.3	17.5
	UMTS	10.3	1.92MHz	4.3			0.9	3.7
[Gome02]	GSM	12.2	200kHz	2.4	0.13 μm	Sampling	1.3	8.8
	UMTS	7.8	2MHz	2.9			3.1	0.2
[Jant97]	IS54/AMPS	11.2	30kHz	130.0	0.8 μm	(Complex filtering)	940.7	0.006
	GSM	10.0	200kHz				315.9	0.008
[Lim06]	CDMA-2000	11.5	615kHz	22.7	0.13 μm	Sampling/Adaptive biasing	6.4	1.1
	WCDMA	10.5	1.92MHz				4.1	0.9
[Mill03]	AMPS	15.0	18kHz	30.0	0.18 μm	Sampling	25.6	3.2
	GSM	13.2	200kHz				8.2	2.8
	CDMA	12.5	625kHz				4.1	3.5
	WCDMA	11.3	1.92MHz	50.0			5.0	1.3
[Salo02]	GSM	11.7	270kHz	56.0	0.35 μm	--- (BP ΣΔM)	31.9	0.3
	CDMA	6.7	3.84MHz				70.9	0.004
[Salo03]	GSM	12.7	270kHz	24.0	0.35 μm	# stages	6.8	2.4
	IS-95	12.2	1.25MHz	37.0			3.0	3.8
	DECT	11.2	1.76MHz				4.6	1.3
	CDMA	7.7	3.84MHz	38.0			24.1	0.02
[Shim05]	GSM	12.1	100kHz	4.0	0.18 μm	Sampling	4.6	2.3
	UMTS	8.6	2.5MHz				2.1	0.5
This Work A	GSM	12.8	200kHz	23.9	0.13 μm	Sampling/# stages/# bits/ Adaptive biasing	8.5	2.1
	Bluetooth	11.6	1MHz	24.5			4.1	1.8
	UMTS	8.6	4MHz	44.5			14.1	0.1
This Work B	GSM	12.1	100kHz	4.6	90nm	Sampling//# stages/ Adaptive biasing	5.3	2.0
	Bluetooth	11.1	500kHz	5.3			2.4	2.3
	GPS	10.6	1MHz	6.2			2.0	1.9
	UMTS	10.2	2MHz	8.0			1.7	1.8
	DVB-H	9.5	4MHz				1.4	1.3
	WiMAX	7.8	10MHz	11.0			2.5	0.2
CT ΣΔ Converters								
[Aria06]	WLAN (Zero-IF)	8.7	20MHz	30.4	0.25 μm	--- (Complex CT filtering)	1.9	0.5
	WLAN (Low-IF)	8.6		32.0			2.1	
[Crom10]	Bluetooth	12.5	500kHz	5.0	90nm	Tunable resonation/ Scalable loop coefficients/ Reconfigurable circuitry/ Adaptive biasing	0.9	16.7
	UMTS	12.0	1.92MHz	6.4			0.4	25.0
		11.2		4.3			0.5	11.7
	DVB	11.3	4MHz	5.5			0.3	24.2
	WLAN	10.5	10MHz	6.8			0.2	15.5
[Ho10]	GSM	13.3	200kHz	8.2	0.18 μm	Sampling (complex CT filtering)	2.1	12.0
	Bluetooth	11.3	1MHz	8.5			1.7	3.7
[Ouzo07]	GSM	13.3	200kHz	1.4	90nm	Programmable resonation and loop coefficients/Adaptive biasing	0.3	73.4
	Bluetooth	12.2	1MHz	3.4			0.4	31.0
	WLAN	7.8	10MHz	7.0			1.5	0.4
[Putt07]	EDGE	14.3	135kHz	2.6	65nm	Sampling/ (Hybrid CT-SC)	0.5	109.3
	CDMA	13.3	614kHz	3.1			0.2	104.7
	UMTS	11.8	1.92MHz	3.7			0.3	34.5
[Silv07]	AM	18.5	3kHz	210	0.18 μm	--- (IF-to-BB)	95.8	9.5
	FM	14.7	200kHz				20.3	3.2
	IBOC	12.7	500kHz				32.3	0.5
[Vadi08]	GSM	12.3	200kHz	2.1	0.18 μm	Sampling	1.0	12.6
	UMTS	9.7	1.92MHz	3.2				2.0
[Veld03]	EDGE	15.0	200kHz	7.6	0.18 μm	Sampling	0.6	139.1
	CDMA2000	13.5	1.228MHz	8.2			0.3	99.6
	UMTS	11.7	3.84MHz	9.0			0.4	22.5

Table 7.4 State of the art in CMOS multi-standard Nyquist-rate and hybrid ADCs

Ref.	Standards/ ADC type	Resolution (bit)	BW	Power (mW)	Technology	Reconfiguration strategy	FoM_1	$FoM_2 x10^5$
Hybrid ΣΔ-Pipeline Converters								
[Gula01]	ΣΔM mode	13.3	9.8kHz	17.7	0.6 μm	Sampling/ # stages/ Adaptive biasing	89.5	0.28
		12.5	9.8kHz	9.4			82,8	0.17
		11.4	9.8kHz	4.3			81.2	0.08
		9.7	9.8kHz	2.2			135	0.015
	Pipeline mode	11	1.3MHz	24.6			4.6	1.1
		6	10MHz				19.2	0.08
[Yurt04]	ΣΔM mode, GSM	10.9	100kHz	5.5	0.25 μm	Sampling	89.5	0.28
	Pipeline mode, UMTS	4.8	3MHz				82.8	0.17
Pipeline Converters								
[Ahme05]	--	8.9	500Hz	15μW	0.18 μm	Sampling/ Adaptive biasing	31.5	0.04
		8.9	25MHz	35mW			1.4	0.8
[Ande05]	--	9.5	5MHz	81mW	0.18 μm	Sampling/ (Cyclic-Pipeline)	11.1	0.16
		9.1	40MHz	94mW			2.2	0.63
[Audo06]	--	9.1	10MHz	9.8mW	0.18 μm	# stages	0.9	1.5
		5.1		3.5mW			3.2	0.04
[Tahe10]	--	10.7	200kHz	0.33mW	90nm	Sampling/# stages/ Adaptive biasing	0.5	9.2
		9.5	22MHz	12.0mW			0.5	4.6
		10.5		22.9mW			0.4	10.4
[Xia06]	Bluetooth	10.3	550kHz	14.7mW	0.18 μm	Sampling/# stages/ (Time-interleaving)	10.4	0.3
	WLAN	9.7	5.5MHz	20.2mW			2.3	0.9

summarized in terms of the signal bandwidth, effective resolution and power consumption for each IC and mode of operation. For the sake of completeness, the employed reconfiguration strategies and the technological process of the corresponding implementations are also included. For comparison purposes, their global performance is quantified through FoM_1 and FoM_2. The following observations can be made from Table 7.3 and 7.4:

- Approximately 75% of reconfigurable ADCs are based on ΣΔ modulation. Moreover, ΣΔ ADCs achieve larger resolutions than the rest for narrowband standards. Reconfigurable pipeline ADCs usually cover a lower range of signal bandwidths and are more appropriate for higher bandwidths, such as those required by WiMAX and WLAN standards.
- SC circuit techniques have been traditionally used in most reported ΣΔMs. However, more and more reconfigurable ICs are being implemented using CT circuits, especially in those applications targeting wideband signals and/or requiring low power consumption.
- Most of multi-standard ΣΔMs only handle two or three standards and very few cover operation modes with BW up to 10 MHz [Chri07, Ouzo07, Crom10] or 20 MHz in [Chri10].
- Although changing the sampling frequency (denoted as "Sampling" in Table 7.3 and in 7.4[1]) has been the reconfiguration strategy most commonly applied, the mandatory use of low oversampling ratios in wideband standards has forced ΣΔMs to achieve the required resolution by resorting to different strategies, including: adaptive high-order shaping [Burg01, Aria06, Ouzo07, Crom10],

[1] The value of the sampling frequency is twice the signal bandwidth in pipeline ADCs. Thus, unless bandwidths are the same in all modes, the sampling frequency will be varied.

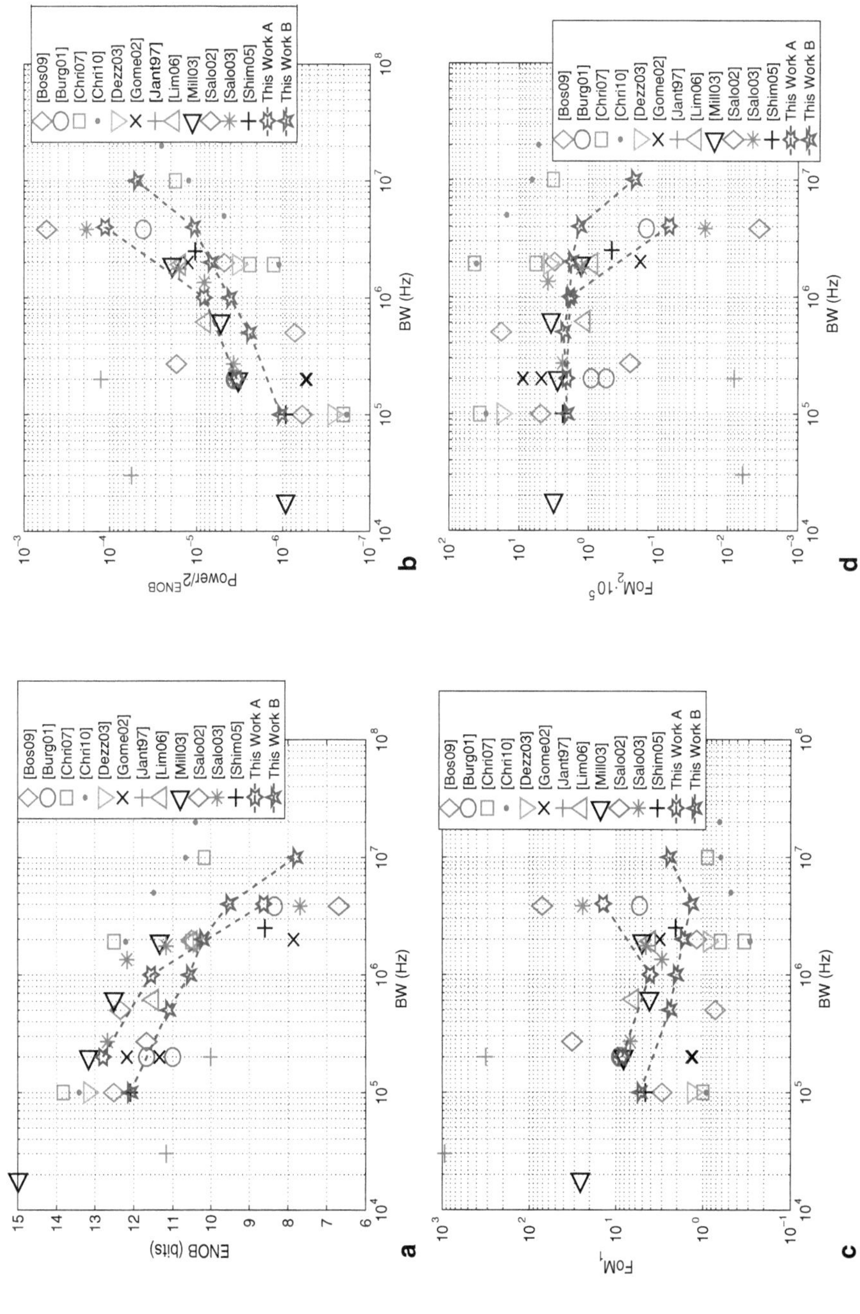

Fig. 7.1 State of the art in SC multi-standard $\Sigma\Delta$Ms: **a** ENOB, **b** normalized power consumption, **c** FoM_1 and **d** FoM_2 versus BW

switchable cascade topologies [Dezz03, Salo03, Chri07, Bos10, Chri10], programmable notches within the signal band [Ouzo07, Chri07, Chri10, Crom10] and adaptive embedded quantizers [Dezz03, Chri10].

Figure 7.1 represents different figures of the state of the art in multi-mode SC $\Sigma\Delta$Ms extracted from the data in Table 7.3. Figure 7.1a and b place the reported performances in the resolution- and normalized power-bandwidth planes (power consumption by number of levels), respectively, whereas Fig. 7.1c and d compare the performance of the SC $\Sigma\Delta$M ICs in terms of the two FoMs, as defined in (2.35) and (2.36).

Prototype 'A', published in 2007 [Morg07b], was one of the first $\Sigma\Delta$M ICs that demonstrated reconfiguration operation, while complying with the requirements of GSM, Bluetooth and UMTS. The covered range of signal bandwidths and the achieved performance made it competitive with the state of the art at that time.

Prototype 'B' was recently published in [Morg10]. This $\Sigma\Delta$M IC covers the requirements of 6 standards (GSM, Bluetooth, GPS, UMTS, DVB-H and WiMAX); i.e., it operates with the largest number of wireless standards, while covering one of the widest regions of signal bandwidths (from 100 kHz to 10 MHz). This region is similar to that achieved in [Chri07] and it is only overpassed by [Chri10] (from 100 kHz to 20 MHz). Note that slightly lower conversion bandwidth ranges can be found in [Ouzo07] (from 200 kHz to 10 MHz) and [Crom10] (from 500 kHz to 10 MHz). Prototype 'B' accomplishes competitive performance as illustrated in Fig. 7.1. This chip achieves the best FoMs for DVB-H standard (with a signal bandwidth of 4 MHz) when comparing only with SC $\Sigma\Delta$Ms and the second best also considering CT $\Sigma\Delta$Ms.

Prototype 'C' can convert up to three modes in parallel. This prototype therefore presents a clear advantage in a comparison with all the reconfigurable ICs in Table 7.3 and 7.4.

The prototypes presented in this book contribute to the state of the art of reconfigurable ADCs with competitive performance, while covering a large number of wireless standards and a wide region in the resolution-versus-bandwidth plane, thanks to the use of many reconfiguration strategies.

Appendix A
Study of Optimum Resonation

Chapter 3 shows different architectures that include resonation strategies. Some of the NTFs zeros are thus shifted from DC to create notches optimally located within the signal band in order to minimize the in-band quantization error power [Schr93]. The effect of the notches in the NTF frequency response will be evaluated in this appendix.

A.1 Optimum Resonation Factors

Chapter 3 provides three different cases of resonation-based NTFs, namely:

- A second-order high-pass filter with a pair of complex conjugate zeros close to the upper limit of the signal bandwidth.
- A fourth-order in-loop filter with two zeros in DC and another pair of zeros in optimal locations within the signal band.
- A 2-2 cascade $\Sigma\Delta$M with all NTF zeros optimally spread over the unit circle is not usually employed in conventional cascade $\Sigma\Delta$Ms—including those made of unity-STF stages—because the digital cancellation logic must compensate one of the two resonation coefficients employed. This results in a more complex implementation of the corresponding digital cancellation logic [Sanc06]. Chapter 3 however presents three architectures that circumvent this issue. The scheme of the first topology is depicted in Fig. 3.37, which exploits the use of local resonation in the two stages of an SMASH $\Sigma\Delta$Ms made up of unity-STF stages. All the four zeros of the NTF corresponding to the last-stage quantizer error can be placed in optimal location in the unit circle, while the contribution of the first-stage quantizer error is removed due to the unity-STF implementation in the last stage. The second and third architectures, respectively shown in Figs. 3.49 and 3.51, use local resonation in the first stage of an SMASH $\Sigma\Delta$M with also global resonation. An optimum location of all zeros of the second-stage quantization error is obtained due to the combined used of both types of resonation, whereas only two of the five (or of the four) zeros of the first-stage quantization error are moved from DC for the architecture in Fig. 3.49 (or Fig. 3.51).

A. Morgado et al., *Nanometer CMOS Sigma-Delta Modulators for Software Defined Radio*,
DOI 10.1007/978-1-4614-0037-0,

The three different cases of resonation-based NTFs initially presented are summarized in Table A.1, where L stands for the $\Sigma\Delta$M order, β stands for the number of complex conjugated zeros spread over the unit circle, ΔSNR stands for the resolution improvement provided by the corresponding NTF[1] notches and f_{zn} stands for the location of the zeros normalized to BW. A set of references are also given in Table A.1 in which a study of the optimum zero locations can be found. The proposed architectures with the corresponding NTFs are also included in the table.

The value of f_{zn} gives the location of the optimum zero in frequency (f_z) as follows:

$$f_z = f_{zn} \cdot \mathrm{BW} = f_{zn} \cdot \frac{f_s}{2\mathrm{OSR}} \tag{A.1}$$

The equivalent angle in the Z unit circle is given by:

$$\theta_z = \frac{2\pi f_z}{f_s} = f_{zn} \cdot \frac{\pi}{\mathrm{OSR}} \tag{A.2}$$

Besides, the optimum resonation factor g_{opt} is defined as follows [Sanc06, Sanc07]:

$$g_{opt} = \cos(\theta_z) \tag{A.3}$$

Therefore, once f_{zn} is known—based, for example, on the information provided by Table A.1—g_{opt} can be obtained from (A.1), (A.2) and (A.3).

A.2 Effect of Spreading the NTF Zeros to Optimal Locations

The first type of resonation shown in Table A.1 is illustrated in Fig. A.1, where the effect of OSR and resonation on the IBE power is shown for a second-order $\Sigma\Delta$M with NTF defined in (3.13)—first row of Table A.1. Note that (A.3) suggests that the larger the OSR, the lower g_{opt} for $\theta_z \ll 1$. This can be observed in Fig. A.1 as well.

Figure A.2 shows the optimal distribution of the NTF zeros in the unit circle for an OSR of 4, 8 and 16 when there are two optimized zeros in a fourth-order quantization noise filtering—see second row of Table A.1. Figure A.3 depicts the effect of this NTF zeros distribution on the noise shaping. The resolution improvement provided by the optimum placement of two NTF zeros is illustrated in Fig. A.4 for an OSR of 4 and embedded quantizers with 4-bit resolution. Let us assume that the cascade $\Sigma\Delta$M employs global resonation as shown in Figs. 3.16 or 3.17. The optimum values of *g/d* for an inter-stage gain d of 1 and 4 therefore are roughly 0.4 and 0.1, respectively. These last values can be respectively implemented with capacitor ratios of 2/5 and 1/10. If there is no inter-stage signal scaling along the cascade $\Sigma\Delta$M ($d = 1$), the resolution is increased by 11 dB [Hamo06] as shown in Table A.1, whereas an additional increase of 12 dB is achieved if $d = 4$ as expected from (3.15).

[1] A factor d is included in the NTFs of the last two rows in the table assuming that there is an inter-stage factor, d, in a 2-2 cascade $\Sigma\Delta$M with resonation. $d = 1$ is assumed in this appendix unless contrary stated.

Table A.1 Optimum resonation factors

L	β	NTF	ΔSNR (dB)	f_{zn}	References	Proposed examples
2	2	$z^{-2}-(2-g)\cdot z^{-1}+1$	3.5	$\pm\frac{1}{\sqrt{3}}$	[Schr93]	Fig.3.10
4	2	$\frac{[z^{-2}-(2-g)\cdot z^{-1}+1]\cdot(1-z^{-1})^2}{d}$	11	$\pm\sqrt{\frac{35}{7}}$	[Hamo06] [Sanc07]	Fig.3.11[a] Fig.3.12 Fig.3.15 Fig.3.17[a] Fig.3.38[a] Fig.3.47
4	4	$\left[z^{-2}-(2-g_1)\cdot z^{-1}+1\right]\times \left[z^{-2}-(2-g_2)\cdot z^{-1}+1\right]/d$	13	$\pm\sqrt{\frac{3}{7}\pm 0.313}$	[Schr93]	Fig.3.37[a] Fig.3.49[b] Fig.3.51[c]

[a]These architectures suffer from timing issues that make their implementation not reliable in practice

[b]There is an additional fifth-order shaped contribution of the first-stage quantization error to the NTF of this cascade ΣΔM [see (3.39)]

[c]In this architecture, the additional contribution of the first-stage quantization error to NTF is fourth-order shaped [see (3.41)]

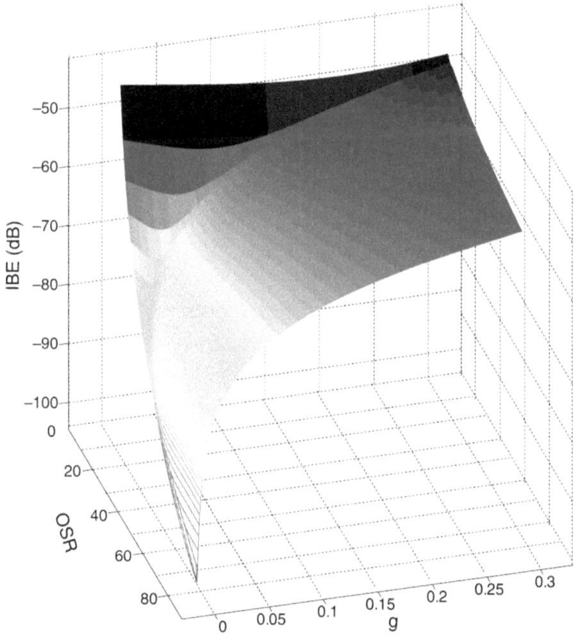

Fig. A.1 Illustration of the effect of resonation and OSR for NTFs defined as (3.13)

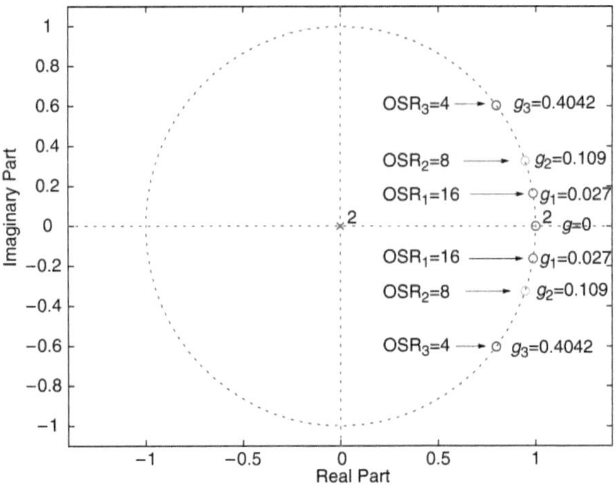

Fig. A.2 Optimum location of two zeros in the unit circle for a 4th-order filtering

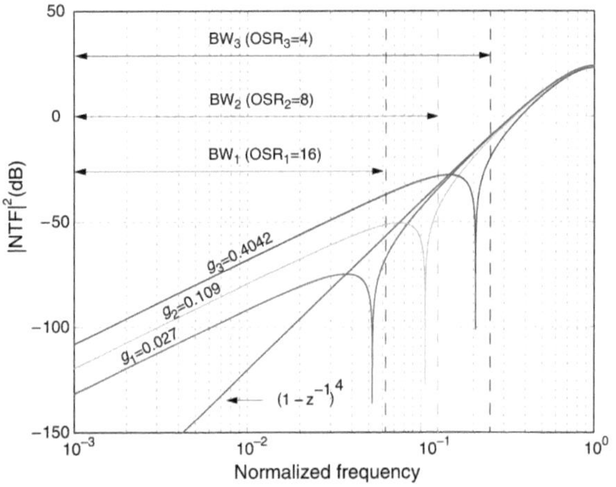

Fig. A.3 NTF amplitude versus frequency for the optimized zeros in (3.15)

The third type of resonation-based NTF considered is that in which all the 4 zeros of a fourth-order NTF are optimally spread from DC to optimal locations. Figure A.5 shows the variation of the IBE power with both coefficients g_1 and g_2. An OSR of 4 and a last-stage tri-level quantizer are assumed in this last case. Both a tri-dimensional and a contour plot are depicted in Fig. A.5a and b, respectively. If the optimal coefficients are chosen (0.0709 and 0.4403), the resolution will improve in approximately 13 dB [Schr93]. Figure A.5 shows that values for the resonation coefficients close to the optimum ones still provide a resolution improvement, the lower the further from the optimum values. This can help to make their electrical implementation of capacitor ratios easier. Let us assume, for instance, resonation coefficients $g_1 = g_2 = 0.25$—that can be easily implemented as 1/4 capacitor ratios—still provide 9- and 2.5-dB resolution improvements compared to not using resonation ($g_1 = g_2 = 0$) and compared to employing $g_1 = 0.25$, $g_2 = 0$ (or vice versa), respectively.

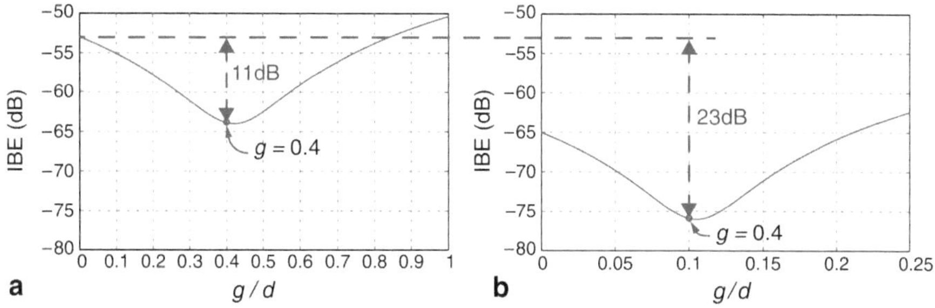

Fig. A.4 Variation of IBE with g/d for: **a** $d=1$, **b** $d=4$ (OSR = 4 and $B_1=B_2=4$)

Fig. A.5 Illustration of the effect of the resonation coefficient values on the IBE power for OSR = 4: **a** IBE versus g_1 and g_2, **b** contour plot

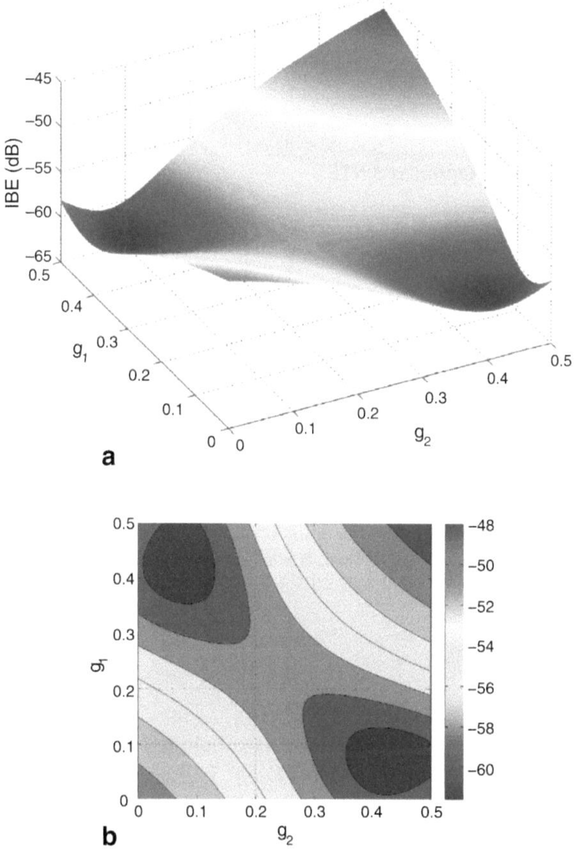

Figures A.6a and b show the optimum zeros location in the unit circle for the respective OSR values of 4 and 8. The optimum coefficients are $g_1=0.0709$ and $g_2=0.4403$ for OSR = 4, and $g_1=0.0178$ and $g_2=0.1133$ for OSR = 8. Figures A.6c and d illustrate the effect of optimum zero placement on both the NTF and the

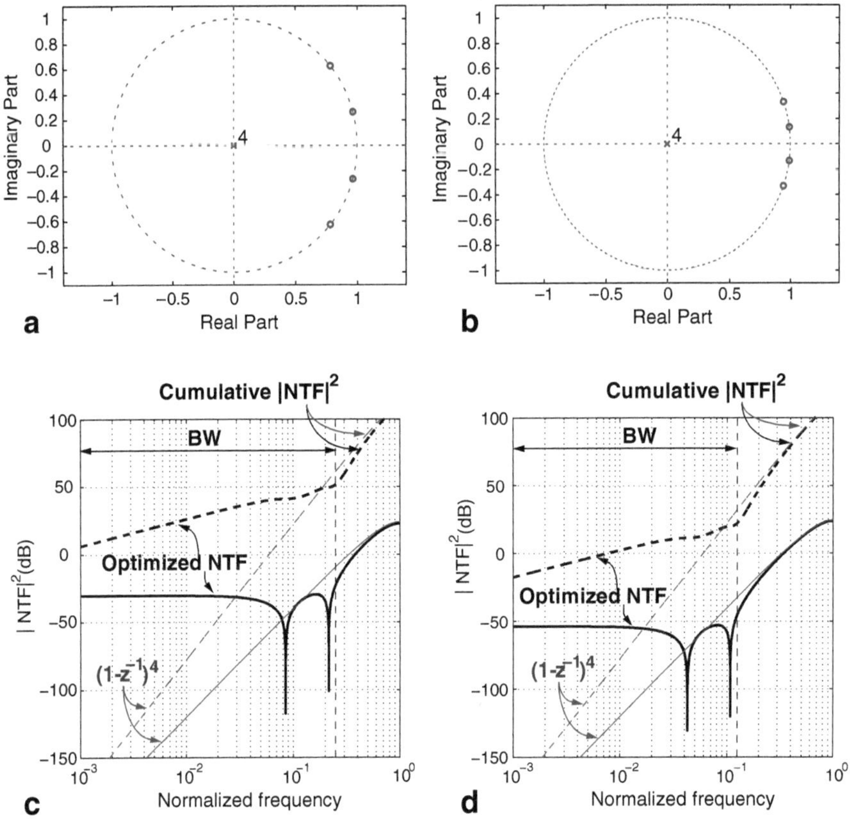

Fig. A.6 Illustration of the optimal spread of all complex conjugated zeros in a fourth-order NTF: **a** pole-zero plot for OSR = 4, **b** pole-zero plot for OSR = 8, **c** NTF amplitude versus frequency for OSR = 4, **d** NTF amplitude versus frequency for OSR = 8

cumulative response of $|\text{NTF(z)}|^2$. The case in which all zeros are located at DC [$\text{NTF(z)} = (1 - z^{-1})^4$] is also included for comparison purposes. Note that the cumulative response of $|\text{NTF(z)}|^2$ is directly proportional to the IBE power. It can be verified in these figures that a resolution improvement of approximately 13 dB is obtained as predicted by [Schr93].

Note that if expandable cascade $\Sigma\Delta$Ms are used (see Sect. 3.1), larger orders and a larger number of optimum zeros can be spread over the unit circle. Analytical expressions can be obtained for these cases by nulling the derivatives of the integrated power of $|\text{NTF(z)}|^2$ with frequency for the corresponding resonation coefficients, as done in [Schr93] and [Sanc06]. By contrast, high-order cascade $\Sigma\Delta$Ms that allow the optimum location of more than two complex conjugated zeros can be difficult to study analytically[2]. A numerical approximation to the problem can be the most effective approach in this case. To this purpose, figures as those depicted in Figs. A.1, A.4 and A.5 can be useful.

[2] Note that the solution when all NTF zeros are optimized (from 1 to 8) within the band of interest can be found in [Schr93].

Appendix B
Design of Electrical Blocks for the Practical Implementation of SMASH $\Sigma\Delta$Ms

In general, SMASH architectures need a digital adder located inside the first-stage loop, which entails an increase of the number of levels in the first-stage feedback DAC. Thus, DEM may be needed in practice when facing high-to-medium resolutions.

As stated in Sect. 3.3, unity-STF SMASH $\Sigma\Delta$Ms only have half of a clock cycle ($z^{-1/2}$) for the sequential timing operation of the quantizer, the digital addition, DEM and the in-loop D/A conversion[1]. The quantizer can work very fast and the digital adder too, since comparators in the quantizer can be based on very fast regenerative latches and the last block is made of a small number of digital gates as illustrated at the end of this appendix. The D/A operation can be divided as follows: first, the digital DAC gates that work very fast and, secondly, the front-end integrator sampling switches that perform the in-loop feedback operation, while being controlled by the aforementioned gates. Therefore, given the small time slot that one half of a clock cycle is, specially for wideband applications, there is no much room left for the DEM scrambling operation.

This appendix presents an implementation of DEM compatible with high-speed operation that allows to relax the timing requirements of the remaining blocks inside the feedback loop. Two different approaches for the implementation of the DAC are also presented and evaluated. Finally, two alternatives for the operation of the digital adder and the corresponding implementations are also shown.

B.1 DEM

Data-weighted averaging (DWA) [Bair95] is probably the most widely used DEM algorithm because of its efficiency. It is based on the sequential selection of unit elements in a DAC, beginning with the first available element which has not been previously used. A first-order noise shaping of the mismatch errors associated to the

[1] Note that, aside from the digital adder, this timing requirement is usually the same for conventional $\Sigma\Delta$Ms with multi-bit embedded quantizers.

A. Morgado et al., *Nanometer CMOS Sigma-Delta Modulators for Software Defined Radio*, 263
DOI 10.1007/978-1-4614-0037-0, © Springer Science+Business Media, LLC 2011

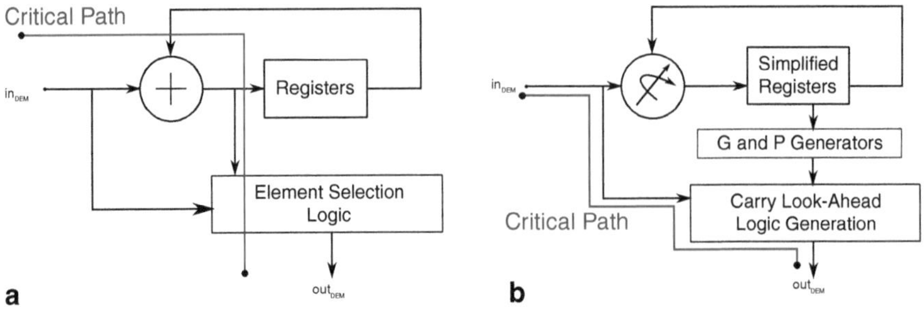

Fig. B.1 DWA implementation: **a** standard scheme, **b** approach proposed in [Lee02]

Fig. B.2 High-speed and low-area scheme proposed for the implementation of DWA

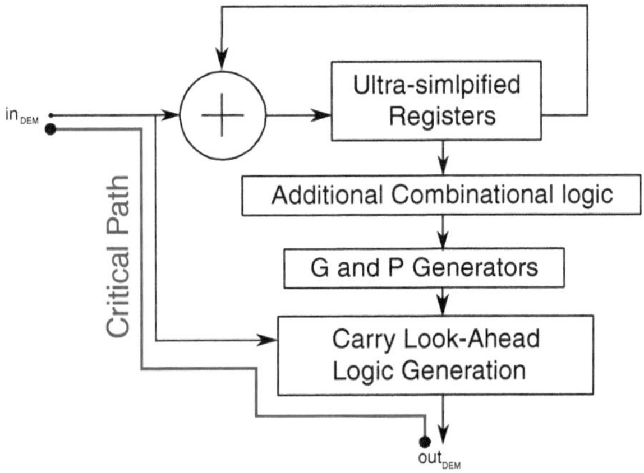

DAC elements is obtained [Hend96] thus pushing the error power to high frequencies, where it will be filtered out by the decimation filter.

Three possible implementations of the DWA algorithm are discussed in this appendix. Figure B.1a shows the one traditionally selected which is not very time-efficient; on the contrary, the approach depicted in Fig. B.1b [Lee02] improves the timing performance. The reason for this is that the first approach performs the full addition in the critical path, while only the carry outputs of the addition are needed for performing the DEM scrambling operation. The second approach takes full advantage of this feature and makes the element rotation, *G* and *P* signals generation—needed for the carry generation—and register loading in a non-critical time slot. The latter topology replaces the full digital adder with an element rotator in order to reduce the number of flip-flops. The approach proposed in this work, depicted in Fig. B.2, consists of changing the rotator in Fig. B.1b with a full adder whose operation, contrary to the structure in Fig. B.1a, is not performed in the critical path. Moreover, an additional advantage of the proposed approach is its low area occupation, since the number of registers can be drastically reduced as explained below.

Let us consider, as an example, a practical implementation of the modulator in Fig. 3.38 which uses 3- and 5-level quantization in the first and second stages,

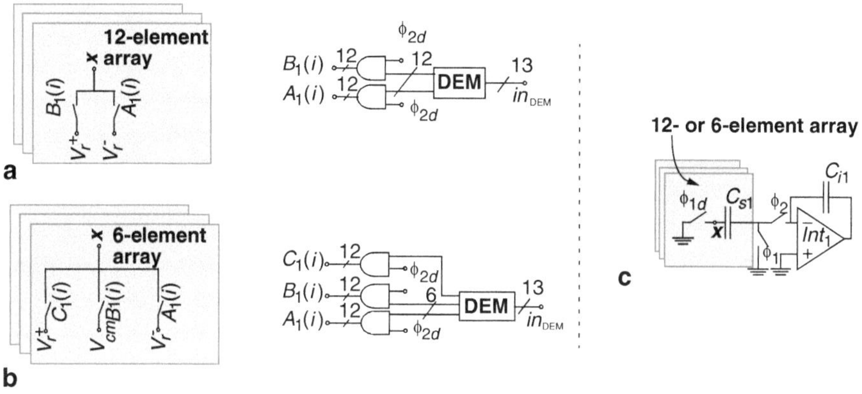

Fig. B.3 DAC implementations: **a** 2-level, **b** 3-level, **c** detail of the DAC connection to the front-end integrator

respectively, and an inter-stage scaling factor of 2 that leads to an in-loop 13-level DAC in the front-end feedback operation. The implementation of the 13-level DEM block following the DWA approach in Fig. B.1a would require 4×13 flip-flops, since 4 bits are needed to codify 13 levels. The implementation presented in [Lee02] would use $1+2+4+8$ flip-flops for 4 bits. Based on the data contained in those registers, it is very simple to obtain that of the remaining elements. However, this can not be directly applied to our specific case because the number of levels of the front-end DAC (13) is not a power of two. A more efficient solution consists in using only one flip-flop per bit, thus reducing the numbers of registers to only 4. The strategy shown in Fig. B.2 is selected in order to diminish the area devoted to registers at the price of a reduced increase in combinational logic and also in delay times for the generation of the G and P signals used for the carry look-ahead operation. Note however that this additional combinational block does not penalize the speed of the DEM since it is located outside the critical path[2].

B.1.1 DAC

Multi-bit DACs in $\Sigma\Delta$Ms are usually implemented by a set of digital gates that enables the feedback of either the positive or the negative reference voltages (V_r^+ and V_r^-) to each feedback capacitor during the integration phase, as shown in Fig. B.3a. An alternative implementation is depicted in Fig. B.3b, in which the common mode (V_{cm}) is used as an additional voltage for the DAC feedback operation. This last implementation reduces, in our specific case, the number of capacitors in the first

[2] Indeed, the delay of the DEM block following the proposed approach after a full digital synthesis in a 90-mm CMOS process shows to be only 360ps.

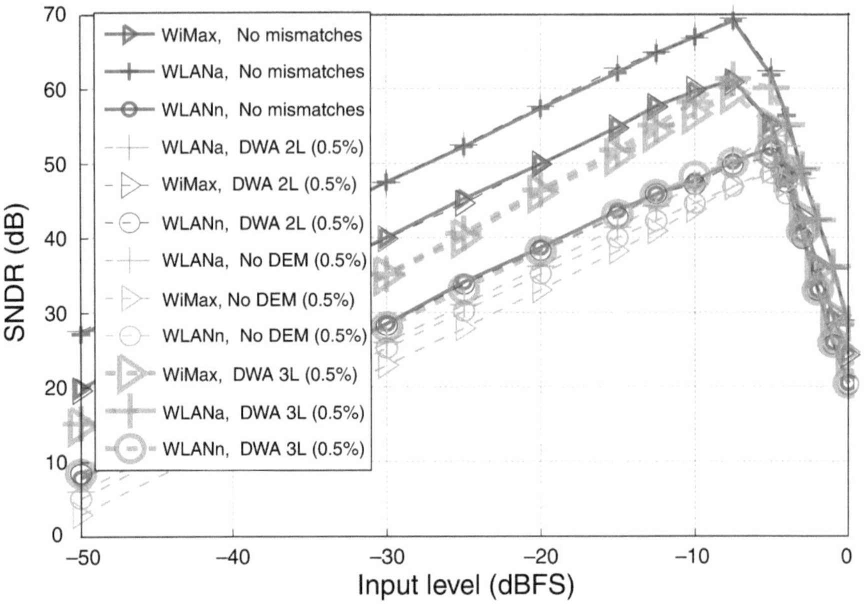

Fig. B.4 SNDR versus input level with DWA and different DAC implementations

integrator from 12 to only 6. These two alternatives for the front-end DAC will be called "2-level" and "3-level" DAC implementations, respectively.

In order to verify the correct operation of the DWA implementation and to investigate the influence of both front-end DAC alternative realizations on the modulator topology in Fig. 3.38, a set of behavioral simulations[3] have been performed for the operation modes that need a front-end multi-bit DAC in the multi-standard case study in Sect. 3.2.3; i.e., for WiMAX, WLANa and WLANn. These simulations also allow to compare the 2- and 3-level implementations of the front-end DAC.

Figure B.4 depicts the modulator SNDR curve when capacitors in the front-end DAC present mismatches with a standard deviation of 0.5%. The cases with no DEM and no mismatches are also depicted for comparison purposes. Note the evident degradation of the performance when no DEM is employed, specially for WLANa. Figure B.4 shows that there is no appreciable SNDR degradation due to mismatch errors as large as 0.5% if the 2-level DAC implementation is used together with a DWA scrambling operation (DWA 2L). However, an SNDR degradation larger than 10 dB comes out for a 3-level implementation of the DAC (DWA 3L) in WLANa mode, being the degradation also highly dependent on the input level. These results are consistent with the fact that the 3-level implementation is capable of scrambling less capacitors—only 6 in this case study—while DWA 2L uses twice that number,

[3] It is worth noting that the DWA algorithm has been included into SIMSIDES as a C-coded block in the SIMULINK platform.

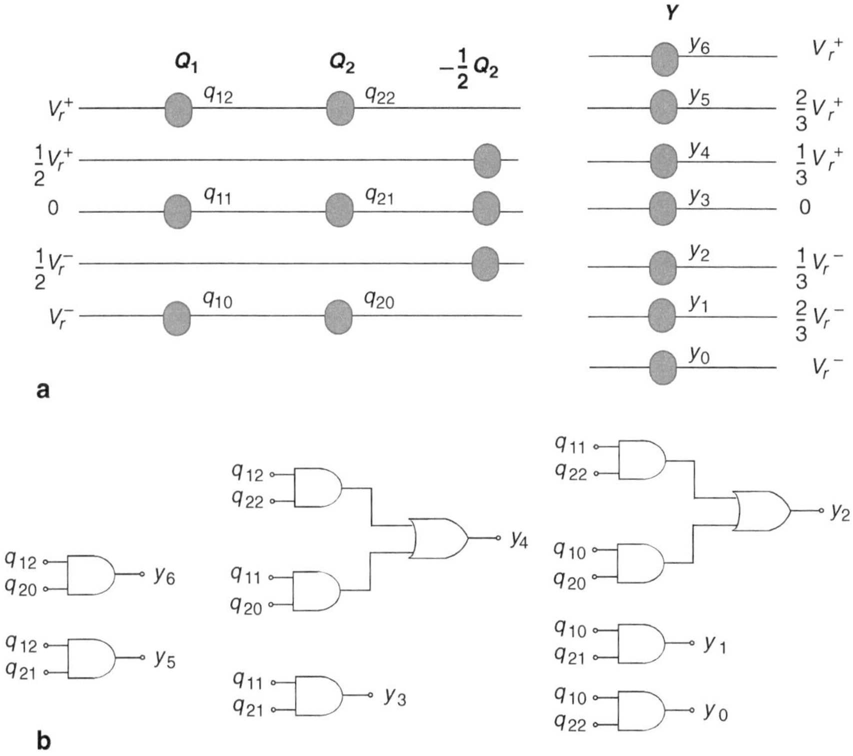

Fig. B.5 Illustration of the digital subtraction of two 3-level quantizer outputs with $d = 2$: **a** thermometric-quantized values and output addition, **b** corresponding schematic

so that a better randomization and, subsequently, a better linearity are expected, as confirmed by these results.

B.2 Digital Adder

All the SMASH architectures proposed in Chap. 3 use an in-loop digital adder. The operation and implementation and logic of the digital adder are illustrated in this appendix by means of two examples. Figures B.5 and B.6 show the digital subtraction of the output of a 3-level first-stage quantizer (Q_1) and that of the second-stage quantizer (Q_2) with an inter-stage gain factor $d = 2$. The resolution of Q_2 gives rise to the examples in Figs. B.5 and B.6 for 3 and 5 levels, respectively. Both figures give a thermometric representation of the digital values after quantization, after applying the inter-stage scaling factor d and after performing the digital subtraction. Note that the equivalent analog voltages of the corresponding DACs are enclosed in the figures

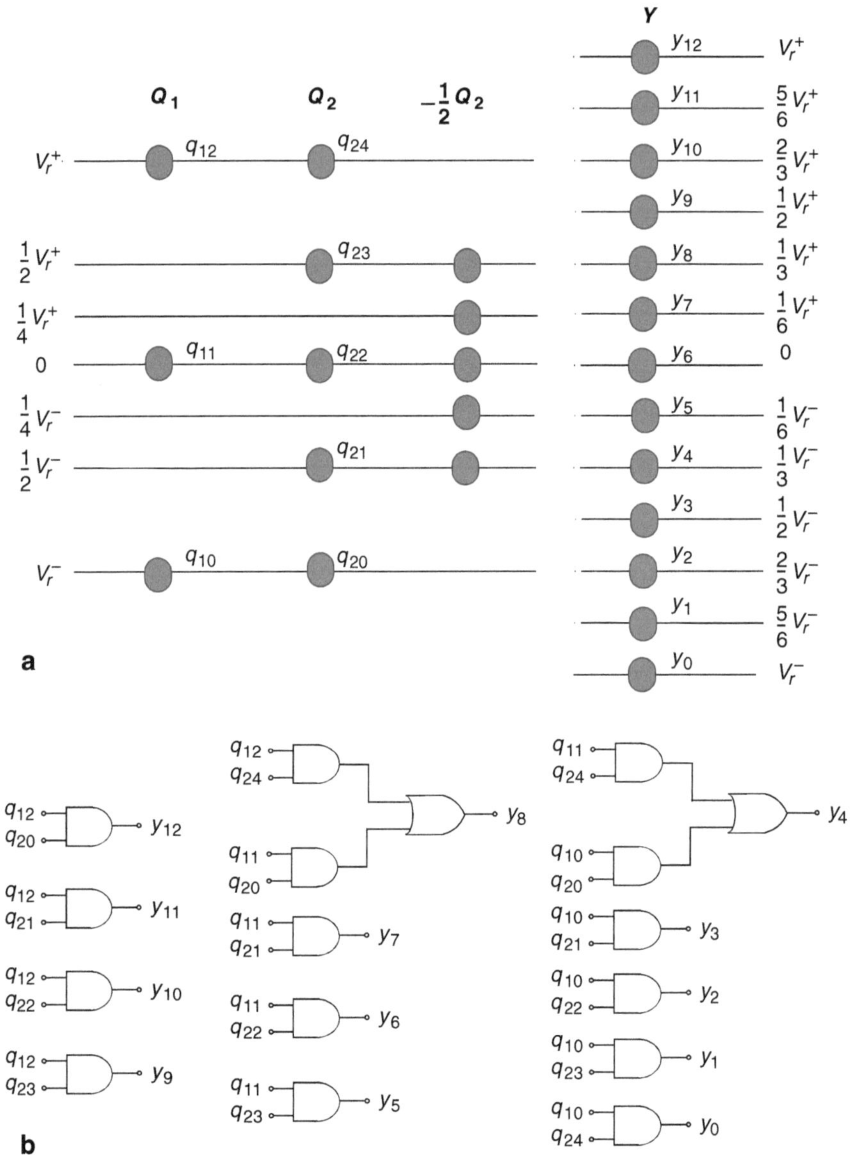

Fig. B.6 Illustration of the digital subtraction of 3- and 5-level quantizer outputs with $d=2$: **a** thermometric-quantized values and output addition, **b** corresponding schematic

as well. Also, Figs. B.5 and B.6 include the gate-level implementation of the digital adder.

The algorithm shown in Fig. B.6a, which can be implemented at the gate level as shown in Fig. B.6b, has been included as a C-coded block in SIMSIDES [Ruiz05] standing for the in-loop digital adder of the modulator in Fig. 3.38. Behavioral simulations show a correct performance of the $\Sigma\Delta M$ as shown in Figs. 3.39, 3.40 and 3.41; i.e., in the case study in Sect. 3.2.3. It is worth noting that the digital subtraction entails the use of a larger number of levels at the modulator output when compared to the resolution of the quantizers. If 3 and 3 levels, and 3 and 5 levels are respectively used for the first- and second-stage quantizers with $d = 2$, the corresponding number of levels of the modulator output will be 7 and 13 as illustrated in Figs. B.5 and B.6. Digital addition also leads to an equivalent reference scaling factor that depends on the full scale of the quantizers and the value of d. For instance, let us assume that the reference voltages of the quantizers are the same and that d equals 1 or 2; the resulting equivalent scaling factors will then be 1/2 and 2/3, respectively. Note that the effective full scale of the $\Sigma\Delta M$ will be scaled by this same factor.

References

[Abid95] A.A. Abidi: "Direct-Conversion Radio Transceivers for Digital Communications". *IEEE J. of Solid-State Circuits*, Vol. 30, pp. 1399-1410, Dec. 1995.

[Abid07] A.A. Abidi: "The Path to the Software-Defined Radio Receiver". *IEEE J. of Solid-State Circuits*, Vol. 42, pp. 954-966, May 2007.

[Adam97] R.W. Adams and R. Schreier, "Stability Theory $\Sigma\Delta$ Modulators", Chapter 4 in *Delta-Sigma Data Converters: Theory, Design and Simulation (S.R Norsworthy, R. Schreier and G.C. Temes, Editors)*. IEEE Press, 1997.

[Agne06] F. Agnelli et al.: "Wireless Multi-Standard Terminals: System Analysis and Design of a Reconfigurable RF Front-End". *IEEE Circuits and Systems Magazine*, pp. 38-59, First Quarter 2006.

[Ahme05] I. Ahmed and D.A. Johns: "A 50-MS/s (35mW) to 1-kS/s (15mW) power scalable 10-bit pipelined ADC using rapid power-on opamps and minimal bias current variation". *IEEE J. of Solid-State Circuits*, Vol. 40, pp. 2446–2455, Dec. 2005.

[Ahol04] R. Ahola et al.: "A Single-Chip CMOS Transceiver for 802.11a/b/g Wireless LANs". *IEEE J. of Solid-State Circuits*, Vol. 39, pp. 2250-2258, Dec. 2004.

[Alex05] M. Alexander: "Power Distribution System (PDS) Design - Using byPass/ Decoupling Capacitors". *Xilinx Application Note: Virtex-IISeries*, Feb. 2005.

[Ande05] M. Anderson et al.: "A Reconfigurable Pipelined ADC in 0.18 μm CMOS". *Dig. Tech. Papers of the Symp. on VLSI Circuits*, pp.326–329, 2005.

[Arag99] X. Aragonès, J.L. Gonzàlez and A. Rubio, *Analysis and Solutions for Switching Noise Coupling in Mixed-Signal ICs*. Kluwer Academic Publishers, 1999.

[Aria06] J. Arias et al.: "A 32-mW 320-MHz continuous-time complex delta-sigma ADC for multi-mode wireless-LAN receivers". *IEEE J. Solid-State Circuits*, Vol. 41, pp. 339-351, Feb. 2006.

[Audo06] W. Audoglio et al.: "A 6-10 bits reconfigurable 20MS/s digitally enhanced pipelined ADC for multi-standard wireless terminals". *Proc. of the European Solid-State Circuits Conf.*, pp. 496–499, 2006.

[Bagh06a] R. Bagheri et al.: "Software-Defined Radio Receiver: Dream to Reality". *IEEE Communications Magazine*, pp. 111-118, Aug. 2006.

[Bagh06b] R. Bagheri et al.: "An 800-MHz-6-GHz Software-Defined Wireless Receiver in 90-nm CMOS". *IEEE J. Solid-State Circuits*, Vol. 41, pp. 2860-2876, Dec. 2006.

[Bahe92] H. Baher and E. Afifi: "Novel fourth-order sigma-delta convertor". *Electronics Letters*, Vol. 28, pp. 1437-1438, July 1992.

[Bair95] R.T. Baird and T.S. Fiez: "Linearity enhancement of multibit $\Delta\Sigma$ A/D and D/A converters using data weighted averaging". *IEEE Trans. on Circuits and Systems - II*, Vol. 42, pp. 753-762, 1995.

[Bale99] F. Balena, *Programming Microsoft Visual Basic 6.0.* Microsoft Press, 1999.

[Balm00] P. Balmelli, Q. Huang and F. Piazza: "A 50-mW 14-bit 2.5-MS/s Σ—Δ modulator in a 0.25 μm digital CMOS technology". *Dig. Tech. Papers of the Symp. on VLSI Circuits*, pp. 142-143, 2000.

DOI 10.1007/978-1-4614-0037-0,

[Basc06] A. Baschirotto et al.: "Baseband Analog Front-End and Digital Back-End for Reconfigurable Multi-Standard Terminals". *IEEE Circuits and Systems Magazine*, First Quarter, pp. 8-28, 2006.

[Behz07] A. Behzad et al.: "A fully integrated MIMO multiband direct conversion CMOS transceiver for WLAN applications (802.11 n)". *IEEE J. Solid-State Circuits*, Vol. 42, pp. 2795-2808, Dec. 2007.

[Bena93] P. Benabes, A. Gauthier and D. Billet: "New wideband sigma-delta convertor". *Electronics Letters*, Vol. 29, pp. 1575-1577, Aug. 1993.

[Bena95] P. Benabes, A. Gauthier and R. Kielbasa: "New high-order universal $\Sigma\Delta$ modulator". *Electronics Letters*, Vol. 31, pp. 8-9, Jan. 1995.

[Bena96] P. Benabes, A. Gauthier and R. Kielbasa: "A multistage closed-loop sigma delta modulator (MSCL)". *Analog Integrated Circuits and Signal Processing*, Vol. 11, pp. 195-204, Nov. 1996.

[Benn48] W.R. Bennett, "Spectra of quantized signals". *Bell System Technical J.*, vol. 27, pp. 446–472, July 1948.

[Blue03] H. M. Bluethgen, C. Grassmann, W. Raab and U. Ramacher: "A programmable platform for software-defined radio". *Proc. of the International Symp. on System on Chip*, pp. 15, 2003.

[Bolz17] B. Bolzano: "Rein analytischer Beweis des Lehrsatzes dass zwischen je zwey Werthen, die ein entgegengesetztes Resultat gewaehren, wenigstens eine reele Wurzel der Gleichung liege". Prague, 1817. English translation in S.B. Russ, "A translation of Bolzano's paper on the intermediate value theorem". *Historia Mathematica*, 7, pp. 156–185, 1980.

[Bos10] L. Bos, G. Vandersteen, P. Rombouts, A. Geis, A. Morgado, Y. Rolain, G. van der Plas and J. Ryckaert: "Multirate Cascaded Discrete-Time Lowpass $\Sigma\Delta$ Modulator for GSM/Bluetooth/UMTS". *IEEE J. of Solid-State Circuits*, Vol. 45, pp. 1198-1208, June 2010.

[Bose88] B.E. Boser and B.A. Wooley: "The Design of Sigma-Delta Modulation Analog-to-Digital Converters". *IEEE J. of Solid-State Circuits*, Vol. 23. pp. 1298-1308, Dec. 1988.

[Boyd06] J. Boyd: "Sneak Peek at Cellphone Future". *IEEE Spectrum*, pp. 15, June 2006.

[Bran91] B.P. Brandt and B.A. Wooley: "A 50-MHz Multibit Sigma-Delta Modulator for 12-b 2-MHz A/D Conversion". *IEEE J. of Solid-State Circuits*, Vol. 26, pp. 1746–1756, Dec. 1991.

[Bran05] M. Brandolini, P. Rossi, D. Manstretta and F. Svelto: "Toward Multi-standard Mobile Terminals – Fully Integrated Receivers Requirements and Architectures". *IEEE Trans. on Microwave Theory and Techniques*, Vol. 53, pp. 1026–1038, March 2005.

[Bult92] K. Bult and G. Geelen: "An inherently linear and compact MOST-only current division technique". *IEEE J. of Solid-State Circuits*, Vol. 21, pp. 1730-1735, 1992.

[Burg01] T. Burger and Q. Huang: "A 13.5mW 185-Msample/s $\Delta\Sigma$ Modulator for UMTS/GSM Dual-Standard IF Reception". *IEEE J. of Solid-State Circuits*, Vol. 36, pp. 1868-1878, Dec. 2001.

[Cabr06] D. Cabric et al.: "Spectrum Sharing Radios". *IEEE Circuits and Systems Magazine*, pp. 30-45, Second Quarter 2006.

[Cade08] Cadence Design Systems, *Virtuoso Spectre Circuit Simulator Reference*, June 2008.

[Cand85] J.C. Candy: "A Use of Double Integration in Sigma-Delta Modulation". *IEEE Trans. on Communications*. Vol. 33, pp. 249-258, March 1985.

[Carl97] L.R. Carley, R. Schreier and G.C. Temes, "Delta-Sigma ADCs with Multibit Internal Converters", Chapter 8 in *Delta-Sigma Data Converters: Theory, Design and Simulation (S.R. Norsworthy, R Schreier and G.C. Temes, Editors)*. IEEE Press, 1997.

[Chal07] A. Chalvatzis, E. Gagnon, M. Repeta and S. Voinigescu: "A low-noise 40 GS/s continuous-time bandpass AZ ADC centered at 2 GHz for direct sampling receivers". *IEEE J. of Solid-State Circuits*, Vol. 42, pp. 1065-1074, May 2007.

[Chan97] P.J. Chang et al.: "A CMOS Channel-Select Filter for Direct-Conversion Wireless Receiver". *IEEE J. of Solid-State Circuits*, pp. 722-729, May 1997.

[Chao90] K.C.-H. Chao, S. Nadeem, W.L. Lee and C. Sodini: "A Higher Order Topology for Interpolative Modulators for Oversampling A/D Converters". *IEEE Trans. on Communications*, Vol. 37, pp. 309-318, March 1990.

[Char01] E. Charbon, R. Gharpurey, P. Miliozzi, R.G. Meyer and A. Sangiovanni- Vincentelli, *Substrate Noise: Analysis and Optimization for IC Design.* Kluwer Academic Publishers, 2001.

[Cher00a] J.A. Cherry and W.M. Snelgrove, *Continuous-Time Delta-Sigma Modulators for High-SpeedA/D Conversion.* Kluwer Academic Publishers, 2000.

[Cher00b] J. Cherry, W. Snelgrove and W. Gao: "On the design of a fourth-order continuous-time LC delta-sigma modulator for UHF A/D conversion". *IEEE Trans. on Circuits and Systems - II*, Vol. 47, pp. 518-530, Jun. 2000.

[Chri07] T. Christen, T. Burger and Q. Huang: "A 0.13 μm CMOS EDGE/UMTS/ WLAN Tri-Mode ΣΔ ADC with -92dB THD". *Dig. Tech. Papers of the Intl. Solid-State Circuits Conf.*, pp. 240-241, 2007.

[Chri10] T. Christen and Q. Huang: "A 0.13 μm CMOS 0.1-20MHz Bandwidth 86-70dB DR Multi-Mode DT Delta-Sigma ADC for IMT-Advanced". *Proc. of the European Solid-State Circuits Conf.*, 2010.

[Chou89] W. Chou, P. Wong and R. Gray: "Multi-Stage Sigma-Delta Modulation". *IEEE Transaction on Information Theory*, Vol. 35, pp. 784-796, July 1989.

[Chun06] J.H. Chun, *ESD protection circuits for advanced CMOC technologies.* Ph.D. Thesis, Stanford university, 2006.

[Colo02] F. Colodro and A. Torralba: "Multirate ΣΔ modulators". *IEEE Trans. on Circuits and Systems - II*, Vol. 49, pp. 170-176, March 2002.

[Cran07] J. Craninckx et al.: "A Fully Reconfigurable Software-Defined Radio Transceiver in 0.13μm CMOS". *Dig. Tech. Papers of the Intl. Solid-State Circuits Conf.*, pp. 346–347, Feb. 2007.

[Crol97] J. Crols and M. Steyaert, *CMOS Wireless Transceiver Design.* Kluwer Academic Publishers, 1997.

[Crom09] P. Crombez, *Fully Reconfigurable Analog Baseband Circuits for Multimode Radios.* Ph.D. Thesis, Katholieke Universiteit Leuven (KUL), Sep. 2009.

[Crom10] P. Crombez et al.: "A Single-Bit 500 kHz-10 MHz Multimode Power- Performance Scalable 83-to-67 dB DR CT ΔΣ Modulator for SDR in 90 nm Digital CMOS". *IEEE J. of Solid-State Circuits*, Vol. 45, pp. 1159-1171, June 2010.

[Dara05] H. Darabi et al.: "Dual-Mode 802.11b/Bluetooth Radio in 0.35-μm CMOS". *IEEE J. of Solid-State Circuits*, Vol. 40, pp. 698-706, March 2005.

[Dedi94] I. Dedic: "A Sixth-Order Triple-Loop ΣΔ CMOS ADC with 90dB SNR and 100kHz Bandwidth". *Dig. Tech. Papers of the Intl. Solid-State Circuits Conf.*, pp. 188-189, 1994.

[Delg03] M. Delgado-Restituto et al., "PGAs and Filters", Chapter 14 in *CMOS Telecom Data Converters (A. Rodríguez-Vàzquez, F. Medeiro and E. Janssens, Editors).* Kluwer Academic Publishers, 2003.

[Deso06] G. Desoli and E. Filippi: "An Outlook on the Evolution of Mobile Terminals: From Monolithic to Modular Multi-Radio, Multi-Application Platforms". *IEEE Circuits and Systems Magazine*, pp. 17-29, Second Quarter 2006.

[Dezz03] A. Dezzani and E. Andre: "A 1.2-V Dual-Mode WCDMA/GPRS ΣΔ Modulator". *Dig. Tech. Papers of the Intl. Solid-State Circuits Conf.*, pp. 58-59, 2003.

[Dias93] V.F. Dias and V. Liberali: "Cascade Pseudomultibit Noise Shaping Modulators". *IEE Proceedings-G*, Vol. 140, pp. 237-246, Aug. 1993.

[Domi03] R. Domínguez-Castro, M. Delgado-Restituto, A. Rodriguez-Vàzquez, J.M. de la Rosa and F. Medeiro, "CMOS Comparators", Chapter 4 in *CMOS Telecom Data Converters.* Kluwer Academic Publishers, 2003.

[Enz96a] C.C. Enz and G.C. Temes: "Circuit techniques for reducing the effects of Op- Amp imperfections: Auto-zeroing, correlated, double sampling and chopper stabilization". *Proceedings of the IEEE*, Vol. 84, pp. 1584-1614, Nov. 1996.

[Enz96b] C.C. Enz and E. Vittoz: "CMOS low-power analog circuit design". *Proceedings of the IEEE Designing Low Power Digital Systems, Emerging Technologies*, 1996.

[Evan00] B.G. Evans and K. Baughan: "Visions of 4G". *Electronics and Communication Engineering J.*, pp. 293-303, Dec. 2000.

[Fata90] J.W. Fattaruso, M. De Wit, G. Warwar, K.-S. Tan and R. K. Hester: "The effect of dielectric relaxation on charge-redistribution A/D converters". *IEEE J. of Solid-State Circuits*, Vol. 25, pp. 1550-1561, 1990.

[Figu09] P. Figueiredo and J. C. Vital, *Offset Reduction Techniques in High-Speed Analog-to-Digital Converters- Analysis, Design and Tradeoffs*. Springer, 2009.

[Frat06] S. Frattasi, H. Fathi, F.H.P. Fitzek, R. Prasad and M.D. Katz: "Defining 4G Technology from the User's Perspective". *IEEE Network*, pp. 35-41, Jan-Feb 2006.

[Fuji00] I. Fujimori, L. Longo, A. Hairapetian, K. Seiyama, S. Kosic, J. Cao and S.-L. Chan: "A 90-dB SNR 2.5-MHz Output-Rate ADC Using Cascaded Multibit Delta-Sigma modulation at 8x Oversampling Ratio". *IEEE J. of Solid-State Circuits*, Vol. 35, pp. 1820-1828, Dec. 2000.

[Fuji06] Y. Fujimoto, Y. Kanazawa, P. Lore and M. Miyamoto: "An 80/100MS/s 76.3/ 70.1dB SNDR $\Delta\Sigma$ ADC for Digital TV Receivers." *Dig. Tech. Papers of the Intl. Solid-State Circuits Conf.*, pp. 201-210, 2006.

[Gagg04] R. Gaggl, M. Inversi, A. Wiesbauer: "A Power Optimized 14-Bit SC $\Delta\Sigma$ Modulator for ADSL CO Applications". *Dig. Tech. Papers of the Intl. Solid- State Circuits Conf.*, pp. 82-83, 2004.

[Gazi05] V. Gazis, N. Alonistioti and L. Merakos: "Toward a Generic Always Best Connected Capability in Integrated WLAN/UMTS Cellular Mobile Networks (and Beyond)". *IEEE Wireless Communications*, pp. 20-29, June 2005.

[Geer02] Y. Geerts, M. Steyaert and W. Sansen, *Design of Multi-Bit Delta-Sigma A/D Converters*. Kluwer Academic Publishers, 2002.

[Geig90] R.L. Geiger, P.E. Allen and N.R. Strader, *VLSI Design Techniques for Analog and Digital Circuits*. McGraw-Hill, 1990.

[Ghar06] A. Gharbiya and D.A. Johns: "On the implementation of input-feedforward delta-sigma modulators". *IEEE Trans. on Circuits and Systems - II*, Vol. 53, pp. 453-457, June 2006.

[Gian07] V. Giannini et al.: "Flexible Baseband Analog Circuits for Software-Defined Radio Front-Ends". *IEEE J. of Solid-State Circuits*, Vol. 42, pp. 1501-1512, July 2007.

[Gian09] V. Giannini et al.: "A 2-mm2 0.1-5GHz Software-Defined Radio Receiver in 45-nm Digital CMOS". *IEEE J. of Solid-State Circuits*, Vol. 44, pp. 3486–3498, Dec. 2009.

[Gibs02] J.D. Gibson, *The Communications Handbook*. CRC Press, 2002.

[Giel05] G. Gielen and E. Goris: "Reconfigurable Front-End Architectures and A/D Converters for Flexible Wireless Transceivers for 4G Radios". *Proc. of the 2005 IEEE CAS Symp. on Emerging Technologies: Circuits and Systems for 4G Mobile Wireless Communications*, pp. 13-18, June 2005.

[Giel06] G.G.E. Gielen: "Design methodologies and tools for circuit design in CMOS nanometer technologies". *Proc. of the IEEE Solid-State Device Research Conf.*, pp. 21-32, 2006.

[Gome02] G. Gomez and B. Haroun: "A 1.5 V 2.4/2.9mW 79/50dB DR $\Sigma\Delta$ Modulator for GSM/WCDMA in 0.13μm Digital Process". *Dig. Tech. Papers of the Intl. Solid-State Circuits Conf.*, pp. 242-490, 2002.

[Good96] F. Goodenough: "Analog Techniques of all Varieties Dominate ISSCC". *Electronic Design*, Vol. 44, pp. 96–111, Feb. 1996.

[Goth02] A. Gothenberg and H. Tenhunen: "Improved cascaded sigma-delta noise shaper architecture with reduced sensitivity to circuit nonlinearities". *Electronics Letters*, Vol. 38, pp. 683-685, July 2002.

[Gula01] K. Gulati and H.S. Lee: "A Low-Power Reconfigurable Analog-to-Digital Converter". *IEEE J. of Solid-State Circuits*, Vol. 36, pp. 1900-1910, Dec. 2001.

[Gupt02] S. K. Gupta and V. Fong: "A 64-MHz Clock-Rate ΣΔ ADC with 88-dB SNDR and -105-dB IM3 Distortion at a 1.5-MHz Signal Frequency". *IEEE J. of Solid-State Circuits*, Vol. 37, pp. 1653-1661, Dec. 2002.

[Hair91] A. Hairapetian, G.C. Temes and Z.X. Zhang: "Multibit sigma-delta modulator with reduced sensitivity to DAC nonlinearity". *Electronics Letters*, Vol. 27, pp. 990-991, May 1991.

[Hamo06] A. A. Hamoui and F. Maloberti, "Delta-Sigma Modulators for Power-Efficient A/D Conversion in High-Speed Wireless Communications". *Proc. of the IEEE Intl. on Electronics, Circuits and Systems*, pp. 123-127, 2006.

[Hash03] H. Hashemi, *Integrated Concurrent Multi-Band Radios and Multiple-Antenna Systems*. Ph.D. Thesis, California Institute of Technology, 2003.

[Hend96] R. K. Henderson and O. Nys: "Dynamic element matching techniques with arbitrary noise shaping function". *Proc. of the IEEE Intl. Symp. Circuits and System*, pp. 293-296, 1996.

[Ho10] C. Y. Ho, Y. Y. Lin and T. H. Lin, "Dual-Mode Continuous-Time Quadrature Bandpass ΔΣ Modulator with Pseudo-random Quadrature Mismatch Shaping Algorithm for Low-IF Receiver Application". *Proc. of the IEEE Intl. Symp. Circuits and System*, pp. 25-28, 2010.

[Holl01] T. Hollman et al.: "A 2.7-V CMOS Dual-Mode Baseband Filter for PDC and WCDMA". *IEEE J. of Solid-State Circuits*, pp. 1148-1153, July 2001.

[Hui03] S.Y. Hui and K.H. Yeung: "Challenges in the Migration to 4G Mobile Systems". *IEEE Communications Magazine*, pp. 54-59, Dec. 2003.

[Inge10] M. Ingels et al.: "A 5mm^2 40nm LP CMOS 0.1-to-3GHz multistandard transceiver". *Dig. Tech. Papers of the Intl. Solid-State Circuits Conf.*, pp. 458-459, 2010.

[Inos62] H. Inose, Y. Yasuda and J. Murakami: "A Telemetering System by Code Modulation – Δ-Σ Modulation". *IRE Trans. on Space Electronics and Telemetry*, Vol. 8, pp. 204-209, Sep. 1962.

[Isma07] M. Ismail and D. Rodríguez de Llera (Editors), *Radio Design in Nanometer Technologies*. Springer, 2007.

[Jant97] S.A. Jantzi et al.: "Quadrature Bandpass ΔΣ Modulation for Digital Radio". *IEEE J. of Solid-State Circuits*, Vol. 32, pp. 1935-1950, Dec. 1997.

[John97] D.A. Johns and K. Martin, *Analog Integrated Circuit Design*. John Wiley and Sons, 1997.

[Jung04] Y.J. Jung et al.: "A 2.4-GHz 0.25-μm CMOS Dual-Mode Direct-Conversion Transceiver for Bluetooth and 802.11b". *IEEE J. of Solid-State Circuits*, Vol. 39, pp. 1185-1190, July 2004.

[Kana06] Y. Kanazawa, Y. Fujimoto, P. Lo Ré and M. Miyamoto: "A 100-MS/s 4-MHz Bandwidth 77.3-dB SNDR ΔΣ ADC with a Triple Sampling Technique". *Proc. of the IEEE Custom Integrated Circuits Conf.*, pp. 53-56, 2006.

[Kare90] T. Karema, T. Ritoniemi and H. Tenhunen: "An oversampled sigma-delta A/D converter circuit using two-stage fourth order modulator". *Proc. of the IEEE Intl. Symp. on Circuits and Systems*, Vol. 4, pp. 3279-3282, 1990.

[Kim09] H. Kim et al.: "Adaptive Blocker Rejection Continuous-Time ΣΔ ADC for Mobile WiMAX Applications". *IEEE J. of Solid-State Circuits*, Vol. 44, pp. 2766–2779, Oct. 2009.

[Kits09] M. Kitsunezuka et al.: "A Widely-Tunable, Reconfigurable CMOS Analog Baseband IC for Software-Defined Radio". *IEEE J. of Solid-State Circuits*, Vol. 44, pp. 2496–2502, Sep. 2009.

[Koh05] J. Koh, Y. Choi and G. Gomez: "A 66dB DR 1.2V 1.2mW Single-Amplifier Double-Sampling 2nd-Order ΔΣ ADC for WCDMA in 90nm CMOS". *Dig. Tech. Papers of the Intl. Solid-State Circuits Conf.*, pp. 170-171, 2005.

[Kund04] K.S. Kundert, *The Designer's Guide to Verilog-AMS*. Kluwer Academic Publishers, 2004.

[Kuo99] T.-H. Kuo, K.-D. Chen and J.-R. Chen: "Automatic coefficients design for high-order sigma-delta modulators". *IEEE Trans. on Circuits and Systems - II*, Vol. 46, pp. 6–15, Jan. 1999.

[Krau92] J.D. Krauss, *Electromagnetics*. McGraw-Hill, 1992.

[Kwon04] S. Kwon and F. Maloberti: "Op-amp swing reduction in sigma-delta modulators". *Proc. of the IEEE Intl. Symp. Circuits and Systems*, pp. 525-528, 2004.

[Lee85] K.-L. Lee and R.G. Meyer: "Low-Distortion Switched-Capacitor Filter Design Techniques". *IEEE J. of Solid-State Circuits*, Vol. 20, pp. 1103-113, Dec. 1985.

[Lee87] W.L. Lee and C.G Sodini: "A Topology for Higher Order Interpolative Coders". *Proc. of the IEEE Intl. Symp. on Circuits and Systems*, pp. 459-462, 1987.

[Lee02] D.-H. Lee, C.-C. Li and T.-H. Kuo: "High-speed low-complexity implementation for data weighted averaging algorithm". *Proc. of the IEEE Asia-Pacific Conf. on Advanced System Integrated Circuits*, pp. 283-286, 2002.

[Lee04] T.H. Lee, *The Design of CMOS Radio-Frequency Integrated Circuits, 2/E*. Cambridge University Press, 2004.

[Lesl90] T.C. Leslie and B. Singh: "An Improved Sigma-Delta Modular Architecture". *Proc. of the IEEE Intl. Symp. on Circuits and Systems*, pp. 372-375, 1990.

[Lewi92] S.H. Lewis, H.S. Fetterman, G.F. Gross, R. Ramachandran and T.R. Viswanathan: "A 10-b 20-Msample/s analog-to-digital converter". *IEEE J. of Solid-State Circuits*, Vol. 27, pp. 351-358, March 1992.

[Li02] X. Li and M. Ismail, *Multi-standard CMOS wireless receivers: analysis and design*. Kluwer Academic Publishers, 2002.

[Lim06] J. Lim and J. Choi: "A Low-Power Sigma-Delta Modulator for Wireless Communication Receivers using Adaptive Biasing Circuitry and Cascaded comparator scheme". *Analog Integrated Circuits and Signal Processing*, pp. 359-365, Sept. 2006.

[Lim08] H.K. Lim: "The 2nd Wave of the Digital Consumer Revolution: Challenges and Opportunities!". *Dig. Tech. Papers of the Intl. Solid-State Circuits Conf.*, pp. 18-23, 2008.

[Lin09] L. Lin et al.: "A fully integrated 2x2 MIMO dual-band dual-mode direct-conversion CMOS transceiver for WiMAX/WLAN applications". *Dig. Tech. Papers of the Intl. Solid-State Circuits Conf.*, pp. 202-607, 2009.

[Long88] L. Longo and M. Copeland: "A 13 bit ISDN-band oversampled ADC using two-stage third-order noise shaping". *Proc. of the IEEE Custom Integrated Circuit Conf.*, pp. 21.2.1-4, 1988.

[Lu10] C.Y. Lu et al.: "A Sixth-Order 200 MHz IF Bandpass Sigma-Delta Modulator With Over 68 dB SNDR in 10 MHz Bandwidth". *IEEE J. of Solid-State Circuits*, Vol. 45, pp. 1122-1136, June 2010.

[Maed06] T. Maeda et al.: "Low-Power-Consumption Direct-Conversion CMOS Transceiver for Multi-Standard 5-GHz Wireless LAN Systems With Channel Bandwidths of 5-20MHz". *IEEE J. of Solid-State Circuits*, Vol. 41, pp. 375-383, Feb. 2006.

[Magh06] N. Maghari, S. Kwon, G.C. Temes and U. Moon: "Sturdy MASH $\Delta\Sigma$ modulator". *Electronics Letters*, Vol. 42, pp. 1269-1270, Oct. 2006.

[Magh07] N. Maghari, S. Kwon, G.C. Temes and U. Moon: "Mixed-Order Sturdy MASH Δ-Σ modulator". *Proc. of the IEEE Intl. Symp. Circuits and Systems*, pp. 257-260, 2007.

[Mak07] P. Mak, S. U and R.P. Martins: "Transceiver Architecture Selection: Review, State-of-the-Art Survey and Case Study". *IEEE Circuits and Systems Magazine*, pp. 6–25, Second Quarter 2007.

[Malo01] F. Maloberti, *Analog Design for CMOS VLSI Systems*. Kluwer Academic Publishers, 2001.

[Malo09] F. Maloberti: "Technology Scaling and New Challenges in Analog Design." *IEEE Solid-State Circuits Magazine*, pp. 33-34, Fall 2009.

[Mark04] J. Markus and G.C. Temes: "An efficient Delta-Sigma ADC architecture for low oversampling ratios". *IEEE Trans. on Circuits and Systems - I*, Vol. 51, pp. 63-71, 2004.

[Marq98] A.M. Marques, V. Peluso, M.S.J. Steyaert and W. Sansen: "Optimal Parameters for $\Delta\Sigma$ Modulator Topologies". *IEEE Trans. on Circuits and Systems - II*, Vol. 45, pp. 1232-1241, Sep.1998.

[Math08a] The MathWorks Inc., *MATLAB™ User's Guide.* March 2008.

[Mats87] Y. Matsuya, K. Uchimura, A. Iwata, T. Kobayashi, M. Ishikawa and T. Yoshitome: "A 16-bit oversampling A-to-D conversion technology using triple-integration noise shaping". *IEEE J. of Solid-State Circuits*, Vol. 22, pp. 921-929, Dec. 1987.

[Mede99] F. Medeiro, B. Pérez-Verdú and A. Rodríguez-Vàzquez, *Top-Down Design of High-Performance Modulators.* Kluwer Academic Publishers, 1999.

[Mede03] F. Medeiro, R. del Río, J.M. de la Rosa, B. Pérez-Verdú and A. Rodríguez-Vàzquez, "High-Order Cascade Multi-bit $\Sigma\Delta$ Modulators", Chapter 9 in *CMOS Telecom Data Converters.* Kluwer Academic Publishers, 2003.

[Miao98] G. Miao, H.C. Yang and T. Pushan: "An Oversampled A/D Converter Using Cascaded Fourth Order Sigma-Delta Modulation and Current Steering Logic". *Proc. of the IEEE Intl. Symp. on Circuits and Systems*, Vol. 1, pp. 412-415, 1998.

[Mill03] MR. Miller and C.S. Petrie: "A Multibit Sigma-Delta ADC for Multimode Receivers". *IEEE J. of Solid-State Circuits*, Vol. 38, pp. 475-482, March 2003.

[Mito95] J. Mitola: "The Software Radio Architecture". *IEEE Communications Magazine*, pp. 26–38, May 1995.

[Morg07a] A. Morgado, R. del Río and J.M. de la Rosa: "Cascade $\Sigma\Delta$ modulator for low- voltage wideband applications". *Electronics Letters*, Vol. 43, pp. 910-911, Aug. 2007.

[Morg07b] A. Morgado, R. del Río and J.M. de la Rosa: "An Adaptive $\Sigma\Delta$ Modulator for Multi-Standard Handheld Wireless Devices". *Proc. of the Asian Solid-State Circuit Conf.*, pp. 232-235, 2007.

[Morg09] A. Morgado, V.J. Rivas, R. del Río, R. Castro-López, F.V. Fernàndez and J.M. de la Rosa: "Behavioral Modeling, Simulation and Synthesis of Multi- Standard Wireless Receivers in MATLAB/SIMULINK". *Integration, the VLSI Journal*, Vol. 41, pp. 269-280, Feb. 2009.

[Morg10] A. Morgado, R. del Río, J.M. de la Rosa, L. Bos, J. Ryckaert and G. van der Plas: "A 100kHz-10MHz BW, 78-to-52dB DR,4.6-to-11mW Flexible SC $\Sigma\Delta$ Modulator in 1.2-V 90-nm CMOS". *Proc. of the European Solid-State Circuits Conf.*, 2010.

[Mous94] S.M. Moussavi and B.H. Leung: "High-Order Single-Stage Single-Bit Oversampling A/D Converter Stabilized with Local Feedback Loops". *IEEE Trans. on Circuits and Systems - I*, Vol. 41, pp. 19-25, Jan. 1994.

[Muha06] K. Muhammad et al.: "The first fully integrated quad-band GSM/GPRS receiver in a 90-nm digital CMOS process". *IEEE J. Solid-State Circuits*, Vol. 41, pp. 1772-1783, Aug. 2006.

[Murm10] B. Murmann, "ADC Performance Survey 1997-2010" [Online]. Available: http://www.stanford.edu/~murmann/adcsurvey.html.

[Nade08] A. Naderi, M. Sawan and Y. Savaria: "On the Design of Undersampling Continuous-Time Bandpass Delta-Sigma Modulators for Gigahertz Frequency A/D Conversion". *IEEE Trans. on Circuits and Systems - I: Regular Papers*, Vol. 55, pp. 3488-3499, Dec. 2008.

[Nam05] K.Y. Nam, S.M. Lee, D.K. Su and B.A. Wooley: "A Low-Voltage Low-Power Sigma-Delta Modulator for Broadband Analog-to-Digital Conversion". *IEEE J. of Solid-State Circuits*, Vol. 40, pp. 1855-1864, Sep. 2005.

[Neuv04] Y. Neuvo: "Cellular Phones as Embedded Systems". *Plenary Lecture at the 2004 IEEE Intl. Solid-State Circuit Conf.*

[Nors97a] S.R. Norsworthy, R. Schreier and G.C. Temes (Editors), *Delta-Sigma Data Converters: Theory, Design and Simulation.* IEEE Press, 1997.

[Nors97b] S.R. Norsworthy, "Quantization Errors and Dithering in $\Delta\Sigma$ Modulators", Chapter 3 in *Delta-Sigma Data Converters: Theory, Design and Simulation (S.R. Norsworthy, R Schreier and G.C. Temes, Editors).* IEEE Press, 1997.

[OptE90] F. Op't Eynde, *High-Performance Analog Interfaces for Digital Signal Processors.* Ph.D. Thesis, Katholieke Universiteit Leuven (KUL), 1990.

[Ortm06] M. Ortmanns and F. Gerfers, *Continuous-Time Sigma-Delta A/D Conversion.* Springer, 2006.

[Ouzo07] S. Ouzounov, R. van Veidhoven, C. Bastiaansen, K. Vongehrl, R. van Wegberg, G. Geelen, L. Breems and A. van Roermund: "A 1.2V 121-Mode CT Delta-Sigma Modulator for Wireless Receivers in 90nm CMOS". *Dig. Tech. Papers of the Intl. Solid-State Circuits Conf.*, pp. 242-243, 2007.

[Park01] Y.-I. Park, S. Karthikeyaiz, F. Tsay and E. Bartoloine: "A 10 bit 100 MSample/s CMOS Pipelined ADC with 1.8V Power Supply". *Dig. Tech. Papers of the Intl. Solid-State Circuits Conf.*, pp. 130-131, 2001.

[Paul87] J. Paulos, G. Brauns, M. Steer and S. Ardalan: "Improved Signal-to-Noise Ratio using Tri-Level Delta-Sigma Modulation". *Proc. of the IEEE Intl. Symp. Circuits and Systems*, pp. 436–466, 1987.

[Phil00] Philips Semiconductor, *Integrated Circuit Packages Data Handbook*, 2000.

[Pies02] T. Piessens, M. Steyaert and E. Bach: "A Difference Voltage Buffer for $\Delta\Sigma$ Converters". *Analog Integrated Circuits and Signal Processing*, Vol. 31, pp. 31-37, April 2002.

[Pres92] W.H. Press, B P. Flannery, S.A. Teukolsky and W.T. Vetterling, "Bracketing and Bisection", Chapter 9 in *Numerical Recipes in FORTRAN: The Art of Scientific Computing.* Cambridge University Press, 1992.

[Putt07] B. Putter: "A 5th order CT/DT Multi-Mode $\Sigma\Delta$ Modulator". *Dig. Tech. Papers of the Intl. Solid-State Circuits Conf.*, pp. 244-245, 2007.

[Rabi97] S. Rabii and B.A. Wooley: "A 1.8V Digital-Audio Sigma-Delta Modulator in 0.8μm CMOS". *IEEE J. of Solid-State Circuits*, Vol. 32, pp. 783-796, June 1997.

[Rao05] K.R. Rao et al.: "A CMOS RF Front-End for a Multi-Standard WLAN Receiver". *IEEE Microwave and Wireless Components Letters*, Vol. 15, pp. 321-323, May 2005.

[Raza98] B. Razavi, *RF Microelectronics.* Prentice Hall,1998.

[Raza00] B. Razavi, *Design of Analog CMOS Integrated Circuits.* McGraw-Hill, 2000.

[Rebe90] M. Rebeschini, N.R. van Bavel, P. Rakers, R. Greene, J. Caldwell and J.R. Haug: "A 16-b 160 kHz CMOS A/D Converter using Sigma-Delta Modulation". *IEEE J. of Solid-State Circuits*, Vol. 25, pp. 431-440, April 1990.

[Rebe97] M. Rebeschini, "The Design of Cascaded $\Delta\Sigma$ ADCs", Chapter 6 in *Delta- Sigma Data Converters: Theory, Design and Simulation (S.R. Norsworthy, R Schreier andG.C. Temes, Editors).* IEEE Press, 1997.

[Reut02] R. Reutemann, P. Balmelli and Q. Huang: "A 33mW 14b 2.5MSample/s $\Sigma\Delta$ A/ D converter in 0.25μm digital CMOS". *Dig. Tech. Papers of the Intl. Solid- State Circuits Conf.*, Vol. 1, p. 316, 2002.

[Ribn91] D.B. Ribner, R.D. Baertsch, S.L. Garverick, D.T. McGrath, J.E. Krisciunas and T. Fuji: "A Third-Order Multistage Sigma-Delta Modulator with Reduced Sensitivity to Nonidealities". *IEEE J. of Solid-State Circuits*, Vol. 26, pp. 1764-1774, Dec. 1991.

[Rio00] R. del Río, F. Medeiro, B. Pérez-Verdú and A. Rodríguez-Vàzquez: "High- order cascade multibit $\Sigma\Delta$ modulators for xDSL applications". *Proc. of the IEEE Intl. Symp. on Circuits and Systems*, Vol. 2, pp. 37-40, 2000.

[Rio06] R. del Río, F. Medeiro, B. Pérez-Verdú, J.M. de la Rosa and A. Rodríguez-Vàzquez, *CMOS Cascade Sigma-Delta Modulators for Sensors and Telecom – Error Analysis and Practical Design.* Springer, 2006.

[Ritc77] G.R. Ritchie, *High Order Interpolation Analog-to-Digital Converters.* Ph.D. Thesis, University of Pennsylvania, 1977.

[Rodr03] A. Rodríguez-Vàzquez, F. Medeiro, J.M. de la Rosa, R. del Río, R. Tortosa and B. Pérez-Verdú, "Sigma-Delta CMOS ADCs: An Overview of the State-of- the-Art", Chapter 2 in *CMOS Telecom Data Converters (A. Rodríguez- Vàzquez, F. Medeiro and E. Janssens, Editors).* Kluwer Academic Publishers, 2003.

[Rosa09] J.M. de la Rosa, R. Castro-López, A. Morgado, E.C. Becerra-Álvarez, R. del Río, F.V. Fernandez and B. Pérez-Verdú: "Adaptive CMOS analog circuits for 4G mobile

terminals—Review and state-of-the-art survey". *Microelectronics Journal*, Vol. 40, pp. 156–176, Jan. 2009.
[Ru09] Z. Ru et al.: "Digitally Enhanced Software-Defined Radio Receiver Robust to Out-of-Band Interference". *IEEE J. of Solid-State Circuits*, Vol. 44, pp. 3359-3375, Dec. 2009.
[Rube07] R. Rubestein: "Radios Get Smart". *IEEE Spectrum*, pp. 47-50, Feb. 2007.
[Ruiz05] J. Ruiz-Amaya, J.M. de la Rosa, F.V. Fernàndez, F. Medeiro, R. del Río, B. Pérez-Verdú and A. Rodríguez-Vàzquez: "High-Level Synthesis of Switched- Capacitor, Switched-Current and Continuous-Time ΣΔ Modulators Using SIMULINK-Based Time-Domain Behavioral Models". *IEEE Trans. on Circuits and Systems - I*, Vol. 52, pp. 1795-1810, Sept. 2005.
[Ruiz09] J. Ruiz-Amaya, *Methodology and Synthesis of CMOS Pipeline A/D Converters for Telecommunications*. Ph.D. Thesis, University of Sevilla, 2009.
[Ryck09] J. Ryckaert, J. Borremans, B. Verbruggen, L. Bos, C. Armiento, J. Craninckx and G. van der Plas: "A 2.4 GHz Low-Power Sixth-Order RF Bandpass ΔΣ Converter in CMOS". *IEEE J. of Solid-State Circuits*, Vol. 44, pp. 2873-2880, Nov. 2009.
[Ryyn06] J. Ryynänen, S. Lindfors, K. Stadius and K.A.I. Halonen: "Integrated Circuits for Multi-Band Multi-Mode Receivers". *IEEE Circuits and Systems Magazine*, pp. 38-59, Second Quarter 2006.
[Salo02] T.O. Salo, S.J. Lindfors and K.A.I. Halonen: "A 80-MHz Bandpass ΣΔ Modulator for a 100-MHz IF Receiver". *IEEE J. of Solid-State Circuits*, Vol. 37, pp. 798-308, July 2002.
[Salo03] T.O. Salo, S.J. Lindfors, T.M. Hollman, J.A.M. Järvinen and K.A.I. Halonen: "80-MHz Bandpass ΣΔ Modulator for Multimode Digital IF Receivers". *IEEE J. of Solid-State Circuits*, Vol. 38, pp. 464-474, March 2003.
[Sanc06] M. Sánchez-Renedo, S. Patón and L. Hernàndez: "A 2-2 discrete time cascaded ΣΔ modulator with NTF zero using interstage feedback". Proc. of the IEEE Intl. on Electronics, Circuits and Systems, pp. 954-957, 2006.
[Sanc07] M. Sánchez-Renedo and S. Patón: "A Cascaded CT ΣΔ Modulator with NTF Zero and Simple Mismatch Tuning Method using Interstage Feedback". *Proc. of the IEEE Ph.D. Research in Microelectronics & Electronics Conf.*, pp. 77-80, 2007.
[Sand08] M.A.T. Sanduleanu et al.: "Receiver Front-End Circuits for Future Generations of Wireless Communications." *IEEE Trans. on Circuits and Systems – II: Express Briefs*, Vol. 55, pp. 299-303, April 2008.
[Sans06] W. Sansen, *Analog Design Essentials*. Springer, 2006.
[Schr93] R. Schreier: "An Empirical Study of Higher Order Single Bit Sigma Delta Modulators". *IEEE Trans. on Circuits and Systems - II*, Vol. 40, pp. 461-466, Aug. 1993.
[Schr05a] R. Schreier, J. Silva, J. Steensgaard and G.C. Temes: "Design-Oriented Estimation of Thermal Noise in Switched-Capacitor Circuits". *IEEE Trans. on Circuits and Systems - I*, Vol. 52, pp. 2358-2368, Nov. 2005.
[Schr05b] R. Schreier and G.C. Temes, *Understanding Delta-Sigma Data Converters*. John Wiley and Sons, 2005.
[Sedr82] A.S. Sedra and K.C: Smith, *Microelectronic Circuits*. Oxford University Press, 1982.
[Shi02] C. Shi and M. Ismail, *Data Converters for Wireless Standards*. Kluwer Academic Publishers, 2002.
[Shim05] J.H. Shim, I.C. Park and B. Kim: "A Third-Order ΣΔ Modulator in 0-18-μm CMOS With Calibrated Mixed-Mode Integrators". *IEEE J. of Solid-State Circuits*, Vol. 40, pp. 918-925, April 2005.
[Silv01] J. Silva, U. Moon, J. Steensgaard and G.C. Temes: "Wideband low-distortion delta-sigma ADC topology". *Electronics Letters*, Vol. 37, pp. 737-738, June 2001.
[Silv04a] J. Silva, U. Moon and G.C. Temes: "Low-distortion delta-sigma topologies for MASH architectures". *Proc. of the IEEE Intl. Symp. Circuits and Systems*, pp. 1144-1147, 2004.

[Silv04b] J. Silva, *High-Performance Delta-Sigma Analog-to-Digital Converters*. Ph.D. Thesis, Oregon State University, 2004.

[Silv07] K.A.A. Silva et al.: "An IF-to-Baseband $\Sigma\Delta$ Modulator for AM/FM/IBOC Radio Receivers With a 118-dB Dynamic Range". *IEEE J. of Solid-State Circuits*, Vol. 42, pp.1076–1089, May 2007.

[Silv09] A. Silva, J. Guilherme and N. Horta: "Reconfigurable multi-mode sigma-delta modulator for 4G mobile terminals". *Integration, the VLSI Journal*, Vol. 42, pp. 34-46, Jan. 2009.

[Srip77] A.B. Sripad and D.L. Snyder, "A necessary and sufficient condition for quantization errors to be uniform and white". *IEEE Trans. Acoust., Speech, Signal Processing*, Vol. ASSP-25, pp. 442-448, Oct. 1977.

[Stan92] Standford research systems Inc. *FFT Network Analyzer: Model SR770, User's Manual*, 1992.

[Stee07] M. Steer: "Beyond 3G". *IEEE Microwave Magazine*, Vol. 8, pp. 76–82, 2007.

[Tach03] K. Tachikawa: "A Perspective on the Evolution of Mobile Communications". *IEEE Communications Magazine*, pp. 66–73, Oct. 2003.

[Tahe10] M. Taherzadeh-Sani and A. A. Hamoui: "A Reconfigurable 10-12b 0.4-44MS/s Pipelined ADC with 0.35-0.5pJ/step in 1.2V 90nm Digital CMOS". *Proc. of the European Solid-State Circuits Conf.*, 2010.

[Tan93] N. Tan and S. Eriksson: "Fourth-order two-stage delta-sigma modulator using both 1 bit and multibit quantisers". *Electronics Letters*, Vol. 29, pp. 937-938, May 1993.

[Tang07] Y. Tang, S. Gupta, J. Paramesh and D.J. Allstot: "A Digital-Summing Feedforward Σ-Δ Modulator and its Application to a Cascade ADC". *Proc. of the IEEE Intl. Symp. Circuits and Systems*, pp. 485-488, 2007.

[Tenb08] B. Tenbroek et al.: "Single-chip tri-band WCDMA/HSDPA transceiver without external SAW filters and with integrated TX power control". *Dig. Tech. Papers of the Intl. Solid-State Circuits Conf.*, pp. 202-607, 2008.

[Than07] B. Thandri and J. Silva-Martinez: "A 63dB 75-mW bandpass RF ADC at 950MHz using 3.8-GHz clock in 0.25-µm SiGe BiCMOS Technology." *IEEE J. of Solid-State Circuits*, Vol. 42, pp. 269-279, Feb. 2007.

[Tsiv92] Y.P. Tsividis and J. Voorman, *Integrated Continuous-Time Filters: Principles, Design and Applications*. IEEE Press, 1992.

[Vadi08] M. Vadipour, C. Chen, A. Yazdi, M. Nariman, T. Li, P. Kilcoyne and H. Darabi: "A 2.1mW/3.2mW Delay-Compensated GSM/WCDMA Sigma-Delta Analog-Digital Converter". *Dig. Tech. Papers of the VLSI Circuits Symp.*, pp. 180-181, 2008.

[Vars01] U. Varshney and R. Jain: "Issues in Emerging 4G Wireless Networks". *IEEE Computer*, pp. 94-96, June 2001.

[Vass08] I. Vassiliou et al.: "A 65nm CMOS Multistandard, multiband TV tuner for Mobile and Multimedia Applications". *IEEE J. Solid-State Circuits*, Vol. 43, pp. 1522-1533, July 2008.

[Veld03] R.H. Veldhoven: "A Triple-Mode Continuous-Time $\Sigma\Delta$ Modulator With Switched-Capacitor Feedback DAC for a GSM-EDGE/CDMA2000/UMTS Receiver". *IEEE J. of Solid-State Circuits*, Vol. 38, pp.2069-2076, Dec. 2003.

[Vleu01] K. Vleugels, S. Rabii and B. Wooley: "A 2.5-V Sigma-Delta Modulator for Broadband Communications Applications". *IEEE J. of Solid-State Circuits*, Vol. 36, pp. 1887-1899, Dec. 2001.

[Wang90] Z. Wang: "Analytical determination of output resistance and DC matching errors in MOS current mirrors". *Proc. of the IEE Circuits, Devices and Systems*, Vol. 137, p. 397-404, 1990.

[Wang00] Y.-T. Wang and B. Razavi: "An 8-Bit 150-MHz CMOS A/D Converter". *IEEE J. of Solid-State Circuits*, Vol. 35, pp. 308-317, March 2000.

[Will91] L.A. Williams III and B.A. Wooley: "Third-Order Cascaded Sigma-Delta Modulators". *IEEE Trans. on Circuits and Systems*, Vol. 38, pp. 489-498, May 1991.

[Wolf97] C. Wolff, J.G. Kenny and L.R. Carley, "CAD for the Analysis and Design of $\Delta\Sigma$ Converters", Chapter 14 in *Delta-Sigma Data Converters: Theory, Design and Simulation (S.R. Norsworthy, R Schreier and G.C. Temes, Editors)*. IEEE Press, 1997.

[Wu98] J.-T. Wu and K.-L. Chang: "MOS Charge Pumps for Low-Voltage Operation". *IEEE J. of Solid-State Circuits*, Vol. 33, pp. 592-597, April 1998.

[Xia06] B. Xia, A. Valdes-Garcia and E. Sànchez-Sinencio: "A 10-bit 44-MS/s 20-mW Configurable Time-Interleaved Pipeline ADC for a Dual-Mode 802.11b/ Bluetooth Receiver". *IEEE J. of Solid-State Circuits*, Vol. 41, pp. 530-539, March 2006.

[Yama08] K. Yamamoto, A.C. Carusone and F.P. Dawson: "A Delta-Sigma Modulator With a Widely Programmable Center Frequency and 82-dB Peak SNDR". *IEEE J. of Solid-State Circuits*, Vol. 43, pp. 1772-1782, Aug. 2008.

[Yao06] L. Yao, M. Steyaert and W. Sansen, *Low-Power Low-Voltage Sigma-Delta Modulators in Nanometer CMOS*. Springer, 2006.

[Yin94] G. Yin and W. Sansen: "A High-Frequency and High-Resolution Fourth-Order $\Sigma\Delta$ A/D Converter in BiCMOS Technology". *IEEE J. of Solid-State Circuits*, Vol. 29, pp. 857-865, Aug. 1994.

[Yoon04] H.-K. Yoon and M. Ismail: "A CMOS Multi-standard Receiver Architecture for ISM and UNII Band Applications". *Proc. of the IEEE Intl. Symp. Circuits and Systems*, pp. 265-268, 2004.

[Yu99] W. Yu, S. Sen and B.H. Leung: "Distortion Analysis of MOS Track-and-Hold Sampling Mixers Using Time-Varying Volterra Series". *IEEE Trans. on Circuits and Systems - II*, Vol. 46, pp. 101-113, Feb. 1999.

[Yuan05] X. Yuan, A. Gothenberg and X. Wu: "Improved Wideband Low Distortion Cascaded Delta-Sigma Modulator". *Proc. of the IEEE Intl. Database Engineering & Application Symp.*, 2005.

[Yurt04] A. Yurttas et al.: "Data converter for multistandard mobile phones". *Analog Integrated Circuits Signal Processing*, Vol. 38, pp. 53-61, Jan. 2004.

Index

A. Morgado et al., *Nanometer CMOS Sigma-Delta Modulators for Software Defined Radio*,
DOI 10.1007/978-1-4614-0037-0,

D

E

MIX
Papier aus verantwortungsvollen Quellen
Paper from responsible sources
FSC® C105338

If you have any concerns about our products,
you can contact us on
ProductSafety@springernature.com

In case Publisher is established outside the EU,
the EU authorized representative is:
Springer Nature Customer Service Center GmbH
Europaplatz 3, 69115 Heidelberg, Germany

Printed by Libri Plureos GmbH
in Hamburg, Germany